ELECTRONIC AND COMPUTER-AIDED ASTRONOMY: From Eyes to Electronic Sensors

THE ELLIS HORWOOD LIBRARY OF SPACE SCIENCE AND SPACE TECHNOLOGY
SERIES IN ASTRONOMY

Series Editor: JOHN MASON
Consultant Editor: PATRICK MOORE

This series aims to coordinate a team of international authors of the highest reputation, integrity and expertise in all aspects of astronomy. It will make a valuable contribution to the existing literature encompassing all areas of astronomical research. The titles will be illustrated with both black and white and colour photographs, and include many line drawings and diagrams, with tabular data and extensive bibliographies. Aimed at a wide readership, the books will appeal to the professional astronomer, undergraduate students, the high-flying 'A' level student, and the non-scientist with a keen interest in astronomy.

PLANETARY VOLCANISM*
PETER CATTERMOLE, Department of Geology, Sheffield University, UK
SATELLITE ASTRONOMY: The Principles and Practice of Astronomy from Space
JOHN K. DAVIES, Royal Observatory, Edinburgh, UK
THE DUSTY UNIVERSE*
ANEURIN EVANS, Department of Physics, University of Keele, UK
SPACE-TIME AND THEORETICAL COSMOLOGY*
MICHEL HELLER, Department of Philosophy, University of Cracow, Poland
ASTEROIDS: Their Nature and Utilization
CHARLES T. KOWAL, Space Telescope Institute, Baltimore, Maryland, USA
ELECTRONIC AND COMPUTER-AIDED ASTRONOMY
IAN S. McLEAN, Joint Astronomy Centre, Hilo, Hawaii, USA
URANUS: The Planet, Rings and Satellites*
ELLIS D. MINER, Jet Propulsion Laboratory, Pasadena, California, USA
THE PLANET NEPTUNE
PATRICK MOORE
ACTIVE GALACTIC NUCLEI*
IAN ROBSON, Director of Observatories, Lancashire Polytechnic, Preston, UK
ASTRONOMICAL OBSERVATIONS FROM THE ANCIENT ORIENT*
RICHARD F. STEPHENSON, Department of Physics, Durham University, Durham, UK
SURFACE GEOLOGY OF TERRESTRIAL PLANETS AND SATELLITES: Instrumentation, Investigation, Interpretation
YURI A. SURKOV, Chief of the Laboratory of Geochemistry of the Planets, USSR Academy of Sciences, Moscow, USSR
THE HIDDEN UNIVERSE*
ROGER J. TAYLER, Astronomy Centre, University of Sussex, UK
THE ORIGIN OF THE SOLAR SYSTEM: The Capture Theory
JOHN R. DORMAND, Department of Mathematics and Statistics, Teesside Polytechnic, Middlesborough, UK, and
MICHAEL M. WOOLFSON, Department of Physics, University of York,
UKAT THE EDGE OF THE UNIVERSE*
ALAN WRIGHT, Australian National Radio Astronomy Observatory, Parkes, New South Wales, and HILARY WRIGHT

* *In preparation*

ELECTRONIC AND COMPUTER-AIDED ASTRONOMY:
From Eyes to Electronic Sensors

IAN S. McLEAN, B.Sc.(Hons.), Ph.D.
The Royal Observatory, Edinburgh
on secondment to UK Infrared Telescope Unit, Hawaii

ELLIS HORWOOD LIMITED
Publishers · Chichester

Halsted Press: a division of
JOHN WILEY & SONS
New York · Chichester · Brisbane · Toronto

First published in 1989 by
ELLIS HORWOOD LIMITED
Market Cross House, Cooper Street,
Chichester, West Sussex, PO19 1EB, England
The publisher's colophon is reproduced from James Gillison's drawing of the ancient Market Cross, Chichester.

Distributors:
Australia and New Zealand:
JACARANDA WILEY LIMITED
GPO Box 859, Brisbane, Queensland 4001, Australia
Canada:
JOHN WILEY & SONS CANADA LIMITED
22 Worcester Road, Rexdale, Ontario, Canada
Europe and Africa:
JOHN WILEY & SONS LIMITED
Baffins Lane, Chichester, West Sussex, England
North and South America and the rest of the world:
Halsted Press: a division of
JOHN WILEY & SONS
605 Third Avenue, New York, NY 10158, USA
South-East Asia
JOHN WILEY & SONS (SEA) PTE LIMITED
37 Jalan Pemimpin # 05–04
Block B, Union Industrial Building, Singapore 2057
Indian Subcontinent
WILEY EASTERN LIMITED
4835/24 Ansari Road
Daryaganj, New Delhi 110002, India

British Library Cataloguing in Publication Data
McLean, Ian S.
Electronic and computer-aided astronomy
1. Astronomy. Applications of microcomputer systems
1. Title
520′.028′5416

Library of Congress Card No. 88–34702

ISBN 0–7458–0205–2 (Ellis Horwood Limited)
ISBN 0–470–21233–0 (Halsted Press)

Typeset in Times by Ellis Horwood Limited
Printed in Great Britain by The Camelot Press, Southampton

Table of contents

Preface 11

Introduction 15

1 The development of electronic and computer-aided astronomy

1.1 Observational astronomy 19
1.2 The observables 23
1.3 From eyes to electronic sensors 25
1.4 The charge-coupled device 29
1.5 The impact of solid-state imaging 32
1.6 Automation and computers 34

2 Modern astronomical instrumentation

2.1 Electronic cameras 39
2.1.1 Photography and computers 39
2.1.2 Optical photoelectric cameras 40
2.1.3 Non-optical cameras 46
2.2 Spectrometers 48
2.2.1 Single-object spectrographs 54
2.2.2 Multi-object spectrographs and optical fibres 56
2.2.3 The Fabry–Perot interferometer 61
2.3 Polarimeters 63
2.3.1 Polarised light and its measurement 63
2.3.2 CCD polarimeters 64
2.4 Other instruments. 67
2.4.1 Coronagraphs 67
2.4.2 Stellar velocity meters 67
2.4.3 CCD transit survey telescope 88
2.4.4 High-speed CCD photometers. 89
2.4.5 Speckle cameras 70

3 The user-friendly approach
3.1 Common-user instruments .72
3.2 The degree of automation .73
3.3 Getting feedback to the astronomer .74
3.4 Dealing with the growing complexity .76
3.5 Project planning and management .77

4 Charge-coupled devices
4.1 The early years .81
4.1.1 Invention and development .81
4.1.2 The astronomical push .82
4.2 Basic principles .92
4.2.1 Absorption of light in silicon .92
4.2.2 Charge storage, charge coupling and clocking .98
4.2.3 Interline and frame transfer CCDs 103
4.2.4 CCD outputs 104
4.2.5 Buried channel CCDs 106
4.2.6 Two-phase, four-phase and virtual-phase CCDs 108
4.2.7 Backside-illuminated CCDs 109
4.3 The non-astronomy applications of CCDs 111

5 Practicalities of operation
5.1 Clock voltages and basic electrical factors 113
5.2 Dark current and cooling 118
5.3 Cosmic rays and luminescence 121
5.4 Bad pixels and charge transfer efficiency 123
5.5 Readout noise and other noise sources 131
5.6 Uniformity of response 132
5.7 Ultraviolet flashing and quantum-efficiency pinning 136

6 Electronic and computer control systems
6.1 Basic requirements 144
6.1.1 Overall system layout 145
6.1.2 Optical considerations 145
6.1.3 The cooling system 149
6.1.4 The electronics 153
6.2 CCD sequencers and clock drivers 153
6.2.1 Hardwired designs 157
6.2.2 Programmable designs 158
6.3 Signal processing and digitisation 162
6.3.1 Correlated double sampling 162
6.3.2 On-chip binning 170
6.3.3 Overscan 171

7 Image displays and computers
7.1 Data rates, formats and protocols 173
7.2 Computers and image displays — beneficial or bewildering 175
7.2.1 Image processing systems 177

7.2.2 Which operating system? . . . 180
7.2.3 Applications software — a unified approach . . . 180
7.2.4 Software development methodologies . . . 184
7.2.5 Data archives . . . 186
7.3 PCs and microprocessors in astronomy . . . 187
7.4 Automated data reduction . . . 188

8 Characterisation and calibration of CCD instruments
8.1 From photons to microvolts . . . 193
8.1.1 Quantum efficiency . . . 193
8.1.2 Photon transfer function . . . 197
8.2 Noise, bias and dark current . . . 199
8.3 Flat-fielding strategies . . . 202
8.4 Fringes and sky emission . . . 208
8.5 Linearity . . . 208
8.6 Photometry . . . 208
8.7 Spectroscopy . . . 216
8.8 Comparison of CCDs . . . 217

9 Electronic imaging at other wavelengths
9.1 Infrared . . . 220
9.1.1 Optical or infrared — the elusive boundary . . . 220
9.1.2 Infrared detectors . . . 223
9.1.3 The infrared array revolution — *déjà vu* . . . 225
9.2 Ultraviolet . . . 230
9.3 X-rays . . . 236

10 The future
10.1 Detector developments . . . 242
10.2 The Hubble Space Telescope . . . 247
10.3 Very large telescopes (VLTs) . . . 251
10.4 Advances in data handling . . . 256
10.5 Remote observing . . . 257
10.6 Taking stock . . . 258

Exercises . . . 259

Glossary . . . 261

Appendix 1: Physical and astronomical constants . . . 273

Appendix 2: Units of measurement and powers of 10 notation . . . 274

Appendix 3: Astronomical magnitude scale and its relation to lux and watts . . . 276

Appendix 4: Calculating the signal-to-noise ratio of a CCD measurement . . . 279

Appendix 5: Typical CCD clock voltages. 284

Appendix 6: CCD suppliers and camera manufacturers 290

Index 293

for Jan

Preface

Many vistors to a modern international observatory are surprised, initially, by the panoply of complex-looking equipment attached to the telescope. The multitude of electrical cables, banks of whirring computers and electronic racks, plumes of white vapour emerging silently from instrumentation being cooled with liquid nitrogen, flashing multi-coloured computer consoles and TV screens. And often no means of actually looking through the telescope with the human eye! It seems that part of this lack of awareness is due to the fact that we professional astronomers tend to enthuse about the cosmical discoveries themselves, rather than about the techniques used to make them.

The figure captions on the visitor displays refer to 'false-colour' images, 'computer-enhancement', 'electronic imaging' and 'microchips'. Most people expect electronics and computers to be associated with radio astronomy and, of course, with space programmes. And indeed, radio astronomers could quite fairly lay claim to having started the electronic revolution in astronomy. But in fact, modern ground-based astronomy at optical and infrared wavelengths is heavily electronic and computer-aided, and is as exciting and challenging for the professional engineer and applied physicist as it is for the astronomer! And any astronomer wishing to make novel, state-of-the-art, quantitative studies of the universe must reckon on acquiring a wide range of skills, or on working as a member of a team.

The idea for the substance of this book grew out of a desire to explain to others just how much applied physics and engineering goes into the seemingly 'pure' science of astronomy. So, when Felicity Horwood called to invite me to write on this topic I was delighted. Of course, at that point I had no idea of the effort it would take! This book was also stimulated by the remarkable impact which one small 'silicon microchip' — called a CCD — had on the world of conventional astronomy, and on shaping my own career. Together with the rapid pace of developments in the computer industry, the CCD — and its near relatives at infrared wavelengths — have wrought a fundamental change in observational astronomy.

This book is therefore largely the story of the CCD — what it is and how it works, who were the key figures in its development for astronomy, how it is used in astronomical instruments and how the very nature of the approach to the design and development of astronomical apparatus has been affected. The CCD, invented in 1969 and first used on a telescope in 1976, has led to the design and development of ingenious and complex apparatus for the collection and analysis of incredibly faint sources of light; oftentimes those sources are so distant that the light has taken a substantial fraction of the age of the universe to reach the Earth. Without the special features and stability of the CCD such instruments may have been considered — but never implemented!

Of course, it is quite impossible in a single book to describe and explain the principles of operation and calibration of the many complex, computer-controlled CCD spectrographs and electronic imaging systems in use throughout the astronomical world, let alone recount all the marvellous scientific results which have emerged in the last 15 years. As one might expect for a successful invention like the CCD, the body of existing literature is already immense. Instead, I have tried to concentrate on general principles, innovative ideas and solutions, illustrate how progress was made and raise the most vexing issues of the present time, as well as take an historical perspective. My account will undoubtedly be biased and incomplete in some regard, and I apologise for any omissions.

Astronomy has changed in the last 15 years. It has developed a multi-national character and an extensive technological base which often drives industrial manufacturing techniques well beyond normal limits. Intriguingly, the inherent curiosity of scientists and engineers in industry is sparked by the stringent demands of observational astrophysics, and astronomers have been most fortunate to find many willing companies ready to try to 'do the impossible'. In mentioning people and events of the last 15 years I have tried to capture some of this interplay and some of the excitement and fervour of this 'golden age' for astronomy.

The book has been written deliberately on several levels in the hope that a wide range of people will be able to find something in it for them. Chapters 1 through 4 are intended to provide a general introduction to the complex instrumentation of modern practical optical astronomy at the professional level, to develop the theme of the book, which is the reliance on electronics and computers for giving us our best images of the cosmos, and lead up to the basic principle of the ubiquitous CCD and the role that astronomy played in its early development. Mathematical expositions are kept to a minimum in the main body of the book. Readers who are interested in more detailed technical explanations should find the treatments and references given in Chapters 5 through 8 on practicalities, control systems, computers and calibrations a useful starting point. Chapters 9 and 10 which cover other wavelengths — in particular the very exciting 'revolution' occurring in infrared astronomy — and the future of electronic and computer-aided astronomy as we enter the era of very large telescopes, are again general. Much of the book is based on a course of lectures given to Masters students at the University of Edinburgh and on numerous talks and tours given to public audiences. I have tried to bear in mind the barrier of technical 'jargon' which so often can deter any reader. Special terms appear in bold characters in the text. The key words are defined in a Glossary and cross-referenced in the Index. Words not used in their normal literal sense, or contrived words — such as high-tech — are contained in quotes. For all those who are curious about the unlikely marriage of 'high-tech' and astronomy I hope you will find this account enjoyable.

I would like to take this opportunity to thank all who helped and encouraged this work. I am particularly indebted to the Series Editors John Mason and Patrick Moore for their advice and encouragement, and to Felicity Horwood and Beverley Ford of Ellis Horwood Limited for their patience and support. I am grateful to the many people who supplied me with information, reference material and photographs; in particular, I thank Jim Janesick, as well as Jim Westphal, Morley Blouke, Jim Gunn, Brad Smith, Craig Mackay, Jonathan Wright, Tony Tyson, Mike Tompsett, Richard Aikens, Al Fowler, Carol Christian, Pat Waddell, Lloyd Robinson, Bev Oke, John Lowrance, John Geary, Harold Ables, Steve Grandi, Harold Reitsema, Martin Cullum, Claus Madsen, Jeff Hester, Peter Stetson, John McGraw, Paul Jorden, David Thorne, Jasper Wall, John Fordham, Harvey MacGillivray and Fred Watson. I am also grateful to Brian Hadley and Bill Roberton for photographic reproductions, and to Marjorie Fretwell for preparation of diagrams. I am especially pleased to acknowledge Bill Boyle and George Smith for so readily allowing me to interrupt their retirement pursuits with questions about their invention of the CCD back in 1969; it was great fun trying to contact George Smith by 'ham' radio while he cruised the South Pacific! I am also grateful to my students and post-docs, in particular Mark McCaughrean and John Rayner, and to my many colleagues in the Technology Unit of the Royal Observatory, Edinburgh from whom I learned so much; in particular I thank my friend Donald Pettie. I especially acknowledge the trust, support and vision of Prof. Malcolm Longair. Drafts of selected chapters were read by Malcolm Smith, Colin Aspin, Bernard McNally, Mark Casali, Gillian Wright, Carol Christian and Pat Waddell and I am most grateful for their advice. Any remaining deficiencies are entirely my own. Finally, I am most appreciative of the unswerving support of my wife Jan, who also typed the original draft from my notes.

Hilo, Hawaii
August 1988 Ian McLean

Introduction

Having said that modern ground-based astronomy has become more technological, it is perhaps worthwhile to introduce the reader to the subject of this book by means of a review of the procedure which professional astronomers must follow to obtain a typical night of research at a large observatory.

At the world's major optical/infrared observatory complexes such as Mauna Kea in Hawaii, La Palma in the Canary Islands, Kitt Peak and Mt Hopkins in Arizona, the Cerro Tololo Inter-American Observatory and the European Southern Observatory in Chile, and the Anglo-Australian Observatory at Siding Spring in Australia, as well as at most university or privately owned facilities and at radio observatories, access to the telescope is on a competitive basis. To win **observing time** an astronomer must submit in writing a well-argued scientific case for being allowed to carry out his experiment. Deadline dates are set typically twice or three times per year. Selection is done by **peer review** and all of the major telescopes are heavily oversubscribed, so disappointment is a fact of life. It is to overcome such delays in the progress of scientific experiments, or more seriously, in the optimum use of weather conditions, that the astronomical community worldwide is expending considerable effort on technology. As we shall see, this means highly automated observatories with much reliance on well-engineered systems and computers, and it also implies new cost-effective solutions for the design and management of telescopes and measuring equipment.

Observing time on large telescopes is therefore difficult to obtain and is very valuable; it is very important that no time be wasted. Also, since the telescope and instrumentation is complex and valuable, complete control of their operation cannot be relinquished to guest astronomers who may visit only twice per year. To solve this problem all large observatories provide one or more trained personnel to support the visitor. Usually a night assistant/telescope operator is provided; he or she will be responsible for control of the telescope and dome, ensuring efficient operation, keeping an observatory logbook and the preparation of observatory equipment. The night assistant has the final word regarding safety matters such as closing the telescope dome if the windspeed becomes too high. Sometimes a support scientist, who is a professional research astronomer on the observatory staff familiar with the application of the instrumentation in question, is also available to assist first-time or irregular users of the observatory.

A guest observer planning to use a modern, computer-controlled electronic imaging camera or spectrograph at one of these major facilities might encounter the following pattern of work.

The visiting astronomers will probably arrive by air a few days before their allocated time to ensure that they are not travel-weary and to discuss their plans with observatory staff. By mid-afternoon on the day of the first night on the telescope, the observatory staff, which may include the support scientist and an engineer experienced with the system to be used that night, will be in the telescope dome making sure that the telescope control system and the instrument are both functioning correctly. The visiting astronomer(s), often including graduate students receiving training in observational methods or seeking data for a thesis topic, may well elect to be present for these checks, and may wish to practice using the instrument. This may mean becoming familiar with a control panel or with the operation of a computer console on which the various experimental modes can be displayed and changed by typing certain words on a keyboard. To feel confident that they understand the operation of the instrument, the visiting astronomers will carry out some tests of their own such as a **noise** check on the detector, or a **calibration** image or spectrum; all of these terms and procedures will be described later. With everything in readiness for the evening, they return to the observatory residential lodge where a meal is prepared for them. Between one and two hours prior to sunset they go back to the telescope dome, usually in the company of the night assistant, to complete their preparations. This entails discussing the 'observing plan' — which *must* exist — with the night assistant, providing him/her with a list of objects and their positions on the sky (called 'coordinates') which will be typed into the telescope 'control' computer, and taking some important calibration data such as 'dark' frames with the detector covered and 'flat fields' or 'arc lamps' which are exposures using special artificial sources (lamps).

As the twilight fades and the sky becomes dark enough to commence work, the night assistant will 'call-up' the first object on the target list and a computer will instruct electrically driven motors on the telescope's rotation axes to move or **slew** to that position. Using a special TV-type camera on the optical axis of the telescope, the guest astronomer examines the field of view to confirm that the telescope is pointing at the object of interest — usually by reference to an existing star chart. Sometimes, of course, nothing can be seen because the object(s) is too faint and requires a long exposure; in that case the field must be confirmed by checking the pattern of brighter non-target objects in the vicinity. When the object is correctly centred the observation begins. Having put the camera or spectrograph to the required settings by typing command words or letters into the 'instrument' computer (rather than the telescope control computer), usually all that is required is to issue a 'start' command. The total time for which the measurement lasts is called the **integration time** and this may be anything from a fraction of a second to hours depending on the brightness of the object, the efficiency of the instrument plus detector, the wavelength, and the nature of the experiment. If the integration time is fairly long then it will be essential to ensure that the telescope is accurately tracking the object. This can be done manually by viewing the object or a nearby star with the TV camera and pushing buttons on a hand-set, which is electronically connected to the telescope drive motors, in order to counteract any drift of the image. More likely this act, known as **guiding**, is performed automatically by the computer which digitally samples the image of the guide star on the TV screen, computes any motion and issues a correction to the drive motors of the telescope. When the exposure is complete the image or spectrum will be displayed on a computer screen, an adjustment to the setting of the instrument might be made and another exposure started as some rapid analysis of the first result is undertaken using the observatory computing facilities. Again as we shall see in this book, analysis and display of the measurements the moment they are obtained is crucial to the optimum use of telescope time. The same pattern of work is used repeatedly throughout the night.

The night can be long, from before dusk until well after dawn — typically 12 hours. Considerable concentration and often a degree of patience is required — sometimes the latter is in relation to the other observers rather than with the experiment — and so some sustenance or 'night lunch' might be taken 'on the job'.

Depending on how smoothly the experiment has proceeded or on what has been found, tactical decisions may be required to optimise the use of the night. And certainly, as dawn approaches, an extra effort is made to get the most out of the remaining time. A golden rule of observing, which many newcomers forget to follow, is to assume that 'every night is your last' and never leave a crucial measurement or calibration until tomorrow. Finally, with the last exposure complete, the night assistant is given the go-ahead to close the **dome**, a few more calibration frames are made and then the **mirror covers** are closed and the telescope is returned to its vertical **zenith-park** position. At last the equipment is shut-down or placed in standby mode, logbooks or fault reports are filled out, and the weary group rally round for the walk or drive back to the lodge. The observers sleep until early afternoon and then rise to prepare and review for the next night. Several days of this activity constitute the **observing run**. The visitors will then spend a few days at the offices of the observatory to obtain copies of their data on magnetic tapes, discuss their observations with observatory staff and then fly back to their home institute to analyse the astrophysical content of their data in detail and write a scientific paper.

Throughout each evening the night assistant and the visiting astronomers will remain in a comfortably warm **control room** rather than in the cold, dark telescope dome while electronic and computer systems gather the data. Occasionally, forays are made into the cold mountain air outside to check on the weather or, as in my own case, simply to look up at the star-studded canopy of the night sky, marvel at its awe-inspiring beauty and remind oneself that *this* is what it is all about!

1

The development of electronic and computer-aided astronomy

1.1 OBSERVATIONAL ASTRONOMY

Observing and charting celestial phenomena is an activity of mankind which probably dates back to the Stone Age when cave-dwellers scratched pictures on the walls of their caves. Those drawings were of the Sun, of the Moon through its different phases, and of the stars. Observational astronomy was also one of the earliest scientific activities. Not until the invention of the telescope in the early 1600s, however, were Galileo Galilei and others able to increase effectively the sensitivity of the only light detector available to them, the human eye, and to resolve details such as craters on the Moon, the satellites of Jupiter and the rings of Saturn. By making careful drawings (Fig. 1.1) of what their eyes could detect during moments of

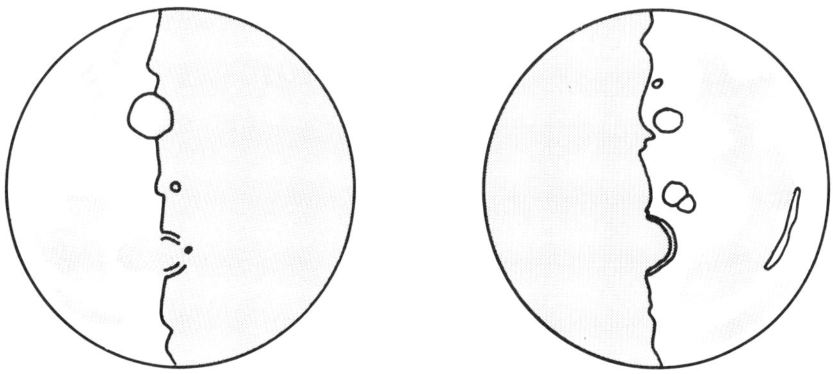

Fig. 1.1 — A sketch such as would have been made by Galileo Galilei of the features on the moon resolved by his telescope. A pictorial representation which can be shared with others.

minimum atmospheric turbulence, what astronomers today call moments of 'good seeing', the early-seventeenth-century astronomers were able to convey pictorially those observations to others.

Better telescopes led to more discoveries. Then, with the invention of the prism spectroscope by Joseph Fraunhofer (1787–1826), the chemical constitution of the Sun and stars became amenable to physical study. In Fraunhofer's early experiments a beam of sunlight was passed through a narrow slit and then through a glass prism to produce a coloured spectrum (Fig. 1.2) which was examined by means of a small

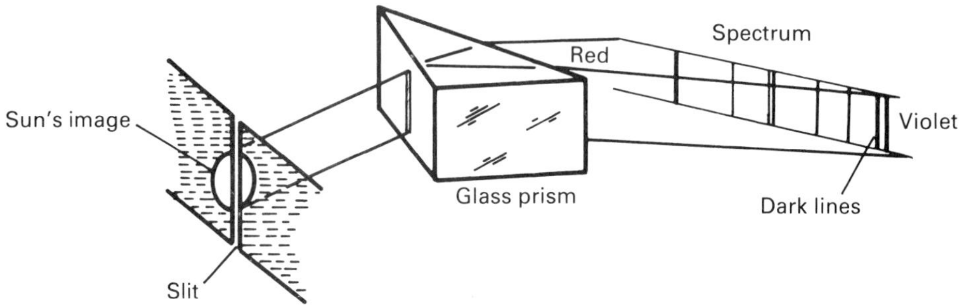

Fig. 1.2 — The principle of Joseph Fraunhofer's spectroscope and the dark lines in the spectrum of the Sun which now bear his name. This instrument, combined with the photographic plate rather than the human eye, opened the way to a physical understanding of the universe.

telescope mounted on an accurate turntable. Initially, the detector was still the human eye. Fraunhofer found that the normal band of colours from violet to red was crossed by numerous dark vertical lines. Eventually the pattern of these 'lines' (actually images of the slit partially devoid of light) were shown to be characteristic of individual chemical elements. The elements hydrogen, calcium, sodium and iron were recognised in the spectra of the Sun and, later, the stars. But spectroscopic observations of the Sun soon led to the discovery of an unknown element which we now know to be a major constituent of the universe; this new element, helium, was named after the Greek word for the Sun, helios. When dry, gelatine-based photographic emulsions became routinely available in the late nineteenth century, astronomers such as Henry Draper (1880) lost no time in putting them to use to catalogue the appearance and properties of a wide range of objects in the night sky. The photographic process was unarguably more accurate and more sensitive than the keenest human eye and the most artistic hand. From planets to stars to galaxies, the new observational tools were applied. Even larger telescopes were constructed, each a technical feat for its era, reaching 100 inches or 2.5 metres in 1917 with the completion of the Hooker Telescope on Mount Wilson by George Ellery Hale. Just one of the great discoveries which followed was the **expansion of the universe** by Hubble and Humason in 1929.

The history of astronomy is marked by such sporadic progress. Each improvement in scientific apparatus, each development in technology, helps to provide answers to old questions, but, inevitably, the new methods uncover a host of new questions, which in turn drive the quest for even better measuring equipment!

Progress in studying the universe has always been related to 'deeper' surveys of the cosmos reaching to ever-fainter objects, higher resolution yielding more and more fine detail, larger statistical samples from which generalisations can be made and broader spectral response to sample ALL the energy forms passively collected by the Earth. That trend has continued since the Renaissance of the sixteenth century to the present day in a kind of ever-increasing spiral, with new tools or technologies leading to new discoveries which in turn drive the development of better tools. A key feature of observational astronomy has been record-keeping: maintaining archives of observations, usually in some pictorial form, for future investigators to compare and consider.

The great advantage of the photographic plate, which in terms of its ability to convert light into a measurable quantity is actually less sensitive than the eye, is that it can build up a picture of a faint object by accumulating light on its emulsion for a long time. It is an **integrating detector**. The eye cannot do this. Moreover, a permanent record is obtained which may be saved for future comparison and study by others.

By using a photographic plate as the recording device in a spectrometer, as well as in direct imaging cameras, astronomers could extend their investigations effectively and efficienctly into the domain of quantitative astrophysics. Initially, of course, the flood of photographic material was analysed by human eye, and those eyes were mostly those of a dedicated group of female assistants hired by the director of the Harvard Observatory College, Edward Charles Pickering, toward the end of the last decade of the nineteenth century. Over 40 women were employed by the observatory during the period of Pickering's tenure as director, and their efforts in handling the torrent of new astronomical data laid the foundations of modern astrophysics. Stellar spectral classifications led to the understanding that the colour range of stars was largely a temperature sequence and that stars shine by the energy released in thermonuclear fusion reactions brought about spontaneously by the enormous temperatures and pressures at their centres. Among the most well-known of the Harvard ladies is Henrietta Leavitt whose work on the class of stars called *Cepheid variables*, which pulsate in brightness with a period which is proportional to their true or absolute average brightness, led to a distance-estimator and an appreciation of the true size and shape of our galaxy. During the first half of the twentieth century, these tools inevitably led to more discoveries (per year) and also to a massive increase in the **data rate**, that is, the amount of information being collected, scrutinised and archived for posterity. But these advances were only the beginning.

Optical astronomy was revitalised in the 1960s, partly by the construction of several new major observatories on excellent mountain sites in different parts of the world, partly by the exciting 'new look' at the universe which accompanied the rise of radio astronomy and the successful developments of the Soviet and American space programmes and partly through the introduction of a variety of **photoelectronic devices**; photocells and photomultiplier tubes and sensitive 'night-vision' TV cameras came first, but the steep rise of consumer micro-electronic products through the sixties and seventies was to accelerate the changes rippling through astronomy. Even the telescopes themselves could be improved by the use of electronically encoded computer-controlled drive systems, thereby enabling much faster set-up times and more reliable tracking across the sky. The new radio and optical telescopes were remotely controlled, and the concept of converting measurements into an electronic form readily suitable for a computer to accept became standard practice. And so the caricature of the 'gentleman astronomer' peering through the eyepiece of some giant telescope was no longer even approximately true. Astronomy became more technological!

Even if we restrict ourselves to optical and near-optical astronomy, the explosion in data-gathering in the last 15 years has been enormous. This flood of quantitative information is due to strides in detector sensitivity. It is the impact of **semiconductor** or **solid-state** electronic sensors attached to the new generation of telescopes over the past 15 years which has had an effect as dramatic as was the introduction of the photographic plate itself. But the new light detectors on the world's largest telescopes are now attaining limits of sensitivity close to theoretical maxima at some wavelengths! Any improvements in sensitivity will require telescopes with much larger collecting mirrors and/or observations in space.

There can be little doubt that we are living in a time of rapid technology development. In some historical accounts the 1950s are known as the Atomic Age and the 1960s and 1970s as the Space Age, but the latter part of the twentieth century will almost certainly go down in history as the Semiconductor Age — the age of the 'microchip'. Semiconductor technology, of which the **silicon chip** is by far the most widely known example, has touched almost every aspect of our daily lives from children's toys, to robotic production lines for the manufacture of cars, from washing machines to personal computers, from bank autotellers to ballistic missiles.

What is a semiconductor? A semiconductor is a crystalline material with some of the properties of a good conductor of electricity (like copper metal), and some of the properties of an electrical insulator (like glass for example). Because of its crystalline (solid-state) structure, a slab of such material behaves the same at all points. Semiconductor crystals can be 'grown' in a controlled way from a melt, and, moreover, the electrical properties can be tailored by introducing so-called impurity atoms into the crystal structure at the atomic level, so that by microscopic sculpting of the semiconductor material all sorts of tiny electrical components and circuits can be constructed. The final piece — often not much larger than a thumbnail — is referred to as an **integrated circuit** or more commonly, as a 'chip'. Besides silicon, there is germanium, gallium arsenide, indium antimonide, and several other materials with these properties. Semiconductors can be used to manufacture a host of low-power microelectronic components including amplifiers, counters, all sorts of logic units, computer memory, very complex chips called **microprocessors** capable of many computer functions, and tiny imaging devices of remarkable sensitivity. Silicon is, so far, the most well-developed semiconductor, but even for silicon the potential for yet smaller and smaller microchips still exists.

Astronomy too has benefitted in this semiconductor revolution, because apparatus for scientific experiments and complex calculations which were impossible before have become viable with the aid of new electronic imaging devices and powerful, extremely compact, high-speed electronic computers.

Almost all modern astronomical research is carried out with **photoelectronic** equipment, i.e. instrumentation which converts weak radiation signals (such as light) from distant objects in space into electrical signals which can be **digitised** into numerical form for a computer. Often highly automated and sometimes remotely controlled, these instruments and the telescopes to which they are attached necessarily rely heavily on electronic computers. Computers play an equally crucial role in helping astronomers assimilate, analyse, model and archive the prodigious quantity of data from the new instruments. The miniaturisation of computers and the ever-increasing availability of large amounts of relatively cheap computer 'memory', means that astronomers can employ fairly complex electronic and computer systems at the telescope which speed up and automate data-gathering and reduction. As a result, those astronomical facilities — which may be costly — and the data they produce can be available to a wider range of scientists than might otherwise be possible. Surprisingly to some, a modern observatory requires an enormous breadth of engineering, scientific and managerial skills to operate efficiently and produce the very best results.

In this book I will try to describe the important impact which solid-state (or semiconductor) imaging devices have had on ground-based astronomy over the past 15 years, and the consequent dependence on computers and software.

1.2 THE OBSERVABLES

Let us start with a very basic look at the practical aspects of astronomy. Astronomy is an observational science. Unlike in a laboratory experiment, the conditions cannot be changed. That is, we on Earth are passive observers (so far) in almost all astronomical experiments, and we can do nothing other than intercept the various forms of energy which reach the Earth from the depths of space. Of course, there have been a few exceptions for solar system studies involving manned and unmanned spacecraft. For ground-based astronomy however, those energy forms can be summarised as:

electromagnetic radiation (gamma-rays through radio waves),
cosmic rays,
neutrinos,
gravity waves.

Of these, the study of electromagnetic radiation which, as shown by James Clerk Maxwell in 1865, includes visible light, is still the most dominant. The nature of electromagnetic radiation is essentially characterised by oscillations of the electric and magnetic energy which gives the radiation the property of a **wave motion**. Different regions of the spectrum correspond to different **wave lengths**. As shown in Fig. 1.3, all the well-known forms of radiant energy are part of this sepctrum.

The measurements which can be made on electromagnetic radiation are also limited. Basically we can determine

the direction of the radiation,
the intensity or energy spectrum,
the polarisation or degree of alignment of the electric and magnetic fields in the radiation, and sometimes
the phase or relation between waves.

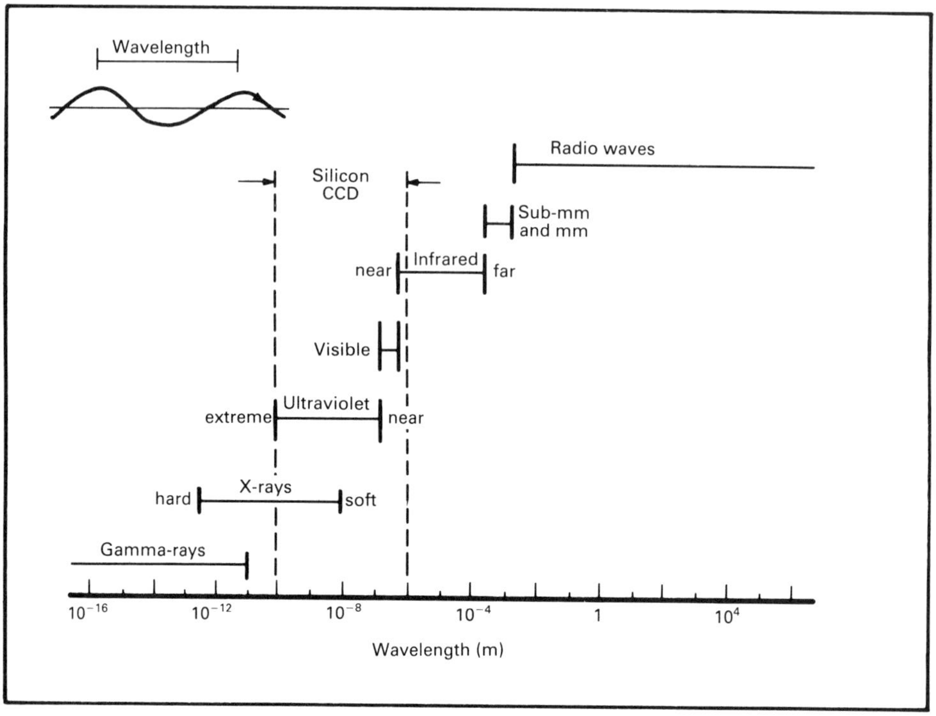

Fig. 1.3 — The electromagnetic spectrum. X-rays, light and radio waves are all different forms of electromagnetic radiation. In the vacuum of empty space each of these forms of radiation travels in straight lines with the same speed — the speed of light.

All that we know about the universe must be extracted from measurements of these energy forms.

Naturally, astronomy began as an **optical** science because human beings have optical sensors (our eyes) which detect a form of electromagnetic radiation which we call **visible light**. Since much of our technology is concerned with enhancing human vision, it is hardly surprising that optical astronomy expanded rapidly with the advent of photography. Photoelectric devices and television cameras accelerated the pace, and now tiny solid-state imaging systems have pushed optical astronomy even further.

The 'new astronomy' subjects of radio, infrared and X-ray have all developed rapidly in the last 30 years, stimulated in part by military and other applications, but largely also the result of the same wave of semiconductor technology. The reliance on electronics and computers holds true for radio and X-ray astronomy also, but a full discussion of those fields deserves separate volumes in their own right. Likewise it is not appropriate to discuss neutrino astronomy or the search for gravity waves. In this book we will be concerned mainly with the detection of electromagnetic radiation from the ultraviolet to the infrared.

1.3 FROM EYES TO ELECTRONIC SENSORS

The blackening of silver halides when exposed to light gave birth to photography well before the beginning of this century, and today photography is still a vital component in the astronomers' arsenal of observational tools. Modern, specially prepared photographic emulsions are sensitive to a wider range of wavelengths than the human eye, from the ultraviolet to the very near infrared. Because it is difficult to relate accurately the sensitivity to light of one part of a plate to another, compared to the corresponding exercise for electronic imaging devices, brightness cannot usually be measured to high accuracies. In addition, the ever-present, albeit faint, light from the night sky eventually **saturates** the emulsion rendering it insensitive. Very clever techniques, called hypersensitisation, have been developed for treating photographic plates; for example Kodak IIIa plates are 'soaked' in nitrogen and hydrogen gas, in order to detect much fainter objects, and highly automated computer-processing of the very large photographic plates used in Schmidt telescopes (see section 2.1) has enabled astronomers to carry out significant statistical studies of many classes of objects, but especially **quasars**. Nevertheless, it is the very low efficiency with which light is converted to blackened grains that limits the sensitivity if not the utility of photographic plates in astronomy.

From the seventeenth century, when Christiaan Huygens and Isaac Newton carried out fundamental experiments on the nature of light, evidence existed that light was a form of energy which tended to interact with matter in a way which suggested that the energy was transported as a 'wave'. Eventually however, situations in which light behaved more like a stream of particles were encountered. Although Newton did develop a 'corpuscular' theory of light, it was based more on his prejudice for mechanics than on evidence. It was not until the end of the nineteenth century with the work of Max Planck that it was understood that light can carry energy only in specific amounts as though it came in individual packets; Planck called these packets **quanta**.

In the early 1900s Einstein's work on the phenomenon called the **photoelectric effect** showed that a quantum of radiant energy (now called a **photon** of light) could eject an electron from certain materials, and that the photon of light behaves like a particle — with energy and momentum — and yet has a 'wavelength' associated with it (see Appendix 4 for the relation between energy and wavelength of a photon). It was also discovered that a beam of electrons impacting on the surface of certain materials could cause the ejection of photons of light (or phosphorescence) and these developments led to the invention of cathode ray tubes, valve amplifiers, radio and ultimately to television (by the Scottish inventor John Logie Baird). In the late 1940s and early 1950s, a photoelectric device called a **photomultiplier tube** became widely available for accurate brightness measurements. In such a device, a photon or quantum of radiation strikes the surface of a certain type of material which responds by emitting an electron. Within the glass-encapsulated photomultiplier tube, that negatively charged electron is accelerated by an electric field and made to impact on another photoemissive surface. That surface generates two or three new electrons which are all accelerated and directed at yet another photoemissive surface where each electron generates two or three more. This cascade process is continued several more times until a final surface collects what is now a huge pulse of electricity generated by just one photon of light (Fig. 1.4). In some cases, the efficiency of

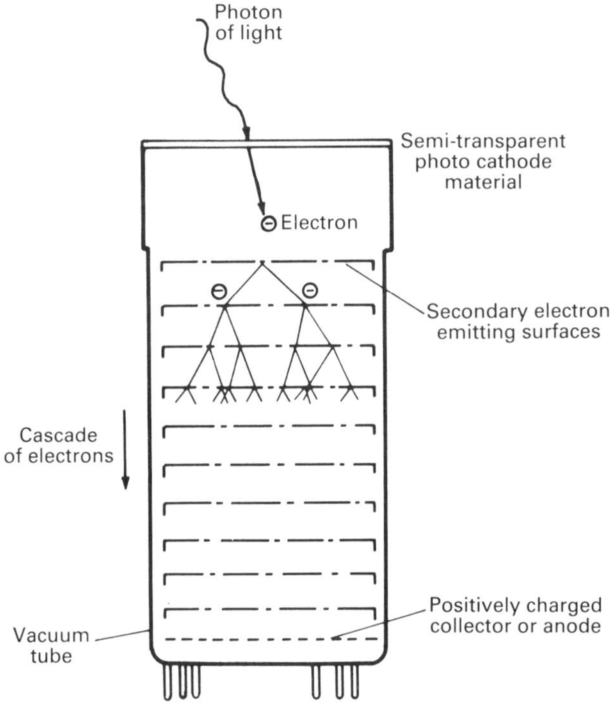

Fig. 1.4 — The photomultiplier tube produces a large cascade of electrical current when illuminated with faint light. It employs the photoelectric effect in which very thin slabs of certain materials emit a subatomic negatively charged particle called an electron.

photomultiplier tubes can be 10 times that of a photographic emulsion and indefinitely long measurements (equivalent to exposures) can be used to detect extremely faint objects. The drawback, of course, is that only one tiny patch of the sky can be observed at a time; a photomultiplier tube has no **panoramic** or **multiplex** advantage — it is like a single cell on the retina of the eye.

Alongside this development came the TV tube and work on **electronic image intensifiers** for very low light level applications. The aim was to combine the attributes of accuracy and unlimited exposure time of the photomultiplier tube with the extended field of view of the TV camera. Military applications for ultra-low light level camera systems helped to stimulate this technology and an astonishing variety of complex television-type image-intensified schemes were proposed and tested for astronomy in the 1960s and 1970s. Most of the devices begin with the light striking a photoemissive surface to release an electron, but a variety of ingenious methods are then employed to amplify the flow of electrons. Some image intensifier stages use electric fields (electrostatic amplification), others use magnetic fields and some use intermediate generation of (brighter) light by using phosphorescent screens. Detection of the amplified stream of charged electrons or the amplified light emission is by some form of TV tube. The characteristic feature of all the various types of TV imaging devices is the use of a beam of electrons to 'scan' across a region called the 'target' on which there is an electrical charge pattern representing the optical image. In essence, the electron beam completes the circuit and provides a source of current to reset the target after exposure to light; the amount of current required for this action is the 'video signal'. TV tubes themselves were greatly improved during this time with the introduction of a range of new variants called the SEC, SIT and silicon vidicons and the Plumbicon or lead-oxide vidicon; the latter is still the primary 'studio' camera. All of the TV-based image-intensified camera systems were complex, physically large, high-voltage (tens of kilovolts) systems. Some found instant success as faint-object TV track-and-guide cameras on large telescopes, but most turned out to be difficult to use for accurate brightness measurements in astronomical cameras and spectrometers. Problems with the stability and reproducibility of the detector, critical for scientific research, often seemed to preclude effective techniques for the removal of systematic errors in long exposures. The most successful of the image-tube devices were those which appealed to **photon-counting** rather than attempting to produce an amplified copy or 'analogue' image of the scene. Two of the best known of these systems are the Wampler Scanner and the Boksenberg Image Photon Counting System (IPCS). In the Wampler Scanner, developed at Lick Observatory, the amplified light emission from the intensifier stage is fed to the unit called an 'image dissector' which has a photocathode and an electromagnetic focussing mechanism which directs photoelectrons to a small rectangular aperture behind which is an electron multiplier stage just as in a normal photomultiplier tube. The area of the photocathode seen by the electron multiplier section can be 'scanned' by controlling the electromagnetic focussing coils. The Wampler Scanner is no longer used and was not truly photon counting. On the other hand, the IPCS (Fig. 1.5) and related systems are still in use and, as we shall see in later chapters, are benefitting from advances in solid-state detectors. Especially well-suited to faint-object spectroscopy, every fraction of a second a frame of the TV output of the IPCS can be digitised and sent to a computer which immediately, or in the jargon 'in **real-time**', could analyse the pattern and intensity of blips in the image and decide which blips should be rejected as **noise** — like the interference or 'snow' on your TV screen when your neighbour is using his unsupressed electric drill!

Over the past 30 years, TV cameras and night vision systems have been developed to a fine art. Tiny colour TV tubes can be enclosed in hand-carried packages, and exceptional real-time sensitivity can be obtained, but for precision scientific measurements all that was not enough. Ironically, this perfection in TV tube construction was reached just about the time that the major breakthrough came in semiconductor devices.

Things began to take a different trend when in 1958 the first multi-transistor **integrated circuit** was demonstrated only 10 years after the invention of the transistor itself at Bell Labs in 1948. In the first transistors a small crystal of germanium was

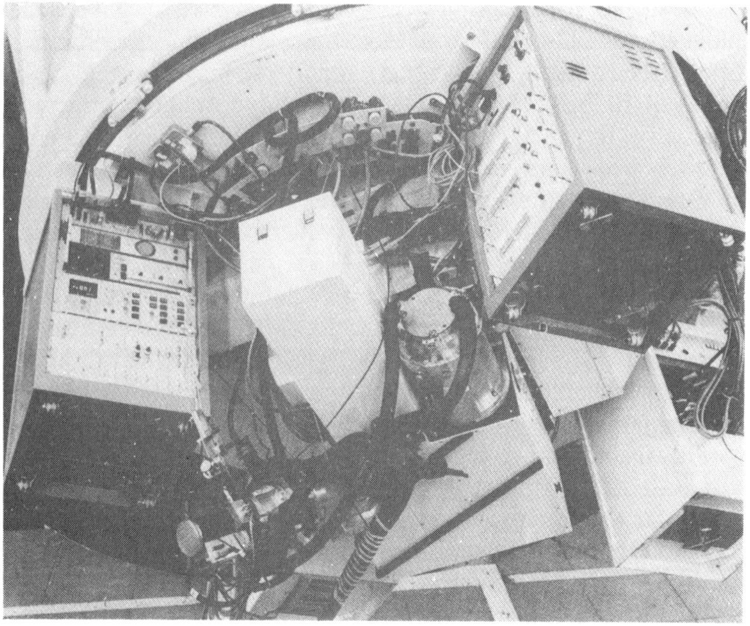

(a)

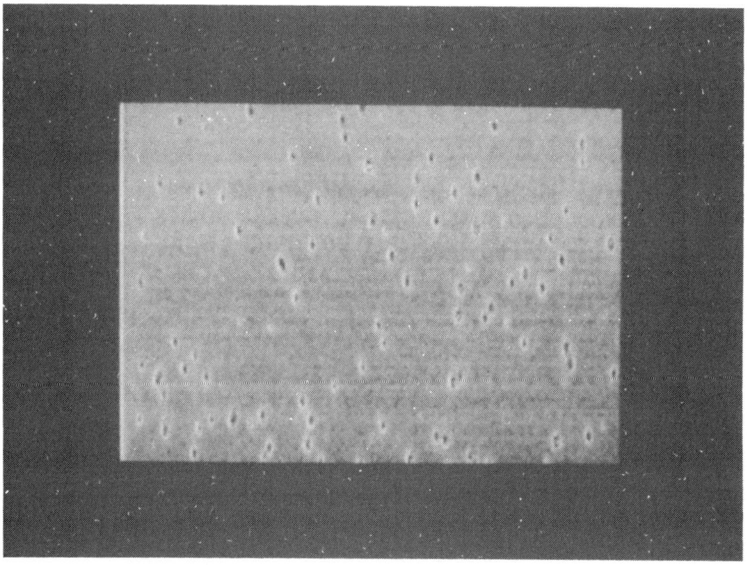

(b)

Fig. 1.5 — (a) The Image Photon Counting System or IPCS (bottom of picture) attached to spectrograph on the Isaac Newton Telescope. Photo courtesy Royal Greenwich Observatory. (b) The IPCS uses the photoelectric effect of light to produce an electron which is then accelerated to impact a phosphor screen which glows brightly enough to be recorded with a TV camera tube. Several photo events are visible in this reproduction. Photo courtesy of John Fordham, University College London.

used; by adding other materials into the crystal structure it was possible to create a small sandwich which amplified electrical currents just like a (much larger) glass-encapsulated tube or valve.

Soon, silicon surpassed germanium as the best material to build transistors, the devices themselves got smaller as fabrication methods improved and then in 1958 the integrated circuit was demonstrated. During the sixties better and better manufacturing processes led to smaller and smaller integrated circuit units still capable of electrical functions. Today, silicon chips the size of a thumbnail can be constructed which contain over 100 000 transistors (amplifiers). Dozens of these silicon chips could fit *inside* the glass vacuum casing of a single valve amplifier!

The photoelectric effect in silicon was put to use in photodiodes: devices in which the size of the electric current flowing from them was proportional to the amount of light falling on them. Silicon photodiodes had the immense benefit of very high **quantum efficiency**, that is 80–90% of the incoming photons of light were converted to electrical charge. Compare this to 1–2% for photographic emulsions (perhaps 4% with hypersensitisation) and 10–20% efficiency for the tube-type systems. Moreover, by the nature of the properties of silicon, this kind of sensor could be used far into the red, reaching in fact to a wavelength of 1.1 micrometres (1 micrometre (μm) is one millionth of a metre; the visible spectrum extends from 0.35 to 0.70 μm). As imaging devices, silicon photodiodes were limited primarily by structural awkwardness in making a large densely packed two-dimensional (i.e. having width and length) assemblage or **array** of photodiodes with no gaps between them. Nevertheless, 'linear arrays' of diodes (just one or two lines wide), used directly and behind image intensifiers, have found applications in astronomical spectrometers, and several systems like this came into operation in the 1970s.

But within the last 15 years a new device has challenged and virtually vanquished all contenders. It is the charge-coupled device or CCD (Fig. 1.6). The CCD is an array of microscopic square-shaped light-sensitive regions arranged in a chequer-board pattern. Tiny light-sensitive squares, usually called **pixels** — for picture elements — are formed directly in a slab of silicon. These pixels are so small that the surface of the CCD looks quite smooth. The introduction of these devices into astronomy has been revolutionary, and has sent repercussions through not only optical astronomy, but infrared and X-ray astronomy as well!

1.4 THE CHARGE-COUPLED DEVICE

Toward the end of 1969, two researchers, Willard S. Boyle and George E. Smith (Fig. 1.7), at the same Bell Labs where the transistor was invented, were investigating new ways of imaging with solid-state silicon methods in an effort to develop a Picturephone! This concept involved having a tiny, inexpensive solid-state camera built into the telephone receiver to enable callers to see each other. Market research later failed to convince Bell Telephone of the worth of the Picturephone concept at that time — though I suspect that now, some 20 years later in the 'video era', the world is ready; Picturephone is a registered trademark of Bell Labs AT&T. As luck would have it, the method devised by the two scientists turned out to be an incredible innovation, destined to change the whole philosophy of imaging away from vidicon-type TV tubes and even from celluloid movie film. Research on magnetic bubble

(a)

(b)

Fig. 1.6 — (a) A typical charge-coupled device or CCD. These tiny imaging devices are made from a small slab of the crystalline form of silicon and are known as solid-state semiconductor detectors; the one shown is manufactured by EEV. Photo courtesy of Royal Greenwich Observatory. (b) A close-up of an early RCA CCD showing how smooth the surface appears to the eye and yet this electronic sensor has 163 840 picture elements. Photo courtesy of NOAO.

Fig. 1.7 — Willard S. Boyle and George E. Smith, inventors of the charge-coupled device, at the Bell Labs AT&T research centre at Murray Hill, New Jersey. Photo courtesy Bell Labs AT&T.

memory, three-phase plasma display panels and silicon vidicons was in full swing at Bell Labs when executive director of the semiconductor division Bill Boyle and department head George Smith — who was in charge of developing a silicon diode array camera tube for Picturephone applications and was a close friend — got together in front of a blackboard one afternoon and began musing about the idea, as George put it, of 'friendly competition' from silicon technology. Bill asked George what about an 'electric bubble'. The obvious analogy to passing a magnetic domain from one site to another was to pass charge from one site to the next. It was already known that charge could be stored by insulating a small metal plate placed on the surface of a silicon crystal, but it was the concept of stringing these storage sites together and using voltage differences between them to pass the charge along that constituted the new idea. Devising the basic concept took only a couple of hours and within a few weeks George had an eight-element device under test! They began their paper, published in the *Bell System Technical Journal* Vol. 49 in April 1970, with the following words 'A new semiconductor device concept has been devised which shows promise of having wide application'. Their invention was of course the **charge-coupled device** or **CCD**, a name chosen by Bill Boyle who recalls the tremendous reception given to the idea following its announcement. As Mike Tompsett, who came to Bell Labs from English Electric Valve (EEV) in the UK where he had worked on the development of the pyroelectric vidicon and himself an author of a classic text on CCDs recalls, 'CCDs were an idea of the time. Within a few months all sorts of applications were listed — and many were actually of relevance to the phone company'. Bill and George recall reactions to their idea as ranging from 'I should have thought of that' to 'it will never work'! But it did work. Experimental verification of the CCD concept was published in the same issue of the *B.S.T.J.* by Gil Amelio, Mike Tompsett and George Smith.

CCDs have a great many advantages over other electronic and photographic imaging devices including:

(a) they are very small, compact, rugged, stable, low-power devices;
(b) they show excellent, near-perfect, sensitivity over a wide range in wavelengths;
(c) they are sensitive from exceptionally low to very high light levels;
(d) they show excellent stability and linearity;
(e) there is no image distortion because all the picture elements in the silicon array are in fixed positions by construction;
(f) they offer the possibility of relatively low cost due to mass production using silicon manufacturing techniques.

Commercial applications of CCDs range from home video cameras to professional broadcast cameras; remote-sensing and surveillance; robotics, medicine and science. For astronomical applications, CCDs are nearly perfect (Fig. 1.8) and represent such a dramatic improvement over all other imaging techniques, except perhaps at the very faintest levels in spectroscopy, that they were promptly embraced by astronomers. Indeed, it is fair to say that astronomers and space scientists were in the forefront of the pioneering days of CCDs. One of those very early developments was an effort by Texas Instruments to provide large CCDs under contract to NASA Jet Propulsion Lab for inclusion in instruments being proposed for the Jupiter Orbiter Probe (later re-named the Galileo mission) and subsequently for the Space Telescope (later renamed the Hubble Space Telescope).

It is quite astonishing that in only 15 years since the first 100×100 pixel device became available CCDs have found such widespread use in science in general and astronomy in particular, and that they should be used for picture-taking — usually called 'imaging' to remove any subconscious reference to photography — with a precision thousands of times more demanding than was ever contemplated in 1970. Today, no major optical observatory in the world lacks a CCD camera!

In the forthcoming chapters I will try to explain how CCDs work and what astronomers have to do in practice to achieve remarkable images, such as those in Fig. 1.9 and others throughout the book, of incredibly faint and distant objects. I will endeavour to dispel at least some of the mystery, interpret the CCD jargon and outline the methodology of CCD use. For more detailed technical discussions on a variety of topics the reader will be guided to publications in the technical literature.

1.5 THE IMPACT OF SOLID-STATE IMAGING

In an introductory paper to the conference on 'Solid state imagers for astronomy' organised by John Geary and David Latham and held on June 10–11, 1981 at the

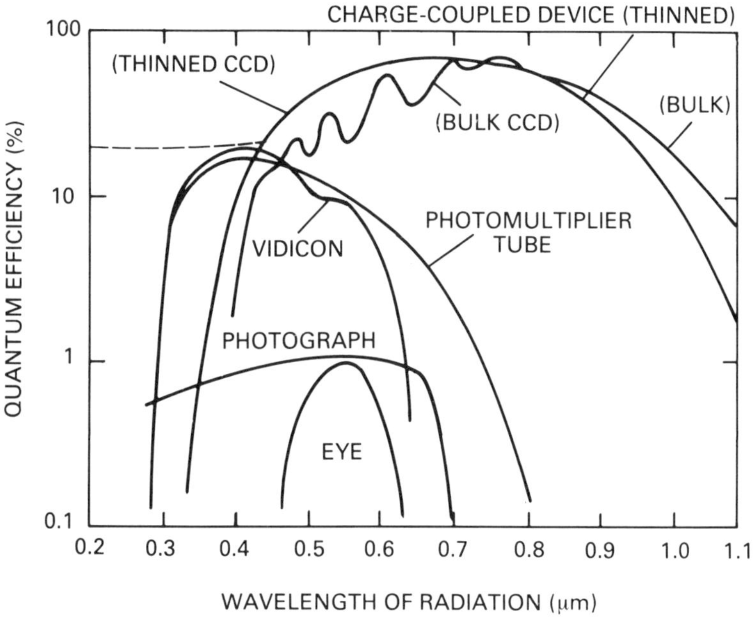

Fig. 1.8 — The sensitivity or quantum efficiency of certain CCDs to light of different wavelengths compared to other forms of light detectors. Note that the scale on the left increases by factors of 10! One of the CCDs many attributes is its extensive range in wavelength coverage.

Harvard-Smithsonian Center for Astrophysics in Cambridge, Massachusetts, Herbert Gursky of that institute, opened the meeting with the following remark: 'There is by now enough information, in the form of laboratory data on a variety of devices and published scientific results, to confirm the original assessment that solid state arrays will become a permanent entry into the repertory of astronomical instruments.' He was right! It took a few more years still, and even today the reliance on manufacturers who themselves are constrained by the rigours of the market place is a fragile one, but he was right.

The impact of CCDs remains phenomenal. For example, at the 5-m (200-inch) Palomar Observatory in California no less than nine CCDs are in use in four different instruments. Even more CCD-based instruments are deployed at the US National Optical Astronomy Observatories (NOAO) on Kitt Peak in Arizona. A 'service' of CCD observations at the prime focus of the famous Anglo-Australian 4-m telescope is offered, and many 1-m telescopes throughout the world are now fitted with CCD cameras. CCDs have been installed in cameras of every description, in spectrometers and in numerous ingenious though more specialised instruments, such as **speckle** cameras, coronagraphs, polarimeters and velocity meters.

Observationally, CCDs have been used for everything from observing the planets, satellites and comets of the solar system, to imaging **gravitational lens** effects in exceedingly distant quasars. Fig. 1.10 shows one rendition of the impact of CCDs on faint-object astronomy. The curve shows how the limiting brightness in faint

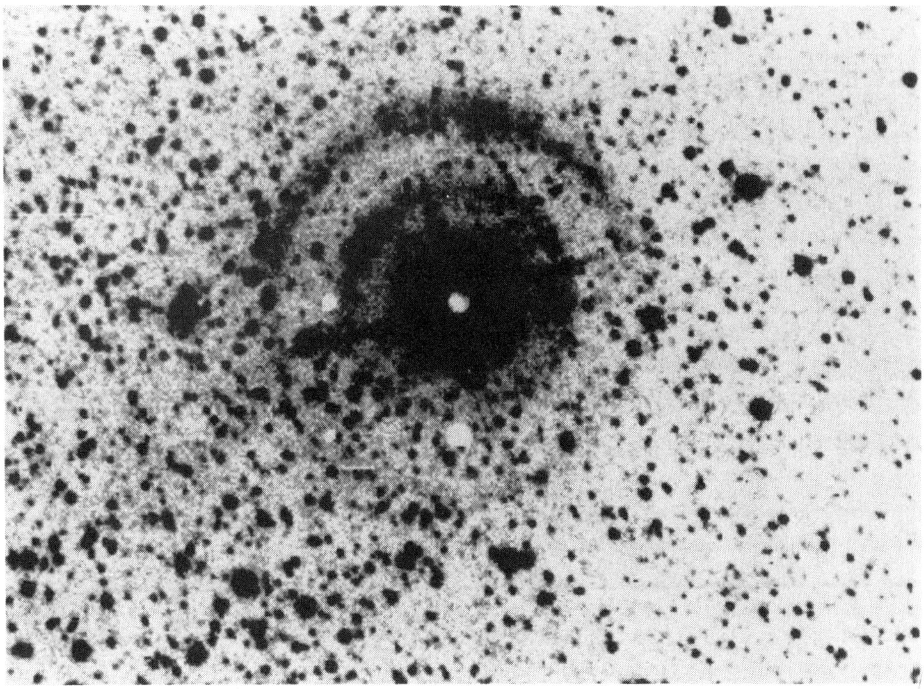

Fig. 1.9 — (a) A deep astronomical image obtained with a CCD camera of a double 'light echo' or reflection from interstellar clouds of the light from the bright supernova explosion of February 23, 1987 in the Large Magellanic Cloud. The light of the supernova was dimmed with a small obscuring disk, seen as a lighter spot in the centre, and the computer-enhanced picture is displayed as a negative. This CCD image was made by Dr Michael Rosa with the ESO 3.6-m telescope in Chile. Photo courtesy of the European Southern Observatory.

galaxy surveys improved with the introduction of the photographic plate and then remained fairly steady until the invention of the CCD.

1.6 AUTOMATION AND COMPUTERS

We will see that real CCDs, for all their many advantages, have idiosyncracies which must be dealt with and, moreover, it will become obvious that astronomers place rather stringent demands on the performance of CCDs and diligently persevere to find ways of meeting those demands! But CCDs are only a means to an end, and this book will also describe some of the ingenious astronomical instruments which incorporate them. It is the instrumentation attached to large modern telescopes which makes all the difference. More and more, these instruments are being engineered to exacting standards and they are inevitably computer-controlled, highly automated and very dependent on sophisticated software for control and data-taking. Microprocessor chips with over 100 000 transistors are used to control functions that were once performed manually or by bulky dedicated electronics.

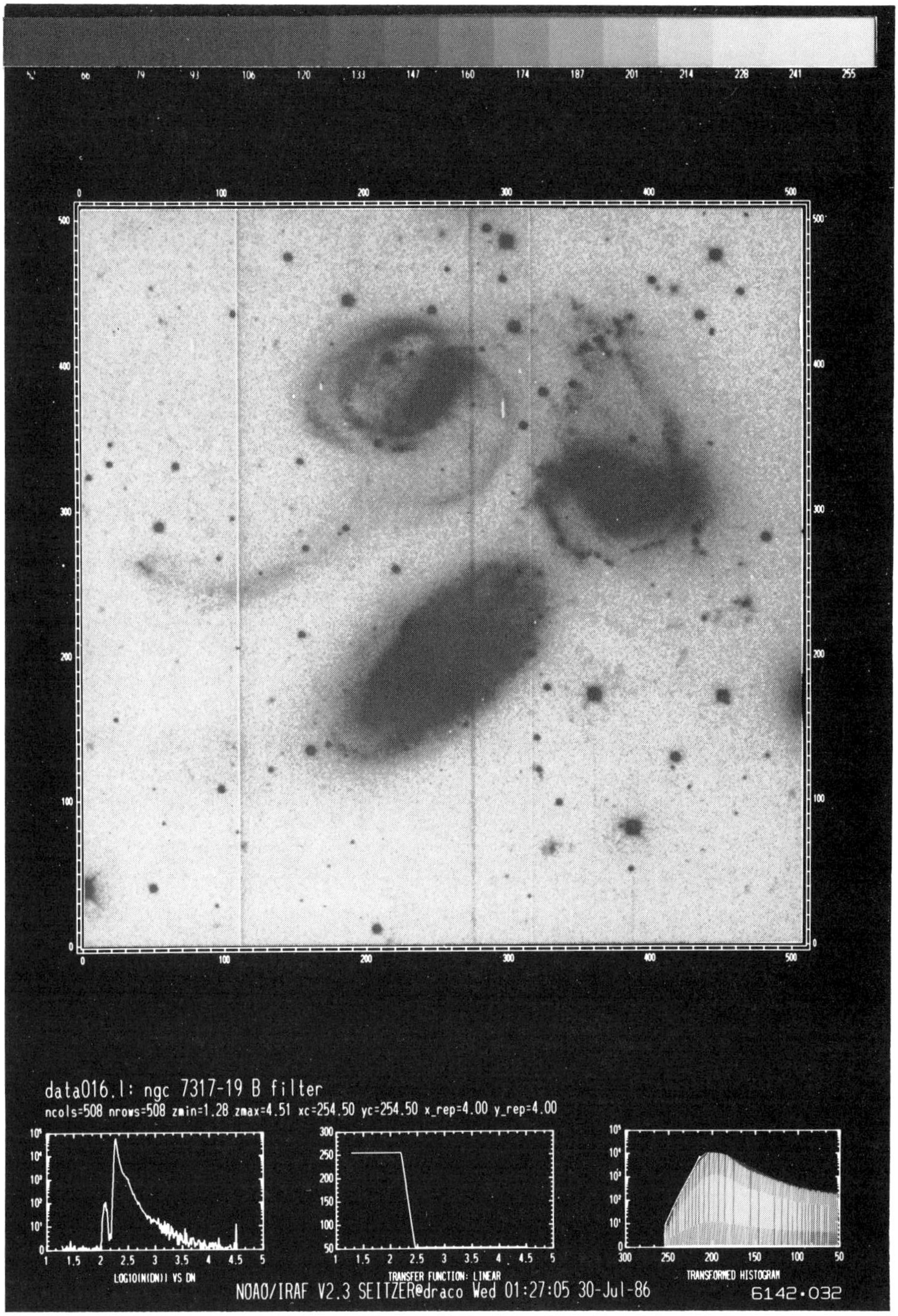

Fig. 1.9 — (b) A deep CCD image of a distant group of interacting galaxies known as NGC 7317-19 obtained by Dr Pat Seitzer of the NOAO in Arizona. Without the incredible brightness range of the CCD it would be virtually impossible to see the tenuous stellar arcs which link up these galaxies. Vertical lines are artifacts (see text).

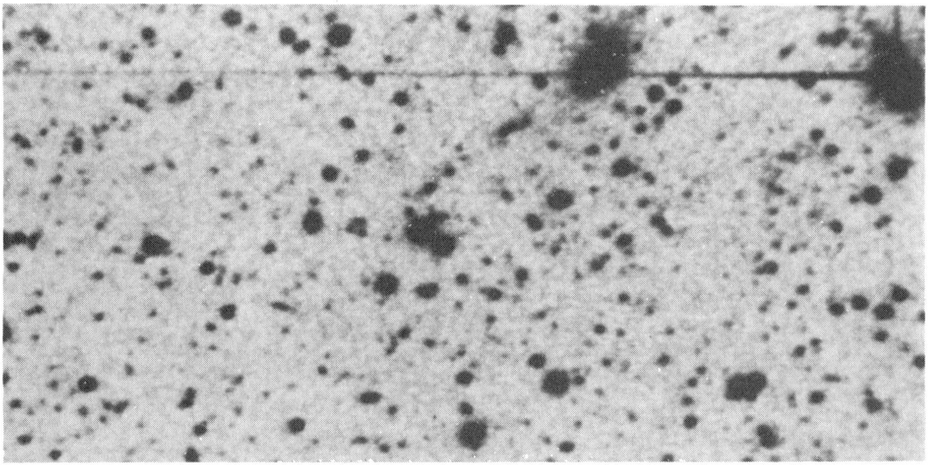

Fig. 1.9 — (c) A very long exposure CCD image (over five hours) showing numerous faint galaxies. The limit of detection is almost 29th magnitude. Courtesy Tony Tyson.

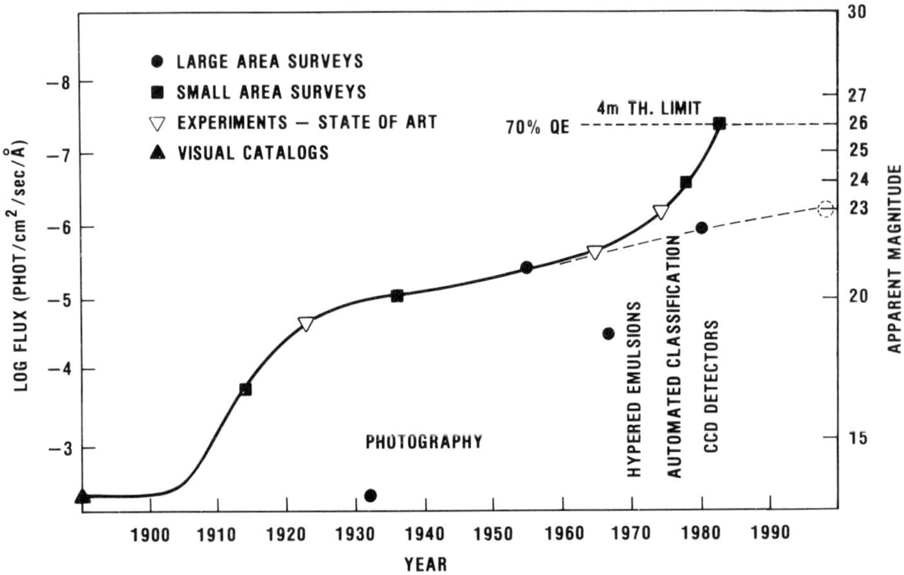

Fig. 1.10 — The history of the faint limit of galaxy photometry has been dominated by advances in detector technology. The diagram shows the faintness levels detected in optical surveys as a function of time. The lower dashed line charts the large-area photographic surveys by Schmidt telescopes, but it is the CCD which has pushed the efficiency per detector area of modern telescopes to their theoretical maximum. Courtesy Tony Tyson, Bell Labs.

Observational sequences can be pre-programmed and linked to the motion of the telescope, and in some cases the entire system can be operated remotely from the other side of the world. Of course, remote operation of satellites and giant radio telescopes has been with us since the late 1950s but it is really only in the last decade and a half that electronic and computer-aided astronomy has become the norm worldwide.

A major problem facing all of astronomy as a result of these technological innovations is **data management**. Ground-based optical and near-optical observational astronomy alone, to say nothing of space and radio astronomy, now produces data, mainly in the form of images or spectra, in such voluminous quantities as to make interactive analysis by human beings difficult.

A great reliance must be placed and is being placed on computers. Unfortunately, the burden of cost on astronomy is considerable and, due to the rapid pace of developments in the computer industry, it is hard to keep up. Some observatories are still under-equipped with suitable computer hardware, assorted peripheral equipment and sufficiently powerful computer programs or **software**. As a result, the 'turn-around' time in getting observations (data recorded at the telescope) into the form of published scientific results in a professional journal can be displeasingly long (one to two years). This need not be so! The same semiconductor technology which has given astronomy the new electronic imaging aids, has also given us incredibly powerful and compact computers easily capable of handling the CCD data, provided they are used optimally.

It is not a question of complexity, merely one of commitment. Astronomers are well able to specify the kind of computing systems needed and software engineers are well able to provide it. Already there is more international cooperation in the development, support and sharing of astronomical software. Getting the numbers needed from the experimental apparatus right there at the telescope requires what is known as **on-line** data reduction. This is becoming more feasible, and excellent proposals have been put forward for automated data analysis and data archiving. It seems highly probable that, provided the funding commitments are made, astronomers will not become drowned in a sea of magnetic and video discs!

REFERENCES AND SUGGESTIONS FOR ADDITIONAL READING

Amelio, G. F., Tompsett, M. F. and Smith, G. E. (1970) Experimental verification of the charge coupled device concept, *The Bell System Technical Journal*, Vol. 49, No. 4, 593–600.

Boyle, W. S. and Smith, G. E. (1970) Charge coupled semiconductor devices. *The Bell System Technical Journal*, Vol. 49, No. 4, 587–593.

Computers and the Cosmos (1988) by the editors of Time-Life Books, USA. (The present writer acted as a consultant.)

Gingerich, O. (ed.) (1975) *The nature of scientific discovery. Proceedings of the symposium in honour of Copernicus*. Smithsonian Institution Press, Washington.

Hearnshaw, J. B. (1986) *The analysis of starlight: one hundred and fifty years of astronomical spectroscopy*. Cambridge University Press, Cambridge, England.

Henbest, N. and Marten, M. (1983) *The new astronomy*. Cambridge University Press, Cambridge, England.

Krisciunas, K. (1988) *Astronomical centers of the world*. Cambridge University Press, Cambridge, England.

Learner, R. (1981) *Astronomy through the telescope*. Van Nostrand Reinhold Company, New York.

Snow, T. P. (1983) *The dynamic universe*. West Publishing Co., St Paul, Minnesota, USA.

Tyson, J. A. (1986) Low-light level charge-coupled device imaging in astronomy, *J. Opt Soc. Am.*, Vol. 3, No. 12, 2131–2138.

2

Modern astronomical instrumentation

We have reviewed the development of astronomy from eyes to electronic sensors, and we have followed a visiting astronomer through a night at a large national observatory where advanced, computer-controlled electronic imaging apparatus attached to the telescope provided the observational data for his/her research. In this chapter we will examine more closely the range of scientific instrumentation available to astronomers.

Broadly speaking, there are two classes of instruments used in astronomy, namely direct imaging cameras and spectrometers. Electronic cameras using CCDs are also referred to as 'area photometers' because their performance and stability enables accurate brightness measurements to be obtained over the entire area of view. All other instruments are basically derivatives of these two classes. For ease of discussion however, I have subdivided modern astronomical instrumentation into four groups, namely electronic cameras, spectrometers, polarimeters and 'other'. The last class contains specialised derivatives from the other three.

2.1 ELECTRONIC CAMERAS

Almost all astronomy, even the 'invisible' astronomy of X-rays and radio, depends on mapping in pictorial form for the human mind to visualise the pattern or distribution on the sky of celestial sources emitting radiation at the wavelength of interest. This is done not only to locate precisely the position of the source — a practice called *astrometry* — but also to provide information on its form or structure, and that of its environment, since much can be learned from the appearance or 'morphology' of certain objects. Images of the sky at different wavelengths are essential for classification of objects, and accurate brightness measurements at different wavelengths can also yield basic physical information. At optical and near-optical wavelengths the appropriate tool is some form of electronic camera.

2.1.1 Photography and computers

Although the classic photographic plate is not an electronic detector and cannot match the CCD for sensitivity, it would be wrong to ignore its utility in modern astronomy. Specially designed telescopes, called Schmidt telescopes after the optical designer who pioneered the technique, such as the 1.83-m UK Schmidt Telescope (now operated by the Anglo-Australian Observatory) at Siding Spring in Australia or the older 1.2-m Palomar Schmidt Telescope in California, have been constructed to accept very large glass-based plates about 14 inches on a side. These telescopes were designed to provide images of very large areas of sky (6.5×6.5 degrees). Recall that the full moon is only 0.5 degrees across. The product of the number of pixels and the quantum efficiency of a typical (hypersensitised) Schmidt plate such as Kodak IIIa-J emulsion is 1000 times better than that of a typical (512×512) silicon CCD simply because the CCD chip is so small. Therefore when area-coverage rather than sensitivity is the issue, photographic plates are extremely useful.

Having obtained one of these 14-inch plates containing tens of thousands of images, the next step is to find some way to extract useful statistical information such as the coordinates or positions of objects, their shapes, sizes and orientations, and of course, their magnitudes or brightnesses. It is at this point that even photographic cameras must rely on computers. At a number of establishments throughout the world special 'plate-measuring' machines have been constructed. In such a machine the original photographic plate is 'scanned' under computer control by a special beam of light and the transmitted signal — which depends on how densely the emulsion has been blackened — is detected by a photoelectric sensor to produce an electrical signal as shown as Fig. 2.1. This signal is converted to a numerical value by some electronics and stored in the memory of a computer — the plate is said to have been 'digitised'. Since these plates are so large, an immense amount of digital data is produced and stored or 'archived' on magnetic tape. The digital data can now be analysed by a high-speed computer to search automatically, for instance, for galaxies rather than stars.

There are several facilities of this kind, one of the largest being at the Royal Observatory, Edinburgh, where plates from the UK Schmidt Telescope are analysed. The Royal Observatory, Edinburgh, operates its plate-measuring facility, called COSMOS for Co-Ordinates Shapes Magnitudes Orientations Sizes, as a service and the computers which control the measurement of the plates and the analysis of the digital data are operated 24 hours per day. There are similar machines elsewhere. One of the most important applications of Schmidt telescopes is the search for relatively bright yet extremely distant quasars. The method involves placing a huge prism plate in the telescope to produce a spectrum of every object in the field; the resolution is sufficient to tell whether or not a bright redshifted emission line is present.

Inevitably, interesting features are found on Schmidt plates, such as candidates for redshift-five quasars — the most distant objects known — or, alternatively, new or peculiar objects may arise as the result of surveys at other wavelengths (X-rays, infrared or radio). In any case, 'follow-up' images of much smaller patches of the sky, but with 100 times more sensitivity, are required. Astronomers call this 'going deep'. To achieve fainter limits we must turn to much larger telescopes and to the more sensitive photoelectric imaging devices such as the CCD.

2.1.2 Optical photoelectric cameras

The most obvious application of the CCD is therefore in a straightforward 'camera' for direct imaging at the focus of a telescope. With the exception of the specialised

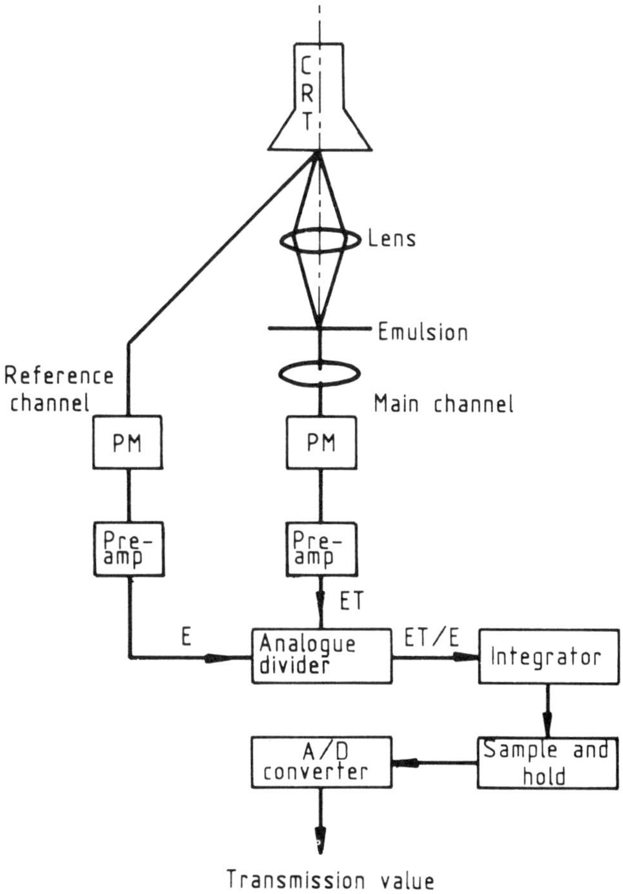

Fig. 2.1 — A schematic drawing of the computer-controlled plate scanning and detection system used by the COSMOS measuring machine to digitise images recorded on the large photographic emulsion plates used with Schmidt telescopes. Photo courtesy of Royal Observatory, Edinburgh.

Schmidt telescopes dedicated to wide-angle photography, CCD cameras have surpassed photographic plates for deep, high-precision imaging of selected objects or regions. The principal advantages of the CCD over photographic plates in this context are:

(a) very high sensitivity — a much larger fraction of the light falling on the CCD is converted to a measurable quantity, in this case, an electronic voltage;
(b) greater coverage of the spectrum — the silicon material of the CCD is sensitive to light of many different colours;
(c) immediate compatibility with computers — no developing or processing of film is required, the 'output' from a CCD is a stream of electronic pulses immediately suitable for transfer to a computer;

(d) instant display of the image.

Compared to TV-type cameras, CCDs offer the additional very important advantages of precision and stability which are essential for quantitative measurements. With care, calibrations of the 'response to light' of a CCD can remain valid for weeks!

Every major observatory in the world now possesses a CCD camera system of some sort. At most of the larger observatories several different kinds of CCDs are offered and CCDs feature in almost all forms of modern instrumentation. Many smaller astronomy groups have transportable or 'travelling' CCD systems. And CCD cameras have even reached the amateur astronomy community (*Sky and Telescope,* Jan. 1985, p. 71; April 1986, p. 407).

As Dr Jasper Wall, astronomer in charge of the United Kingdom's Isaac Newton Group of three telescopes on the island of La Palma where over seven CCD systems are used, has put it, '. . . over half our output, and about 90% of everything done with the 4.2-m William Herschel Telescope to date, is from CCDs. We live and breathe them . . .'.

At the Palomar Observatory, at least nine CCDs are used in four separate systems, including one of the very first prime focus CCD cameras called PFUEI — pronounced 'foo-ee' — for Prime Focus Universal Extragalactic Instrument, a quad-CCD camera called the 'four-shooter', and a double CCD spectrograph. In fact, CCD cameras are in use at all of the well-known large optical telescopes including the 3.9-m Anglo-Australian Telescope (AAT) at Siding Spring, the 3.8-m Mayall Telescope on Kitt Peak (Arizona), the 4-m telescope at Cerro-Tololo Interamerican Observatory in Chile and the 3.6-m Canada–France–Hawaii Telescope (CFHT) on Mauna Kea (Hawaii). An example of a CCD camera system is shown in the accompanying photograph (Fig. 2.2).

Numerous smaller telescopes in the 1-m to 3-m class have been astonishingly revitalised by the addition of CCD cameras or other compact, CCD-based instruments. For example, at the Lick Observatory on Mt Hamilton, Lloyd Robinson reports that over two-thirds of all the nights on their 1-m and 3-m telescopes are scheduled for CCD instruments; Lloyd even uses CCD cameras to replace normal TV tubes for acquiring and guiding on the object of interest (Fig. 2.3).

Most CCD cameras are conceptually straightforward in their design. Usually, CCD cameras will be found at the **prime** or **Cassegrain** foci of a telescope. Prime focus cameras have a larger field of view than Cassegrain cameras, typically about 5 arcminutes (or 300 arcseconds) on a side, and are ideal for faint objects, whereas Cassegrain focus cameras are usually designed to better sample the blurred image of a star (caused by the earth's atmosphere) called a 'seeing disk'. (To gauge this scale remember that the image of a star's seeing disk is about 1 arcsecond in size (at a good site) and the full moon is about 30 arcminutes or 1800 arcseconds across.) In each case the cameras include a rotatable wheel containing a variety of glass filters designed to isolate different regions of the spectrum of light (such as blue, green and red) coming from the object(s) being imaged. Following the filter wheel is a high-speed electrically operated shutter which is placed immediately in front of the transparent window of the small vacuum vessel containing the CCD chip (Fig. 2.4); the CCD is kept in a vacuum because it must be cooled to a very low temperature to

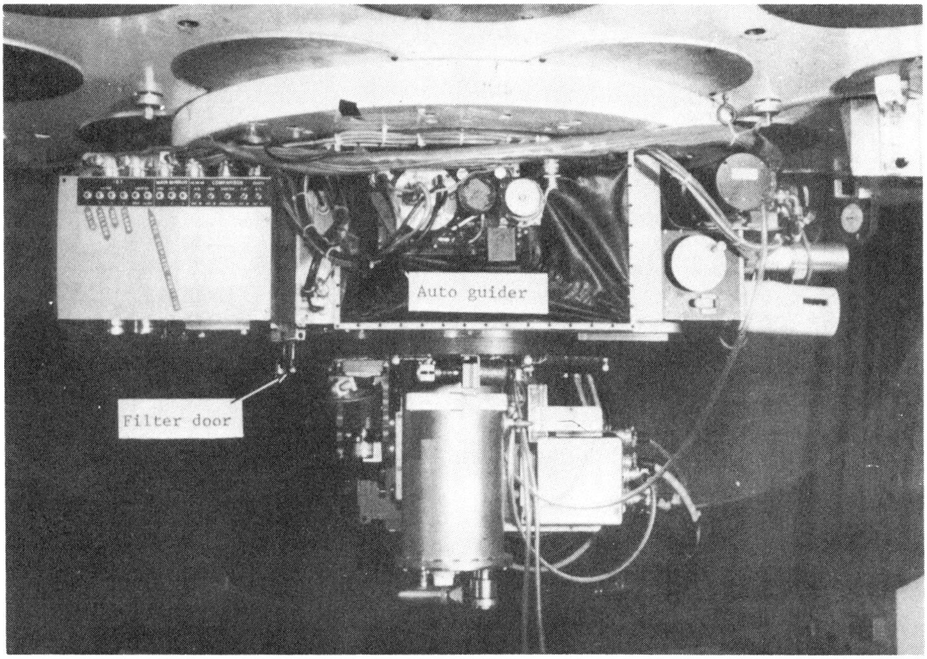

Fig. 2.2 — The CCD camera on the 2.1-m telescope of the NOAO on Kitt Peak. Photo courtesy of NOAO.

allow long exposures by eliminating a property known as **dark current** (section 5.2) and would otherwise become frosted by moisture in the air.

All CCD cameras require a box of electronics with circuits to operate the CCD (see again Fig. 2.4). One of these circuits is called an analogue-to-digital converter (ADC or A/D), and this is the device which converts the voltage signal from the CCD into a number which can be stored in the memory of a computer. Actually, the data are so voluminous that they must be stored on a separate memory device such as a magnetic disk, a magnetic tape or a laser disk. A typical night's imaging or spectroscopy run using a 1024 × 1024 pixel CCD will amass 100–200 megabytes (see Chapter 7) of digitised pictures. These numbers will almost certainly increase by a factor of 10 in the next few years. Recall that the actual physical memory of many modern, high-speed computers is usually about 16 Mbytes, and that each picture or 'frame' obtained with the CCD must be manipulated numerically by the computer to enable the astronomer to extract and display, as accurately as possible, the important astrophysical information.

At the time of its construction in 1980/81, Jim Gunn and Jim Westphal described the Palomar observatory instrument PFUEI as '. . . an order of magnitude [meaning 10 times] more efficient than the best photoelectric instruments . . .'. And there have even been some improvements since then such as the development of the quad-CCD camera or 'four-shooter'! PFUEI, although more or less obsolete now, reached a

Fig. 2.3 — The Lick Observatory CCD television guide camera developed by Lloyd Robinson. Photo courtesy of Lloyd Robinson.

'red' magnitude of $m=25$ (see Appendix 3 for explanation of magnitudes) in about an hour under moderately good 'seeing' conditions in which the image blur due to the atmosphere had an angular size of about 1 arcsecond in diameter. This result, about 100 times better than Schmidt plates, was considered remarkable at the time. The follow-up instrument to PFUEI was named the 'four-shooter' because it used four, rather than one, 800×800 Texas Instruments CCDs simultaneously. A schematic outline of the optical arrangement of the four-shooter is shown in Fig. 2.5(a) and an image of the Horsehead Nebula in Orion is shown in Fig. 2.5(b); the joins between the four CCD images are only barely perceptible. As a complete contrast, Fig. 2.6 shows another CCD image of the Horsehead Nebula but this time using a 30-s exposure on a 6-inch Celestron rather than the mighty 200-inch; the CCD camera is from Photometrics Limited.

Today, a few determined astronomers are within a factor of three or so of reaching the astonishing magnitude limit of $m=30$; one million million times fainter than the star Vega (Alpha Lyrae).

Fig. 2.7 shows just how important CCDs are by comparing a Schmidt photograph and a 4-m telescope prime focus photograph with a CCD image. On deep images of the sky away from the plane of the Milky Way, CCD exposures with large telescopes are now approaching what astronomers call 'confusion-limited'. That is, there is

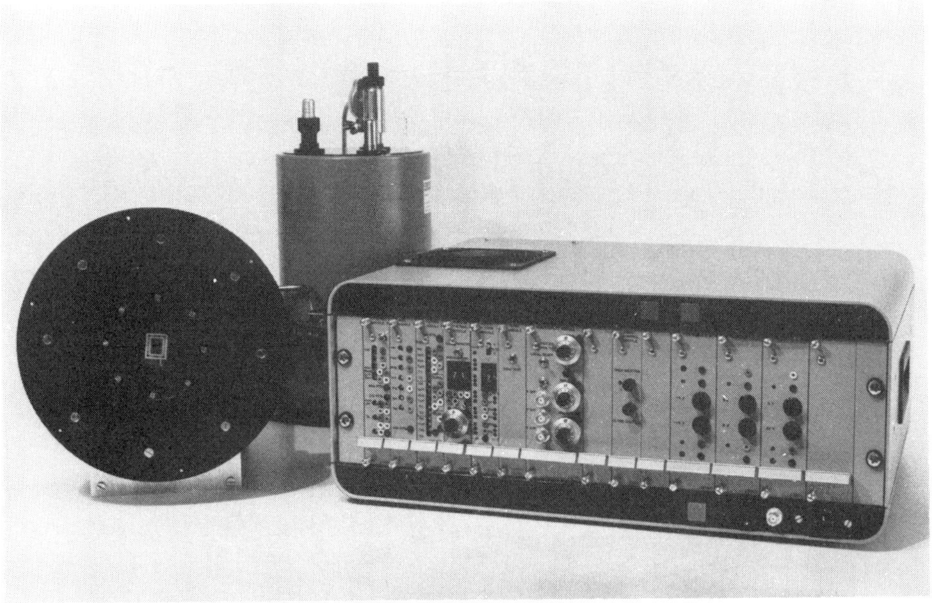

Fig. 2.4 — A liquid-nitrogen-cooled, slow-scan, low-noise CCD camera and electronics from Astromed Limited, England. Photo courtesy of Craig Mackay.

almost no true blank sky on these images; virtually every line of sight from the earth intercepts some incredibly faint, very distant galaxy (Fig. 2.7(c)). Fig. 2.8 on the other hand shows recent images from the 3.6-m CFHT on Mauna Kea, Hawaii — reputed to possess some of the finest seeing conditions in the world. One image shows the familiar planet Saturn with Cassini's Division — a narrow break in the ring structure — clearly visible and the other, a deep CCD image on an excellent night, is of three objects very close together (less than about 1 arcsecond separation). This triplet of images are in fact all of the same object — a distant quasar — but it appears multiple because the gravitational pull of an extremely faint, yet much more massive, intervening galaxy has bent the light rays in a manner similar to a lens; astronomers call this a **gravitational lens**. Such displacement of light waves by gravity is predicted by Einstein's General Theory of Relativity, and the phenomenon was first observed in 1919 during the solar eclipse with the Sun acting as the bending agent for starlight near the Sun's limb.

Plate 1 in the centre spread shows the field of the double quasi-stellar object known as 1146+111B,C obtained with the Bell Labs AT&T camera, developed by Tony Tyson, and used on the Lowell Observatory 1.1-m telescope. This is a modest-sized telescope and yet the faintest objects corresponded to a red magnitude of about 22 in an exposure of about only 30 minutes.

CCDs come in all shapes and sizes as we shall learn in later chapters. One type is the 640×1024 double density device from RCA in which the pixels are only 15 μm square. At the Cassegrain focus of the CFHT each pixel sees only 0.1 arcseconds! But this is ideal on a superb night on which the image blur is 0.5 arcseconds or better. An

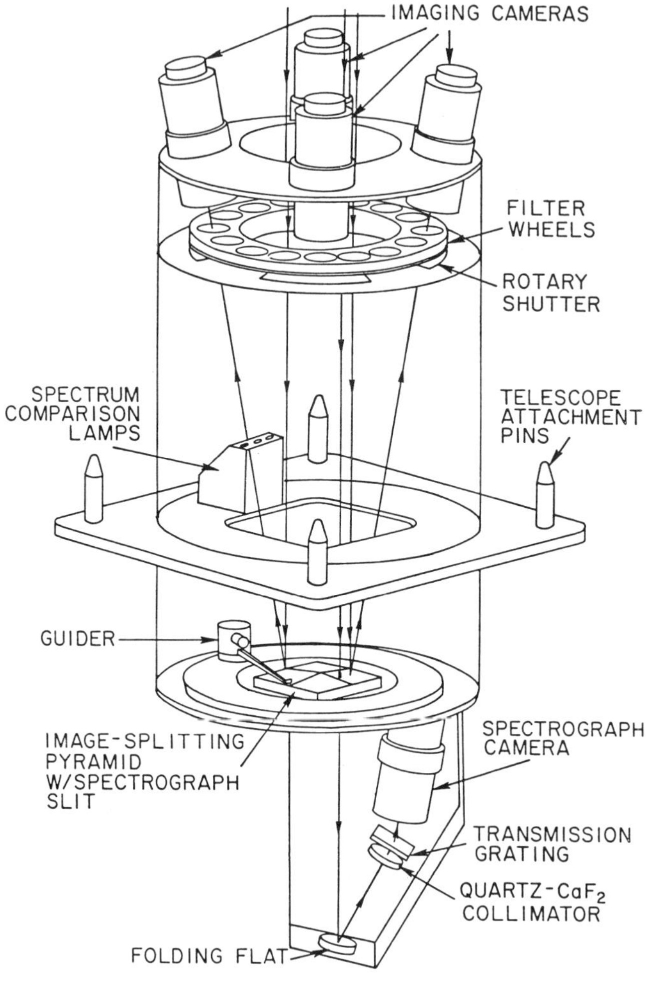

Fig. 2.5 — (a) A drawing of the 'four-shooter', the first quad-CCD camera plus spectrograph option in a single instrument. The four-shooter is used at the Cassegrain focus of the 5-m (200-inch) Hale Telescope on Mt Palomar. Courtesy of Jim Gunn.

amusing picture which demonstrates the quality of the CFHT camera with its RCA chip is shown in Fig. 2.9, entitled 'Fred near the edge of the Universe'.

2.1.3 Non-optical cameras

By 'non-optical' cameras we mean imaging devices working at **ultraviolet** or **infrared** wavelengths. Once again the CCD turns out to be extremely valuable. In Chapters 4 and 5 we will discuss in some detail the extensive wavelength coverage of the electromagnetic spectrum which silicon CCDs can span. CCDs are sensitive to radiation with wavelengths as small as those of X-rays. Unfortunately however, a basic atomic property of silicon prevents it form being sensitive to light with a

Fig. 2.5 — (b) A 2×2 mosaic of the Horsehead Nebula obtained with the four-shooter; shown as a negative. Courtesy of Jim Gunn.

wavelength longer than about 1.1 μm, i.e. 1.1 millionths of a metre; this region is the infrared. Happily, other semiconductors are sensitive in this region and so electronic infrared cameras very similar to optical CCD cameras are available to enable astronomers to observe the invisible 'heat' radiation from distant objects in the universe. Unlike X-ray and ultraviolet studies, the Earths atmosphere is sufficiently transparent to allow infrared astronomy to be pursued from ground-based observatories. Since 'heat'-sensitive cameras also have military applications, it has taken some time for this particular technology to become available for unrestricted use, but infrared astronomy is currently experiencing a 'boom' period.

One of the most versatile of the new infrared astronomical cameras, and the first to be offered as a facility to visiting astronomers, is the IRCAM. IRCAM — shown in Fig. 2.10 — is a computer-controlled infrared imaging system attached to the 3.8-m UK Infrared Telescope (UKIRT) on the 14 000 ft summit of Mauna Kea, Hawaii —

Fig. 2.6 — A CCD image of the Horsehead Nebula in Orion obtained with a Photometrics Limited camera on a 6-inch Celestron telescope. Courtesy of Richard Aikens.

the world's largest telescope dedicated full-time to infrared astronomy. IRCAM's solid-state infrared array detector uses a material called **indium antimonide (InSb)** to convert infrared light into electronic charges which are then digitised to produce pictures at wavelengths two to 10 times *longer* than achievable with ordinary CCDs. Since the obscuration caused by dusty material in the interstellar medium becomes extremely small at these longer wavelengths, a totally different view of the universe is obtained. Compare the optical and infrared images of the famous Orion Nebula shown in Fig. 2.11; the optical image is from a Lick Observatory photograph and the infrared image is from the UKIRT electronic camera. A more detailed discussion of the astounding impact that CCD-like infrared detectors have had on astronomy recently is deferred until to Chapter 9, but almost all of the discussion on electronic imaging with silicon CCDs in the intervening chapters will be relevant in developing an understanding of the basic physics of solid-state detectors and the practicalities of their operation and calibration.

2.2 SPECTROMETERS

When the electromagnetic energy collected by a telescope is finely subdivided by wavelength the result is a spectrum, and since the invention of the spectroscope by

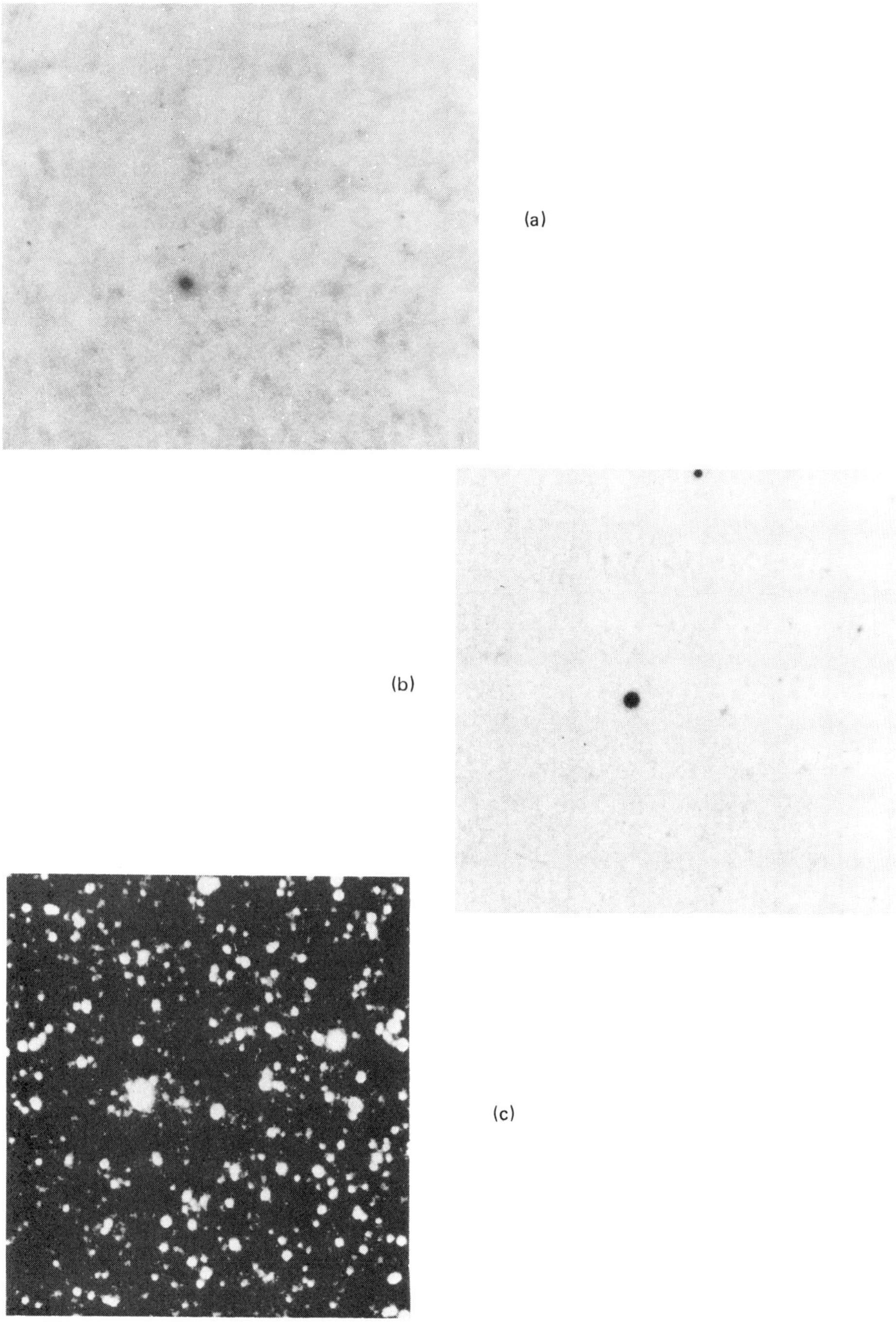

Fig. 2.7 — Three pictures of the same region of sky. (a) A Schmidt telescope photograph; (b) a photograph with a 4-m telescope; (c) a CCD image with a 4-m telescope. The CCD image — shown as a 'positive' — is virtually confused by so many faint objects! Courtesy Tony Tyson.

(a)

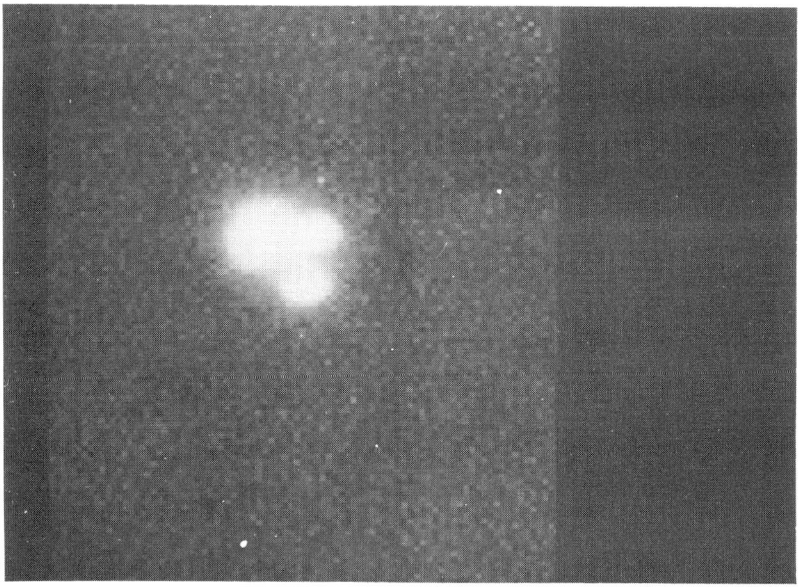

(b)

Fig. 2.8 — (a) A high-resolution image of the planet Saturn obtained with the CCD camera on the Canada–France–Hawaii Telescope (CFHT). (b) A CCD image obtained with the 3.6-m CFHT showing a triple-image gravitational lens effect on a distant quasar. Photos courtesy of Patrick Waddell and Carol Christian.

(a)

(b)

Fig. 2.9 — A test image obtained with the CFHT camera using a very high-resolution CCD from RCA. The picture is entitled 'Fred near the edge of the Universe'. Photo courtesy Patrick Waddell.

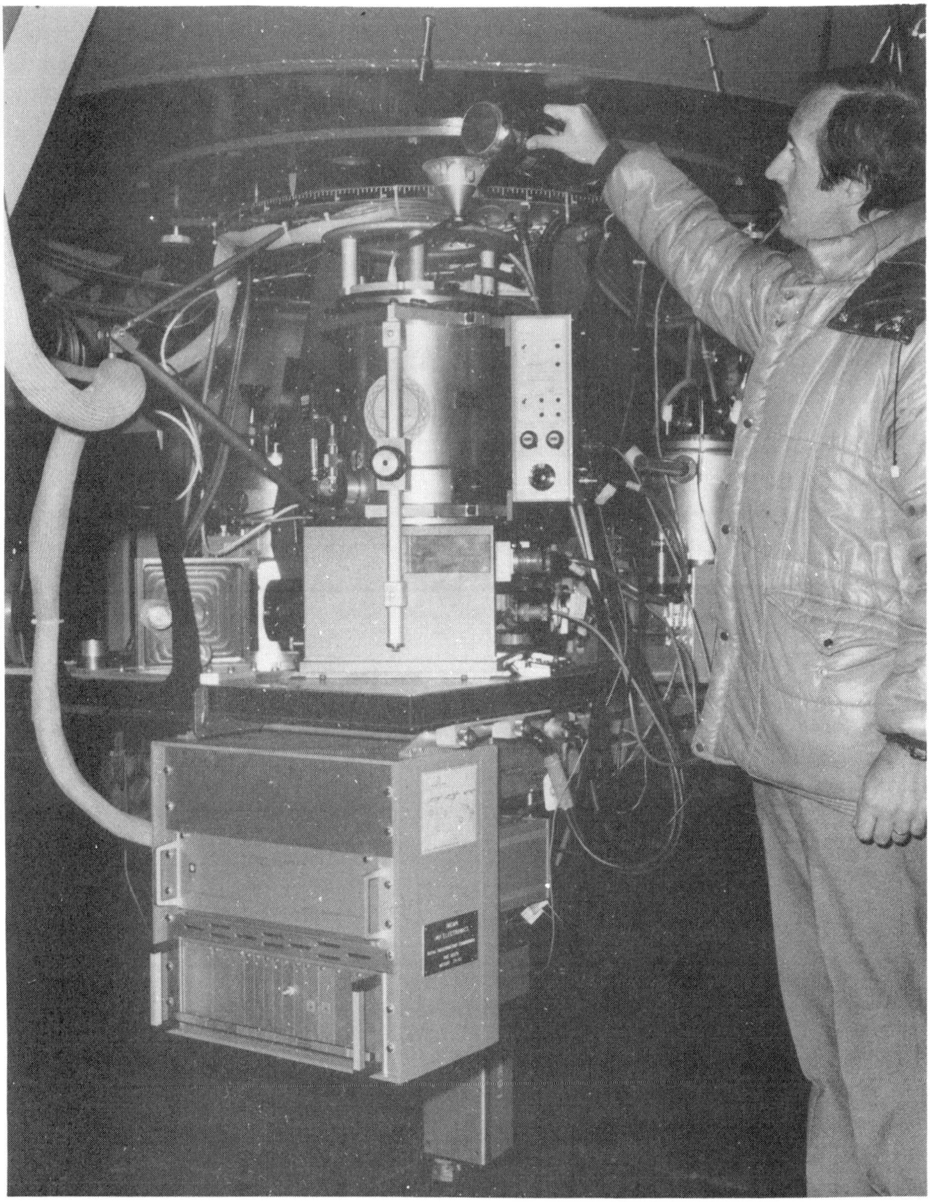

Fig. 2.10 — An electronic camera system called IRCAM, which is sensitive to infrared wavelengths beyond the limits of normal silicon CCDs. The entire instrument is operated at low temperatures and is a 'common-user' facility on the 3.8 m UKIRT on Mauna Kea, Hawaii.

Joseph Fraunhofer (see Chapter 1), astronomers have relied on spectrometers to provide them with fundamental physical information on the chemical constitution, temperatures, densities and velocities of objects in the universe. To record the

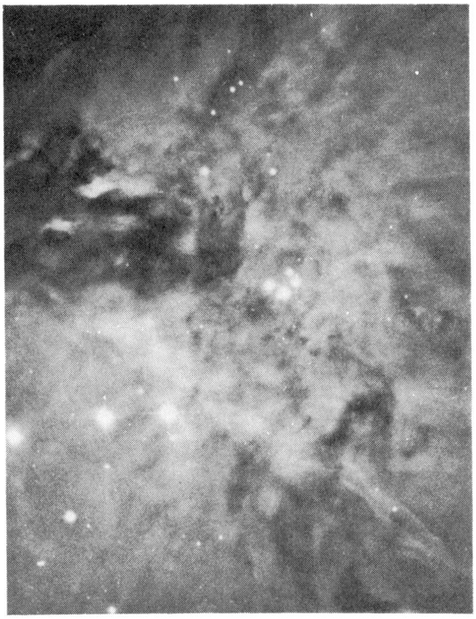

(a)

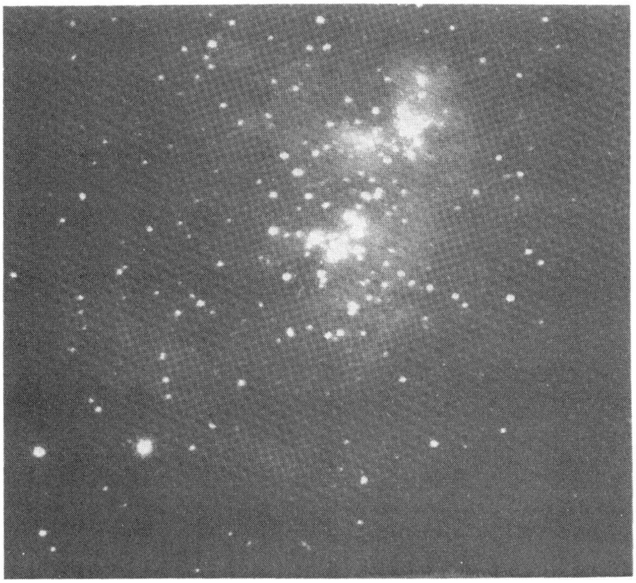

(b)

Fig. 2.11 — A comparison of an optical view and an infrared view of the Great Nebula in Orion. (a) is an optical photograph from Lick Observatory and (b) is a near infrared image obtained by IRCAM of the same region. Because the infrared radiation penetrates the dusty nebula more easily, many more stars *inside* the nebula have been discovered.

spectrum, almost all astronomical spectrometers now contain electronic imaging devices instead of photographic plates and most of these involve CCD detectors. In this context, CCDs are attractive because of their digital nature, their exceptional linear response and especially their good response to very red light since the metallic photoelectric surfaces (or photocathodes) used in photomultiplier-type detectors are less efficient at longer wavelengths. CCDs, because of their sensitivity, are particularly important for spectrographs designed for observations of faint objects. For fainter objects it is often neither necessary nor practical to split or 'disperse' the spectrum into such fine detail; such spectrometers are said to have 'lower resolution'.

Spectrometers — also called spectrographs — fall into several classes depending on the spectral resolution or fine detail to be achieved in the spectrum. These classes are

- faint-object spectrographs — low resolution,
- intermediate dispersion spectrographs,
- high-resolution spectrographs,
- imaging spectrometers.

The more resolution required in the spectrum, the less light there will be on any given pixel of the detector. Virtually all spectrographs which disperse or spread the incident light beam into a spectrum use a **diffraction grating** rather than a glass prism. The principle of the diffraction grating, also pioneered by Joseph Fraunhofer, depends on the wave nature of light. Thousands of very fine parallel lines or grooves are cut in an optical substrate and, because the lines are so narrow, reflected or transmitted light waves 'interfere' to produce a spectrum. The more lines per unit length the better the resolution or detail in the spectrum. But the more the light is spread out the less bright any part of the spectrum becomes, and the more difficult it is for a detector to make an accurate measurement of that brightness. Diffraction gratings can be made quite large, typically 152×254 mm (6×10 in).

2.2.1 Single-object spectrographs

For the highest dispersion spectroscopy on the faintest objects, especially in the blue/violet region of the optical spectrum, there is still a preference for true photon-counting systems such as the Boksenberg IPCS. In the Boksenberg IPCS, the images fall on the photocathode of a high-voltage image intensifier system the output of which is a large phosphor screen. The screen is scanned by a high-quality TV camera which locates the exact centre (or centroid) of the pulse of light and stores this as a count in the correct location in an electronic memory. If another pulse appears before the phosphor screen is scanned again or is coincident with the first then the count is lost, i.e. this kind of photon-counting detector saturates easily and is intended for exceedingly faint sources. The IPCS has no extra electronic noise corresponding to the so-called 'readout noise' of the CCD and is therefore better (more efficient) than the CCD at high dispersion because readout noise is particularly serious when the light collected by the telescope is greatly 'stretched-out' to form a spectrum with very fine detail. In other words, at high dispersion the recorded signal from the astronomical source has poor contrast against the unwanted signals (or noise) associated with the detector itself. Recently, however, very low-noise CCDs with special blue-sensitive coatings (see Chapter 4) have become very competitive with the IPCS, and even the superb Plumbicon scanning TV camera in the IPCS has now been replaced with a CCD detector!

Intermediate dispersion spectrographs are the workhorses of astronomy and are in use all over the world employing blue- and red-sensitive CCDs; sometimes *both* types are utilised in the same instrument. A major fraction of the 'dark time', that is the period of the month around new moon when the sky is darkest, on large telescopes is devoted to obtaining spectra of faint objects to derive their chemical composition, temperatures and densities, and velocities. Many of these objects are of such low brightness that the light of the darkest night sky still dominates the raw picture. Once again, computer processing is required to measure and remove the signature of the background sky from the spectrum of the faint object. This correction is done numerically, after the fact, by taking suitable calibration measurements and it relies greatly on the superb properties of CCD detectors, such as stability of response, linearity and dynamic range — all of which we will discuss in detail later.

An innovative 'double' spectrograph was developed largely by Bev Oke in 1982 for the 5-m Hale Telescope at Palomar. He noted recently that this instrument, which covers the entire available wavelength interval from the ultraviolet at 0.32 μm to the near infrared at 1.00 μm, was used on more than one-third of all nights, that CCDs were used on the 'blue' and 'red' sides 85% of the time and a pulse-counting camera was used on the blue side with a CCD on the red side 15% of the time. Fig. 2.12 shows the blue spectrum of the quasar 1331+170. A trace like this one is produced easily from the raw CCD image of the spectrum by a computer program which can examine and plot the digital data, held as an array of numbers on computer disk memory, corresponding to a row of pixels in the CCD image. The strong emission feature protruding upwards is actually due to triply ionised carbon atoms and occurs at a wavelength of 1550 ångströms (Å) (or 0.155 μm) in the far ultraviolet; it is 'redshifted' by the **Doppler effect** of the quasar's very high velocity into the normal optically visible part of the spectrum.

An even more complex 'triple' spectrograph called ISIS (Fig. 2.13) is undergoing evaluation for the 4.2-m William Herschel Telescope. The ISIS is essentially two intermediate spectrographs (for red and blue wavelengths) but has a quick-change mode to a 'built-in' faint-object low-resolution CCD spectrograph.

The 'faint-object' class of spectrographs is basically a CCD camera capable of imaging a spectrum of a faint object with relatively low resolution of the spectral detail so as not to 'spread out' the available light too much. The less the spectrum is dispersed, the more light there will be on any given pixel of the CCD and therefore the more accurate and rapid the measurment will be. Obviously a compromise is needed which ensures that enough spectral resolution is retained otherwise features or lines in the spectrum which could yield the velocity of the object, for example, will be hopelessly smeared out.

Although designs vary slightly, the usual approach is to use a fixed arrangement like that in Fig. 2.14. Optically, the light rays diverging from the telescope focus pass through a transmission diffraction grating *and* a prism; the prism is oriented to spread or disperse light at right angles to the grating. A Schmidt camera — an optical arrangement of a spherically curved mirror with a special thin lens or corrector-plate — images the spectrum onto the CCD. The entire system is in a vacuum. This

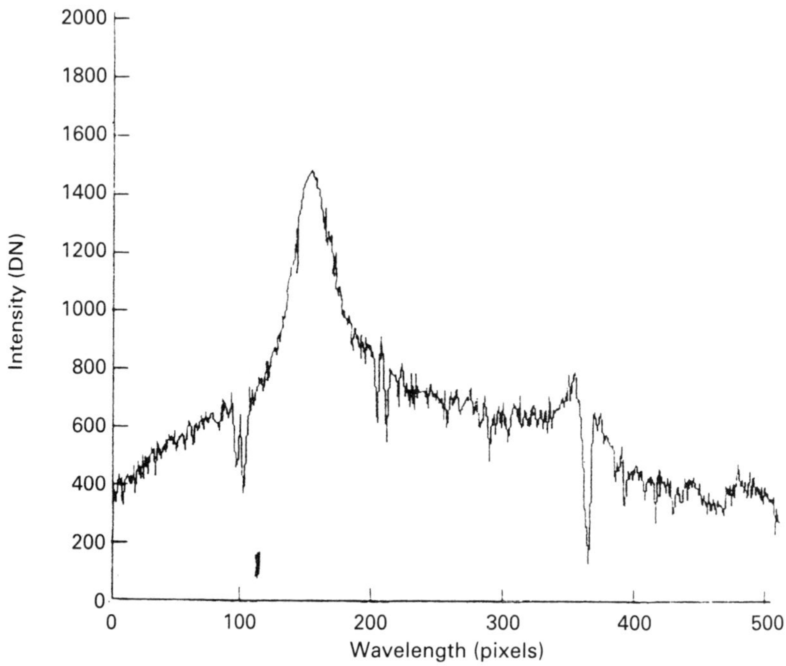

Fig. 2.12 — A blue spectrum of the quasar 1331+170. This drawing is performed by the computer which examines the CCD image of the spectrum and plots the brightness against the pixel number or wavelength along the spectrum.

combination is three times more efficient than the equivalent intermediate dispersion spectrographs. An instrument of this sort is particularly appropriate for the study of objects like quasars which have prominent broad emission lines in their spectra which stand out above the underlying and much fainter general or continuum spectrum. Another use of this kind of spectrograph which demonstrates the range of brightness which CCDs can handle is the study of supernovae, in particular SN1987a which made news in February 1987. Fig. 2.15 shows the spectrum of this object obtained with a red-sensitive faint-object spectrograph called FORS on the 3.9-m Anglo-Australian Telescope; the CCD was made in England by EEV/GEC. Of course, near maximum light, SN1987a was visible to the naked eye and not a 'faint object' at all, but as the years pass it will fade to obscurity and yet the CCD faint-object spectrograph will be able to follow its progress.

2.2.2 Multi-object spectrographs and optical fibres

Spectroscopic measurements can be painfully slow even with very sensitive CCD detectors. This is because the accuracy of the measurement depends on the amount of light (the number of photons) falling on the CCD pixel and, as already stated, the greater the spectral resolution the fewer the number of photons on any pixel. Typically, the signal at any point in the spectrum of an astronomical source will be several hundred times fainter than the signal in a direct image of the source obtained

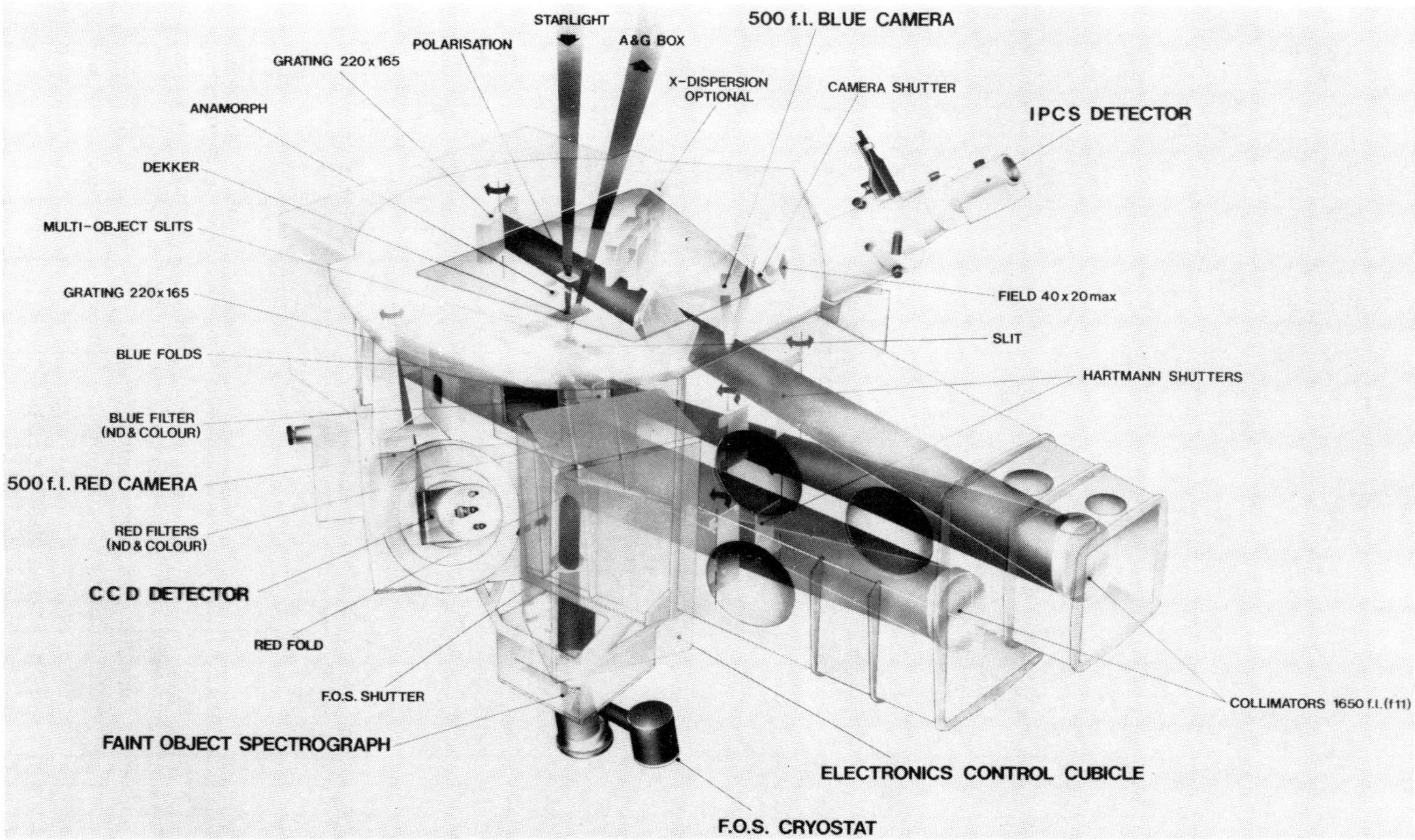

Fig. 2.13 — ISIS: an advanced highly automated spectrograph for the 4.2-m alt–az William Herschel Telescope. ISIS is capable of using CCDs and image photon-counting systems. Courtesy Royal Greenwich Observatory.

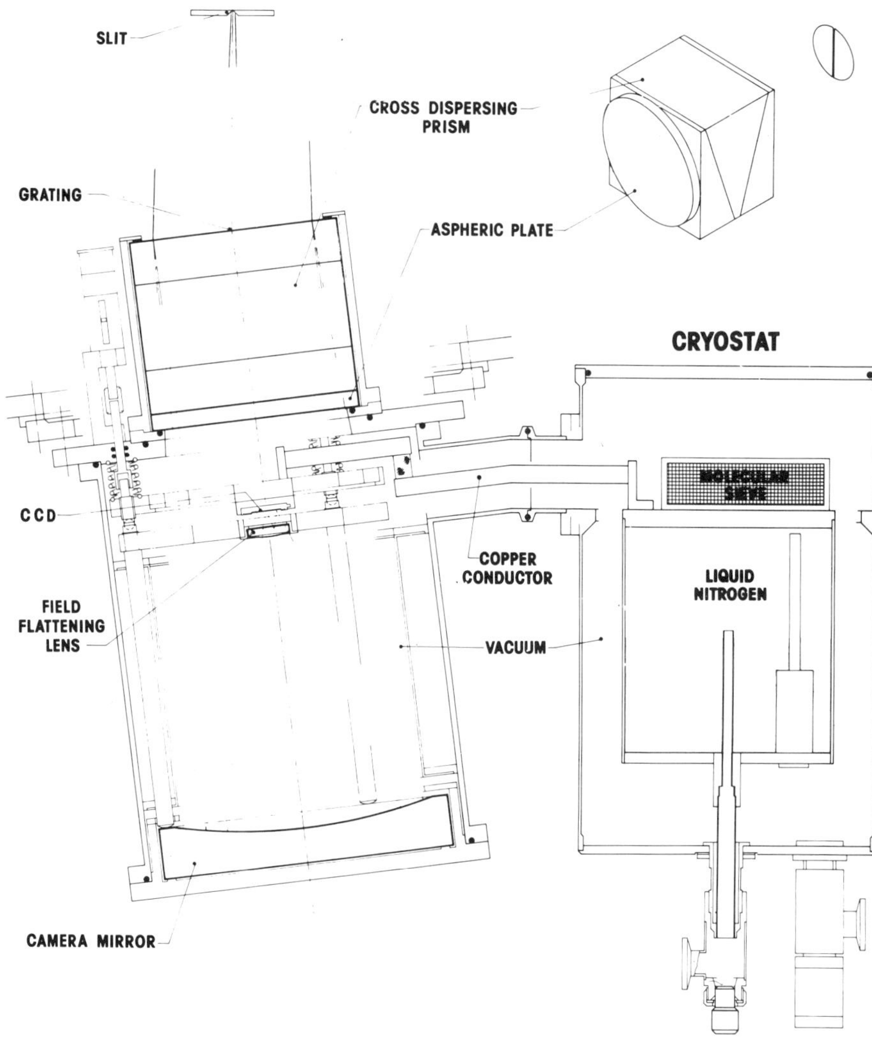

Fig. 2.14 — A drawing showing the basic design of the new generation of faint-object spectrographs (FOSs) which employ a Schmidt camera and corrector plate, together with a CCD detector. The one shown here was designed by Charles Wynne. Courtesy Royal Greenwich Observatory.

with a CCD camera. Thus the possibility of recording spectra from several objects at once is very attractive. One approach to this problem is the multi-object spectrograph which employs an entrance slit composed of multiple subsections which can be positioned by computer to pick up many different objects in the field of view.

Of even greater impact on multi-object spectroscopy has been the new technology of transparent **optical fibres**. Developed mainly for the telecommunications industry, these slim, flexible glass conduits — which resemble normal electrical

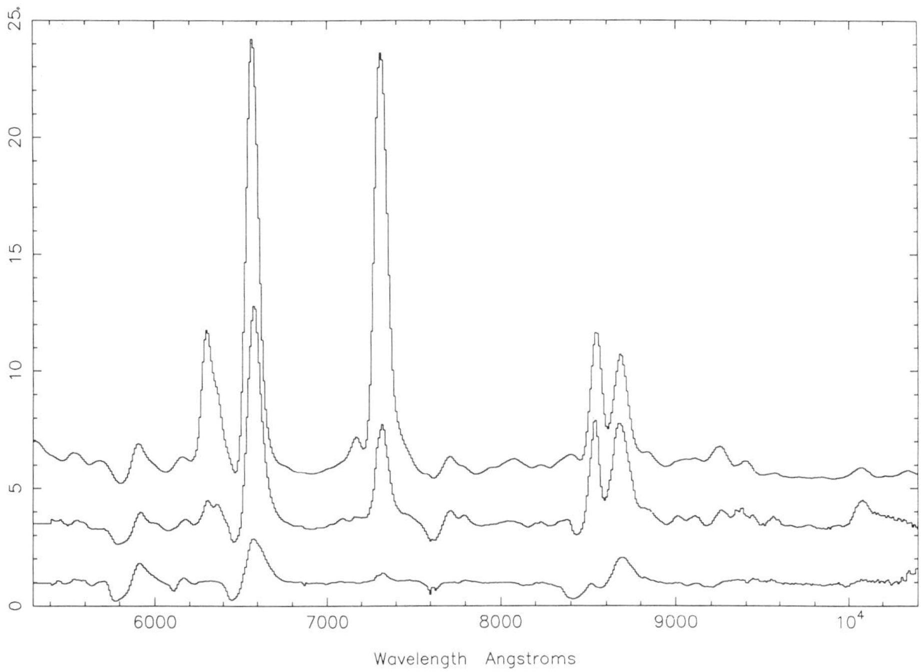

Fig. 2.15 — Plots of the red part of the spectrum of supernova SN1987a taken with the AAT and a faint-object CCD spectrograph. Note the strengthening of emission between July 1987 (lower spectrum) and October 1987 (upper spectrum).

cables — can be used as 'light-pipes' to transmit light over very long distances with only slight losses or attenuation. The basic idea is to position one end of each of several optical fibres in the focal plane of the telescope at points corresponding to interesting objects, such as a cluster of distant galaxies, and to stack up the other ends along the entrance slit of the spectrograph; the spectrum of each source is therefore recorded simultaneously. A multi-fibre spectrograph-feed can be mechanically complex; devices allowing simultaneous acquisition of 30 to 40 spectra have been described and plans for up to 300 simultaneous spectra are proposed for the largest CCDs, such as the Tektronix 2048×2048 device. The first of the fibre-optic-coupled systems — called appropriately enough MEDUSA — was developed at the Steward Observatory in Arizona by John Hill, Roger Angel and coworkers in 1979. Such systems are now fairly common on telescopes of sizes 1 m to 4 m all over the world (see Fig. 2.16), and have drastically transformed many branches of astronomy where statistical properties on many sources are required. More than 50% of all spectroscopic time on the AAT is now multi-object in nature according to Richard Ellis of Durham University, England, who listed 18 distinct observing programmes carried out on the AAT between 1984 and 1987 which used optical fibres. For instance, a complete redshift(distance)-limited survey of quasars and faint galaxies becomes viable thus affording us the opportunity to visualise accurately the universe in three dimensions.

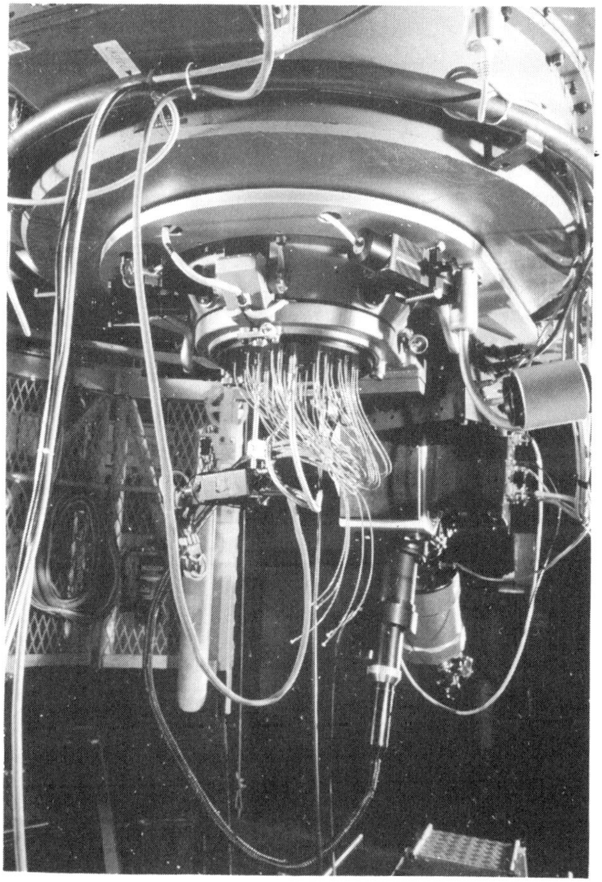

Fig. 2.16 — The photograph shows the fibre-optic spectrograph instrument, Octopus, of the European Southern Observatory. The fibres are led away from the focus to the spectrograph. Courtesy European Southern Observatory.

All of the earlier fibre-positioning schemes were manual 'plug-plate' systems in which an aperture plate is prepared in advance from the astronomer's target list using a milling machine, and then each fibre is inserted by hand — paying careful attention to which fibre goes to which hole! Newer, more automated systems have now been developed such as the Steward Observatory 'MX' system and the Durham/AAT AUTOFIB system. Obviously, these new 'robot-like' systems depend on a control computer to do all the accurate positioning and keep all the necessary logbooks — something which computers do very well.

Computer-controlled optical-fibre multiple spectrographs with CCD detectors have resulted in more consistent sky subtraction and greatly reduced running costs and preparation time.

The optical fibre innovation has also revitalised the photographic Schmidt telescopes in a novel way. At the Anglo-Australian (former UK) Schmidt Telescope in Australia, Fred Watson of the Royal Observatory, Edinburgh has pioneered an instrument called FLAIR which is a fibre optic system which allows simultaneous spectroscopic observations of 40 objects scattered over the huge 6.5×6.5 degree field of view; FLAIR stands for Fibre-Linked Array-Image Reformatter. The location of the fibre ends with the target objects is achieved with a specially made 1-mm-thick positive (rather than the usual negative) glass plate of the field to which the fibres are glued with an ultraviolet-curing optical cement. The bundle of fibre cables is led away from the telescope, stacked one above the other in a vertical line and fed into the slit of a standard grating spectrometer which is set up on an optical bench on the dome floor (Fig. 2.17); the spectrometer is therefore completely stationary at all times which greatly enhances the stability of the apparatus. A cooled CCD camera is used to take images of these spectra.

2.2.3 The Fabry–Perot interferometer

Even the multi-object spectrograph suffers from the fact that in order to record the dispersed spectrum of any object in the field of view of the telescope the remainder of the field, which is most of it, must be masked from the spectrometer; an image of the scene cannot be recorded with high spectral resolution unless the measurements are repeated many times with slight displacements of the telescope. For spectroscopic studies of extended objects such as galaxies and nebulae this is a significant disadvantage.

A fundamentally different type of spectrometer which can address this problem is the Fabry–Perot interferometer. Developed initially by Charles Fabry and Albert Perot as long ago as 1896, it remained dormant until the 1950s because it was considered too difficult to manufacture and use. Technological improvements in fabrication techniques were stimulated by the realisation that the Fabry–Perot 'etalon' — essentially a cavity formed by two face-to-face circular plates of high-reflectivity, low-absorption surfaces held parallel and flat to a tiny fraction (typically 1/200) of the wavelength of light, i.e. a mere 2.5 nanometres (nm) in the mid-visible — has a much greater efficiency or **throughput** for a given spectral resolution than the normally-used diffraction grating. (The throughput of a spectrometer is the 'solid angle' it can accept multiplied by its 'aperture'.) As the spacing or gap between the pair of reflective surfaces is changed, so the wavelength transmitted by the etalon changes. The beauty of the Fabry–Perot etalon lies in the fact that it behaves essentially like an ordinary filter and can be placed directly into the beam of an imaging system to yield a picture of the entire field of view — rather than just a small patch — having the spectral purity and detail that could otherwise only be obtained by a fairly powerful spectrometer.

The Fabry–Perot (or FP for short) can be 'tuned' to a series of different wavelengths by careful adjustment of the gap between the reflective plates. This can be accomplished under computer control and a whole series of images can be stacked up in computer memory corresponding to slightly different wavelengths or, equivalently, slightly different velocities. The resulting collection of images is known in the jargon as a 'data cube' because it has three 'dimensions', namely the ordinary two spatial dimensions of a picture, and a third dimension which is the wavelength at which the picture was recorded. A plot of the signal values along the wavelength dimension for any spatial position yields the spectrum at that point in the source.

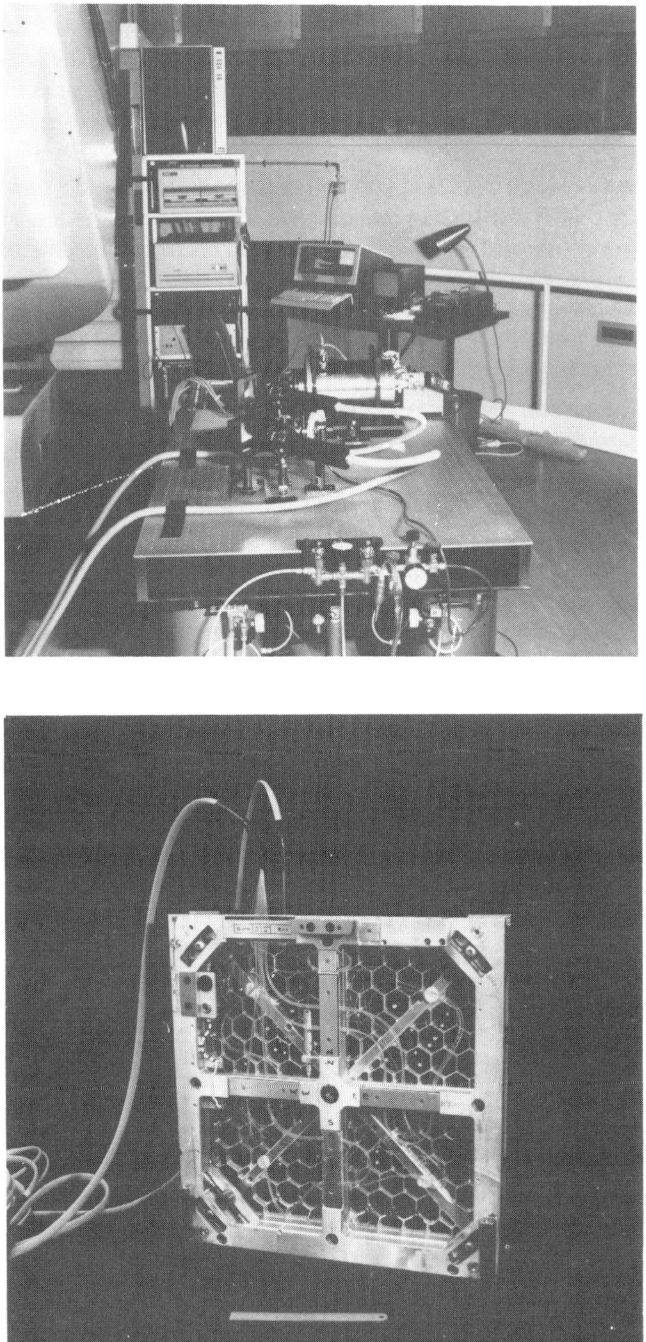

Fig. 2.17 — A CCD spectrograph with a fibre-optical system called FLAIR; the fibres are attached to where the 14-inch photographic plate would be in the wide-field UK Schmidt Telescope. Courtesy Fred Watson.

Controlling a CCD imaging spectrometer employing a Fabry–Perot interferometer, and manipulating the resulting data cube would be virtually impossible without a computer. A 'velocity map' of a galaxy (CenA) was obtained in the light emitted by hot hydrogen gas. The original data cube was obtained on the AAT using a Fabry–Perot instrument called TAURUS devloped by Keith Taylor and Paul Atherton; the etalons were made by Queensgate Instruments (Atherton and Hicks) in England and use very small position sensors whose electrical 'capacitance' is employed to monitor parallelism and spacing of the reflective plates. An electronic control system uses these signals to drive tiny piezoelectric actuators — devices which produce a small movement in response to an electric voltage — to maintain perfect alignment.

Several instruments of this kind now exist for optical astronomy, and very recently the author and his associates have developed a first system for use at near-infrared wavelengths (see Chapter 9).

2.3 POLARIMETERS

CCD cameras constitute good imaging detectors for almost any kind of instrument. Having constructed camera systems to collect light in the form of an image for brightness and position measurements, and instruments for measurements of the distribution of light across the spectrum, it is not difficult to construct a CCD camera system which can determine the polarisation state as well.

2.3.1 Polarised light and its measurement

Polarisation is the term used to describe the phenomenon in which a fraction (or all) of the electromagnetic waves in a beam of light are vibrating in the same plane, such as the vertical or horizontal planes (or any orientation in-between) along the line of sight to the source. If there is no preferred orientation and the beam contains electromagnetic waves having a random jumble of all planes of vibration then the beam is said to be unpolarised. When there is a preferred plane of vibration which does not change no matter at which point in the wave cycle we sample, the beam is said to be **linearly polarised**. On the other hand, if the plane of vibration rotates by 360 degrees through a wave cycle then the beam is said to be **circularly polarised**. The light emitted by an ordinary tungsten lamp is essentially unpolarised, whereas the scattered sunlight which constitutes 'blue sky' is very strongly linearly polarised.

Many important astrophysical phenomena produce polarisation either through the interaction of unpolarised light with matter, or by the generation of polarised light by the atoms themselves. For example, reflection, scattering and absorption by certain materials in the interstellar medium will cause polarisation; atoms emitting light in the presence of a magnetic field (such as accompany sunspots) will suffer the Zeeman effect and the emitted radiation will be polarised; very high energy (relativistic) electrons spiralling in a magnetic field around a neutron star will emit polarised light called synchrotron radiation. Polarisation spectra and polarisation images therefore contain additional information about physical processes and source geometry which cannot be discerned from brightness measurements alone.

Polarimetry depends on *relative* brightness measurements, on the same source, at different 'settings' of a polarisation-sensitive element — often called a **polarisation**

modulator. The modulator can be a rotatable plate of a special optical material (or composite material such as quartz and magnesium fluoride) which exhibits a crystal structure property called **birefringence** which makes the material sensitive to the orientation or plane of vibration of the incident light wave. Essentially, the crystal will distinguish between electromagnetic waves which are vibrating at right angles to one another, and will 'slow-down' one of them. Alternatively, the modulator can be made of a material in which birefringence can be introduced by the application of an alternating electric field — this is called a Pockels cell. Another method of inducing birefringence in certain crystals is by a variable mechanical strain — this is the photoelastic or Kemp modulator. In each case the modulator is followed by a polariser (like Polaroid but usually a glass component rather than a plastic component to improve transmission). Operation of the modulator/polariser combination results in a controlled, periodic variation of brightness of the transmitted light provided the incident light was polarised. These brightness variations are recorded by the CCD camera.

Conceptually, all polarimeters fall into this basic design structure; the light from the telescope must first pass through the polarimeter section comprising the modulator and polariser, then through the camera or spectrometer part to arrive at the CCD detector.

2.3.2 CCD polarimeters

Because of the systematic but unpredictable intensity variations in the optical scene being observed due to poor seeing or poor tracking, it is essential that the rate at which one moves from one setting of the polarimeter module to another — the 'chop rate' — be fairly rapid. Or, alternatively, some means must be found to measure two polarisation positions simultaneously so that both are affected in the same way by any systematic errors. Typically, for relatively faint objects, this chop rate is about 1 per second or a little slower. Even at these rates, however, the small but not quite zero noise of the CCD would degrade the image quality and degrade the detectability of faint sources. Ideally, it would be best to build up the alternating (polarisation) signal on the chip until enough signal was accumulated so that the readout noise would be negligible.

This idea, first mooted to me by Pete Stockman in an imaging context using Pockels cells when we were both at Steward Observatory in Arizona in 1978/79, led me to the idea of developing an instrument which I dubbed the imaging spectropolarimeter (or ISP) based on the unique properties of a three-phase CCD detector.

The ISP, shown in Fig. 2.18 was constructed at the Royal Observatory, Edinburgh, and was an early breed of the general purpose, fully automated, computer-controlled, instruments which were emerging as the result of the new awareness of the power and advantages of electronic and computer-aided methods. Although basically a CCD camera, the ISP could be converted to a spectrometer by placing a 'grism' — a small right-angled prism with a transmission grating deposited on the hypotenuse face — in one of the filter positions, and converted to a polarimeter (imaging or spectro-) by inserting a polarisation modulator in front of the entire optical system. In the spectropolarimetry mode, a special polarizer made of calcite was placed under the slit of the spectrograph which produced two oppositely polarised spectra on the CCD; actually, two tiny slits were used, one for the object

Fig. 2.18 — An automated computer-controlled instrument for obtaining polarisation information from images and spectra called the Imaging Spectropolarimeter or ISP, developed at the Royal Observatory, Edinburgh, by the author. The detector is a CCD camera.

and one for the sky, and so four spectra appeared on the CCD. The three-phase structure of the CCD — described and explained in Chapter 4 — allows so-called 'bidirectional' charge-transfer, therefore alternating images (or spectra) corresponding to opposite polarisation states of the modulator could be stored in the top and bottom thirds of the CCD array while only the central part was used for actual light collection. The collected charge image in the centre is transferred by charge coupling to the appropriate (upper or lower) storage area during the dead-time required to move to the other polarimeter setting. Because the 'transfer time' can be very rapid compared to the 'integration time' little or no smearing occurs. No other detector can perform this trick! The ISP gave much good service, though in practice, this form of operation proved rather stringent for the CCDs of the day (see section 5.4). Any systematic errors remaining were removed by getting the computer to form the ratio of the images produced by the calcite polarising beam-splitter.

The ISP was commissioned in 1981 and used for several years in a number of diverse studies including the first very high-resolution images of the synchrotron polarisation from the Crab Nebula and the consequent disentangling of the nebular and pulsar polarisations (Fig. 2.19 and Plate 3). The ISP was also used to study the

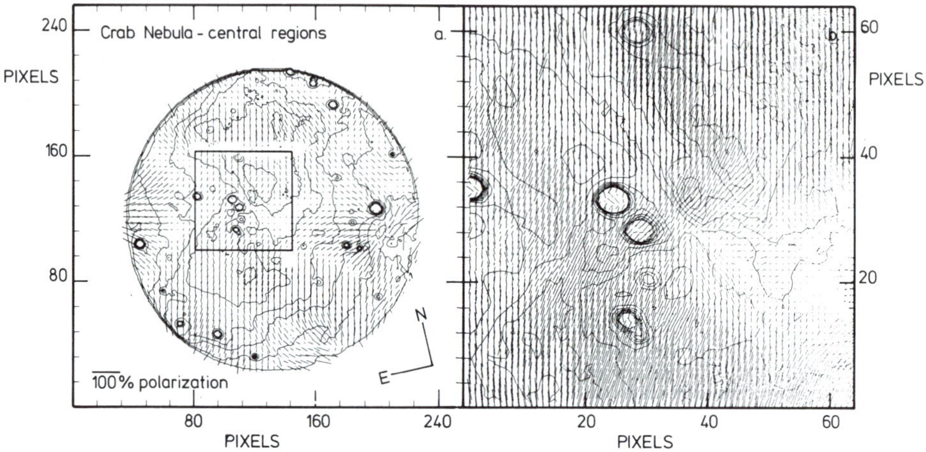

Fig. 2.19 — A contour map of the bright emission associated with the Crab nebula supernova remnant overlaid with tiny line segments which represent the amount of polarisation and the orientation of the magnetic field in the nebula. The data were obtained with the CCD ISP developed by the author and represent a 25-fold improvement in magnetic field resolution over classical photographic images. Plots like this are generated by the computer after intensive manipulation of the digital CCD images.

reflection clouds of gas and dust surrounding peculiar nebulae and, using polarised light, it revealed the first clear evidence of a twisted, disk-like structure around the object Lick H-alpha 233 (Plate 4). In the field of spectropolarimetry, one of its successes was to obtain the detailed polarised spectrum of the active galaxy called NGC 1068 — one of a class of very energetic galaxies named Seyfert galaxies after their discoverer. The ordinary spectrum of the nucleus or power source in this galaxy was diluted by starlight whereas the polarised spectrum revealed hidden sources whose light was 'scattered' in our direction. This result led the way to a totally new interpretation of this galaxy.

Many other CCD imaging polarimeters have been developed such as the compact travelling system designed by Tony Tyson of AT&T Bell Labs, the Durham CCD polarimeter and a new CCD-based spectropolarimeter system — similar to the ISP — which has recently been commissioned by Joe Miller of Lick Observatory. A polarimeter based on a special Pockels cell modulator designed by Gary Schmidt was developed by the author in 1981/82 for use with the Boksenberg IPCS and the Cassegrain spectrograph on the AAT; this system was later upgraded by Jeremy

Bailey and others to utilise a CCD camera — just in time for SN1987a which would have been much too bright for the IPCS! Several national facilities now offer polarimeters and imaging polarimetry has recently been applied to the infrared with the IRCAM on the UKIRT. Of course, in all of these systems computers are used to control the instrument, acquire and store the data from the electronic camera, manipulate the data numerically and display results on a TV-type screen.

2.4 OTHER INSTRUMENTS

2.4.1 Coronagraphs

In all CCD cameras there will be some small amount of light which gets into the system after being scattered from the slender struts or vanes which support the secondary mirror of the telescope, or from the edges of the primary mirror (including its Cassegrain hole). Scattered light is particularly serious when looking for very faint sources close to a very bright source. One way to overcome this is to place a small occulting finger or disc in the focal plane to blot out the bright object in much the same way as the Moon covers up the Sun during a total eclipse to reveal the outer layer or corona of the Sun's atmosphere. To cut down scattering from the secondary mirror support vanes, a blackened mask is constructed with a shape which imitates the size and orientation of the vanes and this mask is placed very accurately inside the camera at a position where there is an optical image of the secondary mirror. If there is no such image then one can be created by means of a 'field lens', i.e. a weak lens placed exactly at the focus of the telescope. In practice the mask (or 'pupil mask') is slightly oversized to ensure that the scattered light is blocked. This is the principal of the CCD coronograph. Beautiful results from an instrument using this technique are displayed in Plate 5; the image is of the disk of proto-planetary material around the star Beta Pictoris. See also Fig. 1.9(a) showing SN1987(a).

2.4.2 Stellar velocity meters

Several astronomers are engaged in the extremely difficult task of searching for evidence of massive planets orbiting around relatively nearby stars and with orbital periods in the range of three to 10 years. Of course, the planets themselves are too faint and too close to their parent stars to be seen directly. The basis of the measurement is the reflex motion on the star being orbited; the star itself must orbit about the common centre of mass — called the 'barycentre' — of the star–planet system. For example, the sun orbits its centre of mass in common with Jupiter with a speed of only 12.7 m/s (metres per second). This speed is many times smaller than the errors of conventional velocity measurements of the Doppler shift from stellar spectra. How can such a small effect be observed?

One system is the University of Arizona Lunar and Planetary Lab's 'stellar accelerometer' developed by Bob McMillan and coworkers from a concept originating with the late Kris Serkowski. Light is fed from the telescope focal plane to the spectrometer entrance aperture by an optical fibre, the spectrum is dispersed by an echelle grating crossed with a normal plane diffraction grating, and the resulting two-dimensional spectrum is focussed onto a CCD detector. Wavelength calibration is imposed on the incoming starlight by a transmitting Fabry–Perot interferometer.

This complex instrument therefore employs a number of techniques all rolled into one instrument.

Other systems obtain precise calibration by allowing the incoming starlight to pass through a chamber containing an almost translucent gas which absorbs a small amount of light at a few very specific places in the spectrum.

Although for some groups the main attraction of these difficult instruments is the search for planets around other stars, one possibly serious confusing effect which is made into a virtue by other astronomers is the fact that the star itself may be undergoing small pulsation patterns. The nature of these pulsation patterns, which also seem to occur in the Sun, contains clues to the internal structure of stars and so has been called, by analogy with studies of the interior of the Earth, 'stellar seismology'. A global oscillation network (GONG) has been set up to monitor the Sun for these effects.

Once again, such experiments could not have been undertaken as little as 10 years ago because of the lack of good detectors and sufficient computing power.

2.4.3 CCD transit survey telescope

A remarkable property of the CCD is that charge can be transferred (along a column of pixels) one row at a time at any desired rate. In particular, the CCD can be operated at the 'sidereal rate', that is, the rate at which the Earth turns on its axis relative to the most distant stars. How is this useful?

John McGraw at Steward Observatory has spotted a very intriguing application. Suppose a sensitive CCD camera is placed at the focus of a moderate-sized telescope which is pointing at the zenith (or anywhere on the meridian), but the telescope drive motors are switched off. Stars will drift across the CCD pixels at the sidereal rate of about 1.5 arcseconds per second of time, producing trails. Now, initiate the electronic process of reading out the CCD charge pattern at the sidereal rate and in the same direction as the star moves, and then open the shutter. There will no longer be star-trails. Instead, the charge image from previous pixels is added to the next one and the current position of the charge pattern will move along the column so as to keep up with the current optical image position (see Fig. 2.20), more and more photons will be collected and, ultimately, the entire column of pixels will be read out and will have contributed to the detection process. Carrying on in this way a huge strip of sky can be surveyed systematically to a deep level without actually moving anything except electrons!

To put this idea into practice a fully automated imaging survey telescope called a CCD Transit Instrument (CTI) has been installed on Kitt Peak Mountain in Arizona (Fig. 2.21). The telescope has no moving parts. The entire system is computer-controlled and operates all night unattended, even shutting down and closing the dome automatically at dawn. In its focal plane are two CCDs aligned east–west. The CCDs are operated in the time-delay and integrate mode at the apparent sidereal rate to produce a strip image of the sky 8.25 arcminutes wide (declination) and an incredible eight hours of time in length. In this way, about 15 square degrees of the sky are surveyed every clear night. After a year, more than 40 square degrees will be surveyed in a continuous strip. Every clear night one CCD produces a V-band (mid-visible) image to a limiting magnitude of about 21, while the other CCD observes in

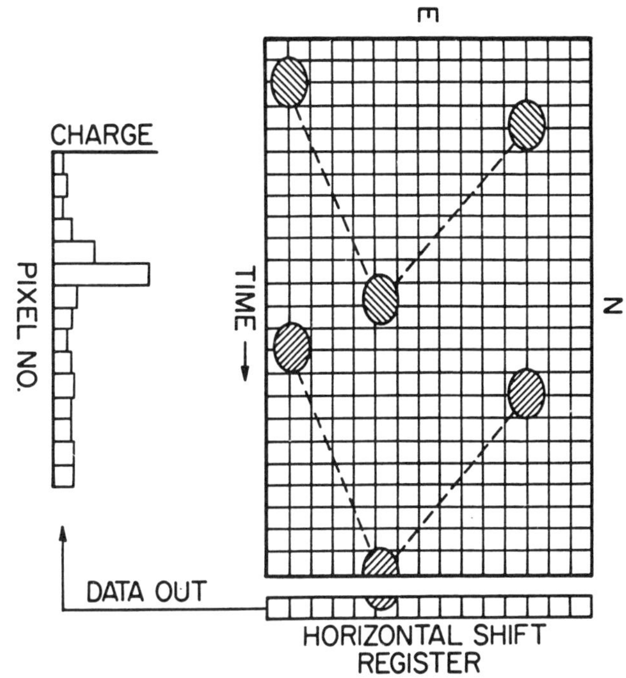

Fig. 2.20 — Showing the principle of operation of the CCD Transit Telescope or 'drift scan'. The star pattern drifts across the CCD because the telescope is stationary, but does not produce a smear because the electronic charges stored in the CCD are moved along at the same rate in the same direction. Courtesy John McGraw.

another filter passband such as ultraviolet (U), blue (B), red (R), or near infrared (I) depending on the sky brightness.

Raw images are calibrated and corrected by computer for artifacts, valid objects are identified and their brightness and position are extracted and stored in a master file and in a history file; the master file contains the list of all objects ever observed by the CTI and maintains a record of average properties such as brightness, whereas the history file is a data base which stores the observed properties for each sighting. The CTI data-handling system is generating and absorbing up to 450 Mbytes of image data per day and updating databases at this rate. This is a lot of data! Data rates are explained and discussed in Chapter 7.

2.4.4 High-speed CCD photometers

One of the great advantages of the photomultiplier tube is its great speed of response to a change in brightness; typically one thousandth of a second is used routinely. There are many useful and important applications of this so-called 'high-speed photometry'. For example, objects such as cataclysmic variable stars and pulsars suffer rapid changes in brightness on short timescales. Also, when stars are occulted by the moon or planets, or satellites of planets are occulted by the planet itself, there

Fig. 2.21 — The University of Arizona's CCD Transit Telescope. There are no moving parts and the space frame is designed to be thermally self-compensating. Courtesy John McGraw.

is a systematic dimming which yields the physical dimensions of the sources. More recently, CCD cameras have begun to be used at rates of about one frame per second for the more slowly varying of these sources and for situations in which it is desirable to refer any brightness changes to another (non-varying) star in the field of view.

2.4.5 Speckle cameras

All stellar objects are so far away as to be practically unresolved by any ground-based telescope. All that should be seen is a compact image surrounded by circular rings — the diffraction image — whose size depends on the wavelength of the light and (inversely) on the diameter of the telescope. In practice, large telescopes cannot achieve this **diffraction limit** because of the motions and turbulence of the earth's atmosphere above the observatory — even at the very best sites — and, of course, the quality of the surface of the mirror. There is one way around this problem however, at least for sufficiently bright objects, and that is to take numerous very short snapshots of such short duration that the image motion which leads to blurring is essentially frozen. A camera with a high magnification is needed. A movie of these snapshots will show a small compact image which changes its position randomly with time; this is called a **speckle pattern**. If a single long integration had been used then the resulting image would be the normal blurred seeing disk. By intensive computerised mathematical manipulation of the speckle pattern it is possible to reconstruct a single image with greatly improved resolution. Currently, there is considerable interest in the use of computer-controlled optical mechanisms — including control of the primary mirror itself — to correct not only for the atmospheric image motion leading to the loss of resolution, but also for imperfections in the mirror surface itself.

In summary, CCDs dominate as the detector of choice in almost all kinds of astronomical apparatus. These instruments are electronic in nature and computer-controlled. In Chapter 3 we consider in more detail the problems facing observatories required to provide and support such equipment, and then in Chapter 4 we look closely at the CCD itself.

REFERENCES AND SUGGESTIONS FOR FURTHER READING

Clarke, D. and Grainger, J. F. (1971) *Polarized light and its measurement.* Pergamon, London.

Crawford, D. L. (ed.) (1986) Instrumentation in astronomy VI, *Proc. SPIE,* Vol. 627.

Eccles, M. J., Sim, M. E. and Tritton, K. P. (1983) *Low light level detectors in astronomy.* Cambridge University Press, Cambridge.

James, J. F. and Sternberg, R. S. (1969) *The design of optical spectrometers.* Chapman and Hall, London.

Robinson, L. B. (ed.) (1988) *Instrumentation for ground-based optical astronomy; present and future.* Springer-Verlag, New York.

Walker, G. (1988) *Astronomical observations; an optical perspective.* Cambridge University Press, Cambridge, UK.

Research and facilities handbooks from: European Southern Observatory, National Optical Astronomy Observatories, Royal Greenwich Observatory, Royal Observatory Edinburgh, Anglo-Australian Observatory.

3

The user-friendly approach

This chapter is a discussion of the impact that electronic imaging and computerisation has had on the way astronomical research is organised, particularly in terms of the development of more sensitive and more sophisticated observational apparatus.

3.1 COMMON-USER INSTRUMENTS

Although many individual university departments throughout the world still actively design and develop astronomical instrumentation which they subsequently transport to various telescopes when they have succeeded in winning 'observing time', there has been a strong trend over the last decade toward shared or **common-user** facilities. Most of the generic instruments described in the last chapter are in this common-user category.

In many ways this arrangement is similar to that encountered in other scientific disciplines, such as elementary particle physics in which experimenters combine resources to construct a single, powerful apparatus for one of the giant accelerators. The situation is similar in radio astronomy and in space astronomy. Apart from making good economic sense by preventing needless duplication, common-user instruments have the advantage that their properties become well-known and more widely known to many people and so questions of systematic errors or optimum methods of calibration become less contentious. Expertise becomes distributed and better scientific results ensue.

Most often, a consortium of university groups and a national laboratory or observatory will get together to define or specify the new instrument. Usually, some effort is made to ensure that the new instrument is 'modular' in design and reasonably flexible to allow innovative additions at a later time. It is particularly important to ensure that new detectors can be readily substituted. To be competitive, each common-user instrument must be very sensitive, easy to use correctly at the telescope and must provide computer-compatible data in a form which can be readily handled by the guest astronomer at his home institute. The instrument must have a high degree of reliability not only because the original builders/designers may not be around to participate in every experiment done with it, but also because such instruments tend to become workhorses and get used night after night on the cold mountain-tops where reside the world's major observatories. Inevitably therefore, common-user instruments must be well-engineered.

3.2 THE DEGREE OF AUTOMATION

One of the most contentious questions about a decade ago was: how much automation? Having invested, for example, in a modern electronic imaging system with a CCD camera, does it make sense to have hand-operated mechanisms for focussing, changing filters or tilting gratings? This was especially a concern at major observatories with large telescopes since the instrumentation clustered at the focus soon becomes quite inaccessible once the telescope moves away from the vertical position (Fig. 3.1). It is also true that there is much interest in 'remote observing' so

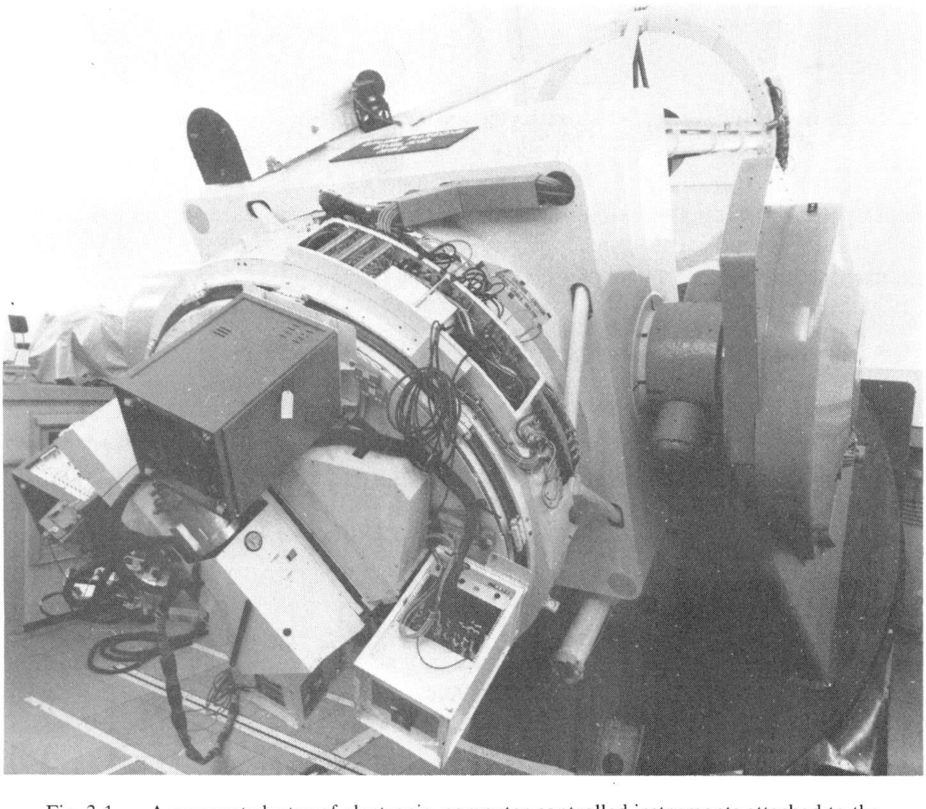

Fig. 3.1 — A compact cluster of electronic, computer-controlled instruments attached to the Cassegrain focus of the 2.5-m Isaac Newton Telescope. Courtesy Royal Greenwich Observatory.

as to permit more participants in a project without incurring large increases in travel costs and logistics problems from having too many collaborators all at the remote site simultaneously.

Given that the CCD camera in question is already under computer control, it is not difficult to ensure that all other mechanisms which are routinely required can also be computer-controlled, and this has been the trend worldwide. For example, stepper motors, dc motors and solenoids can all be used to physically move items, and numerous transducers such as absolute encoders, microswitches, light emitting diode (LED) sensors and Hall-effect switches can be employed to sense electronically that the moveable part is in the correct position.

Most astronomers will have experienced occasions when they have been extremely grateful for a fully automated instrument, not merely because it was more efficient in terms of time saved, but because it is fundamentally safer and more reliable than sending an astronomer out into the cold night to 'ride' in the Cassegrain 'cage' of a large telescope like the AAT or the 5-m Hale Telescope, worse still, trying to reach the instrument by standing in the dark on tall step-ladders when the telescope is pointing at a source well-away from the zenith!

Astronomers have found that automation pays dividends in scientific flexibility also. By providing a modular package of control programs and other software options, modern astronomical instruments can be 'reconfigured' to optimally match the needs of the most ingenious researcher without recourse to manual adjustments. It is this feature of re-programmability which ensures that the extra cost and extra effort which must be expended to design and construct a common-user instrument is worthwhile. Wherever possible then, the trend has been toward having all the key features of a modern, common-user astronomical instrument under remote computer control.

3.3 GETTING FEEDBACK TO THE ASTRONOMER

A major concern with all forms of electronic imaging, whether in cameras or spectrometers, is handling the electronic signals from the detector in such a way that the astronomer can appreciate what is going on while at the telescope. The very earliest CCD cameras, for example, simply recorded the images on magnetic tape for later analysis. With the availability of much more powerful, yet compact, computers such as Digital Equipment's MicroVax series and stand-alone image-processing units such as the SUN Microsystems Inc. work-stations, together with the huge capacity of modern computer disk storage, it is now possible for computer programs to perform basic numerical manipulations, called 'reductions', on the raw data — i.e. the actual untreated numbers digitised by the electronics and stored by the computer — and to display images and spectra on TV screens just minutes after the observation has ended. What kind of manipulations? Almost all detectors will introduce an offset or baseline signal which must be *subtracted* out. Likewise, no matter how good the manufacturing process is, each pixel in the CCD will differ slightly from its neighbours in its response to light and this variation must be cancelled by *dividing* every frame by a calibration image. Performing thousands of repeated subtractions and divisions is just what a computer is good at doing. Sometimes this is not enough, however. In astronomical measurements of faint sources the contrast between the interesting part of the image and the rest of it is often poor. Computers can be used to perform other types of numerical manipulations which help to enhance contrast or pick out the significant information, and present the result in a form more useful to the astronomer such as a brightness profile or cross-section, or a contour map or a picture in 'false-colour' in which shades of grey are replaced by unusual colour hues to make low-contrast features more prominent.

The ability to assess the quality of the data being obtained at the telescope in so-called 'real time' is in no sense a luxury. It requires, of course, considerable provision for computers and associated peripherals at the telescope, and it may require additional staff to support those facilities, but in the end it is more cost-effective and more efficient in the use of telescope time because it gives the astronomer the (only) opportunity to properly control the experiment.

Therefore, at most observatories, the visitor will encounter a telescope 'control room' such as the one shown in Fig. 3.2, containing several TV-type displays and

Fig. 3.2 — The control room of the AAT, typical of many modern facilities of this size. Photo courtesy of the Anglo-Australian Observatory.

several computer screens and keyboards. The status of the telescope and its instrumentation will be on display, together with normal images of the sky seen by 'acquisition and guide' TV cameras — which also may be CCDs — and computer-enhanced images or spectra of the particular region under study.

3.4 DEALING WITH THE GROWING COMPLEXITY

Electronic and computer-aided astronomical equipment, including the telescope itself, is undoubtedly growing in complexity. The technological innovations demanded by modern astronomy are as diverse as they are difficult, and it is interesting to note that, over the years, industrial manufacturers of many sorts have happily participated in these developments for reasons of prestige and subsequent spin-off rather than for the rather small market value of astronomy itself. The CCD is a fine example of a technological innovation driven to near ultimate perfection by the demands of ultra-low light level astronomical studies of distant sources in the cosmos, as much as by the mass market for home video cameras. Yet CCDs, for instance, now find application in fluorescence microscopy, industrial radiology, underwater imaging, high-energy physics and general medical imaging.

Instrumentation developed for a national facility, such as those in mainland North America, La Palma, Hawaii, Chile and Australia, is expected to be used by many different scientists and to be supported by scientists and engineers other than those who designed and built it. To make this truly feasible it is absolutely essential to carefully design and build the instrument to a very high standard of engineering and to a specification agreed with the astronomers. Drawings, documentation, and test procedures need to be excellent. In a phrase, the instrument must be 'user-friendly'. It can be argued, and sometimes is, that it is too costly to attend to such 'details' as accurate drawings and parts schedules, documentation, test procedures, software design and modelling tools such as CAD (computer-aided design). But clearly, if the correct balance is struck, the cost savings in terms of the number of breakdowns and the inefficiency of repair and maintenance once the instrument is finally located at the observatory can be immense — as any observatory director will testify. Astronomers cherish 'telescope time' jealously and are not at all tolerant of inoperative equipment during *their* block of assigned time!

If an instrument, CCD or otherwise, is to be rapidly and successfully deployed it makes sense to deliver it to the telescope with a full suite of computer programs (generally simply referred to as 'software') enabling the instrument to take data and enabling the astronomer to, as conveniently as possible, analyse and archive those data. A common flaw among instrument builders who have not constructed common-user systems before is to overemphasise the hardware development at the expense of software development. Often this is done on the assumption that

(1) the software requirements are too hard to define,
(2) it is up to the astronomer anyway!
(3) software is easier to design and implement than hardware.

On the contrary! It has gradually become accepted that it is the job of the instrument-building team to provide a complete system, including software. The required software *can* be specified and designed alongside the hardware, and it *can* be coded as the hardware is built. Moreover, the software package should provide for data simulations which can test the software, and 'exercise' routines for assisting engineers to check the hardware. When such an approach to software and system management is taken the result can be very powerful. This approach was taken with the development from 1984 to 1986 of the first state-of-the-art, common-user *infrared* electronic imaging system (called IRCAM) by the Royal Observatory, Edinburgh. This new system (described in Chapter 9) went straight into successful operation on the UKIRT on the 14 000 ft summit of Mauna Kea in Hawaii with only a few months of laboratory trials because of a planning technique which emphasised

(1) design reviews — including software,
(2) sub-assembly testing,
(3) system simulation and modelling,
(4) simultaneous development of on-line data reduction and analysis software.

The way to deal with the growing complexity of electronic computer-controlled instrumentation in astronomy is therefore to adopt the highest standards of project planning and management which are already familiar to professional engineers and industrial technologists — especially those in the optoelectronics, computers and aerospace fields.

3.5 PROJECT PLANNING AND MANAGEMENT

The key to the successful construction and commissioning of a complex, modern, common-user astronomical instrument with computer control is 'project planning'. While it is true that a small group of good experimental scientists developing apparatus for their own use can, and do, often avoid the formal project planning approach, it is dangerously cavalier for a national or international facility to take this approach. Actually, in reality there is much to be gained by the small experimental group if they too endeavour to follow some simple rules of project management. For example, they can end up with a more reliable and better documented instrument, win more telescope time and provide better post graduate training.

When setting up a project to develop a complex common-user instrument for astronomy, the most commonly adopted approach is to begin by breaking up the project into clearly identifiable areas requiring specialist expertise. These might be, for example,

(1) optics,
(2) mechanics,
(3) electronics,
(4) cryogenics,
(5) software,
(6) detector physics,
(7) science.

The next step is to further divide each of these areas if required. For example,

(a) structural mechanics, compact mechanisms;
(b) digital electronics, analogue electronics, microcomputers;
(c) control software, data acquisition software, user interface software, data reduction and data analysis software.

Having prepared a detailed scientific case-for-support and a draft specification for the proposed new instrument, it is wise for the scientist to conduct a 'preliminary' design study — sometimes referred to in space-mission jargon as a 'pre-Phase A' study — by consulting professional engineers and instrument physicists with expertise in each appropriate field. In this way the specification can be strengthened, detailed and finalised. These steps are largely self-evident and standard practice in industry. Remember, however, that the specification is not absolutely sacrosanct, compromises may be possible and necessary for purely practical reasons. It is the role of the project scientist to 'point the direction' and to mediate in any design reviews in which there is a request for a change. Once there is a specification, the next step is to obtain a budget and define a project team.

At this stage it becomes essential to have a project engineer, and for very large projects one must also have a project manager, who can deal with day-to-day management, scheduling and financial activities. The rest of the team should be formed so as to cover the required technical areas of expertise (e.g. as outlined above) on as nearly as possible a one-to-one basis. This is important to ensure that a lot of work can proceed in parallel and that a given engineer is not overloaded.

Once the project team has been formed a full design study (Phase A) is carried out in all areas, including software. Before any metal is cut, before any wires are soldered and before any computer code is written it is important to verify that a clear and comprehensive design has emerged. Detailed 'design reviews' must be held to discuss proposals critically. Wherever possible, small demonstration parts can be made to verify a concept and computer simulations used, e.g. for optical ray tracing, mechanical structural stress/deformation analysis, thermal modelling, detector modelling etc.

When the design is complete the next phase (Phase B), that of procurement and manufacture, can begin. Commercial software packages, such as SuperProject Plus, are available for personal computers to help plan project activities in detail and yield graphical displays of the network of overlapping jobs which must be completed in time and in the correct order to complete the project in the most efficient manner. In every project there will be at least one item, one subset of the project, which if not completed on-schedule will seriously delay and disrupt the delivery of the instrument. These items are said to lie on the **critical path**, and it is usually possible to accurately identify items on the critical path. Of paramount importance is Phase C (assembly and test). If care has gone into the design of test procedures and test rigs — as well as the instrument itself — then the design and performance of sub-systems can be verified more easily and any unforeseen problems can be caught at the sub-assembly stage. It is a well-known maxim of the professional engineer that '. . . if you can't test it you can't build it . . .'! A thorough Phase C results in a much easier final phase, namely integration and commissioning (Phase D).

Of course, there are always problems. What seems like an elegant experimental set-up on paper rarely transposes into a straightforward exercise in practice.

Everything is harder than expected, everything costs more and everything takes longer. And, even worse, one actually has to deal with real people to make any progress!

Some of the difficulties encountered in any project are:

- limitations on funding,
- limitations on manpower and technical resources,
- personnel factors such as
  - personality clashes,
  - incompetence,
  - occasional personal matters,
- demarcation rules — who does what,
- conflicts of interest,
- poor communication between personnel.

The last item is certainly an area of responsibility for the project scientist (and/or the project manager) and, at times, can be very trying. If the team is allowed to splinter into non-interacting factions, or if there is no means to ensure that everyone's viewpoint is heard and understood or if the project scientist fails to encourage discussion among the team, then there is a real danger of a design flaw becoming overlooked. Regular monthly 'progress and review' meetings, properly minuted, are essential, as are occasional or ad hoc meetings of special-interest sub-groups. Well-known rules of good management, such as 'catching people in the act of doing well and praising them accordingly', and 'motivating people by your own enthusiasm for their involvement', are just as valid for scientific projects as for any other activity.

All instrument teams have found the need for very substantial reviews called critical design reviews at appropriate junctures before proceeding to procurement and manufacture. Such meetings are very important. A simple example of how a critical design review can reveal an oversight is the following. In a contract to construct a spectropolarimeter based on an electro-optic polarisation modulator called a Pockels cell for the AAT we had built a detailed 'space-model' — a full-size mechanical replica of the correct weight — ready for shipping to Australia. For reasons associated with the available space in the AAT acquisition and guide box, the unit had a 'handedness', that is, it could only be fitted one way up. Due to a failure of communication, a mechanical designer misinterpreted an AAT engineering drawing as a 'view from above' rather than a 'view from below'. Consequently our test unit was of the wrong handedness! Just as the meeting was about to adjourn happily, it occurred to me to ask the obvious 'silly' question '. . . are you quite sure this is the right way round? . . .'.

So, we have learned that there have been changes in the way astronomers carry out their observational research and changes in the way that observatories responsible for supporting a large community of 'users' must react and plan. We have also learned that at the heart of almost all modern astronomical instrumentation is an electronic imaging system, and that in all but a few special cases the imager is a CCD. In the next chapter we begin our detailed look at the tiny device which has had so much impact on astronomy.

SUGGESTION FOR FURTHER READING

Blanchard, K. and Johnson, S. (1982) *The one minute manager*. William Morrow, USA.

4

Charge-coupled devices

4.1 THE EARLY YEARS

4.1.1 Invention and development

The charge-coupling principle was invented in 1969 by Willard Boyle and George Smith and demonstrated in a simple one-line eight-pixel device by Gil Amelio, Mike Tompsett and George Smith at the Bell Laboratories in New Jersey, USA. Larger image-forming devices of 100×100 pixels were not introduced until 1973 and Boyle and Smith received the basic patent at the end of 1974. The first devices were called *surface-channel* CCDs, because the electrical charges formed by the absorption of light were stored and manipulated near the surface of the silicon. Difficulties with this kind of CCD structure led to a more complex architecture, called **buried-channel**. Similarly, to minimise the random variations in the electrical signals coming out of the CCD — the 'noise' — techniques had to be developed to make better transistor amplifiers on the CCD itself, to invent electronic filtering circuits to 'condition' the CCD signals, to improve the quality and stability of power supplies, and, most important of all, to find ways of manufacturing CCDs reliably in large quantities free of microscopic defects which could affect hundreds of otherwise good pixels.

These developments did not happen overnight. Even today there is considerable scope for further development. Moreover, there was a dreadful period of recession and depression from about 1982 to 1987 during which it seemed as though scientific grade CCDs could not be made to live up to their expectations. It was not the CCD concept which was the stumbling block but the negligible profit margin in scientific imagers compared to hand-held colour TV cameras for the home video market.

Initially it was envisioned that the uses of CCDs would span the entire field of electronics, from computer memories to signal processing and from logic circuits to imaging. Their applications to computer memory systems were quashed by the invention of cheap and fast, solid-state random-access memory chips.

From the original small arrays available around 1973 CCDs have come a long way. The most readily available CCDs are now 512×512 pixels, with 2048×2048 pixels† on the way. A marvellous montage of 22 solid-state imaging devices, mostly CCDs, is shown in Fig. 4.1 from a collection put together over the years 1970 to 1987

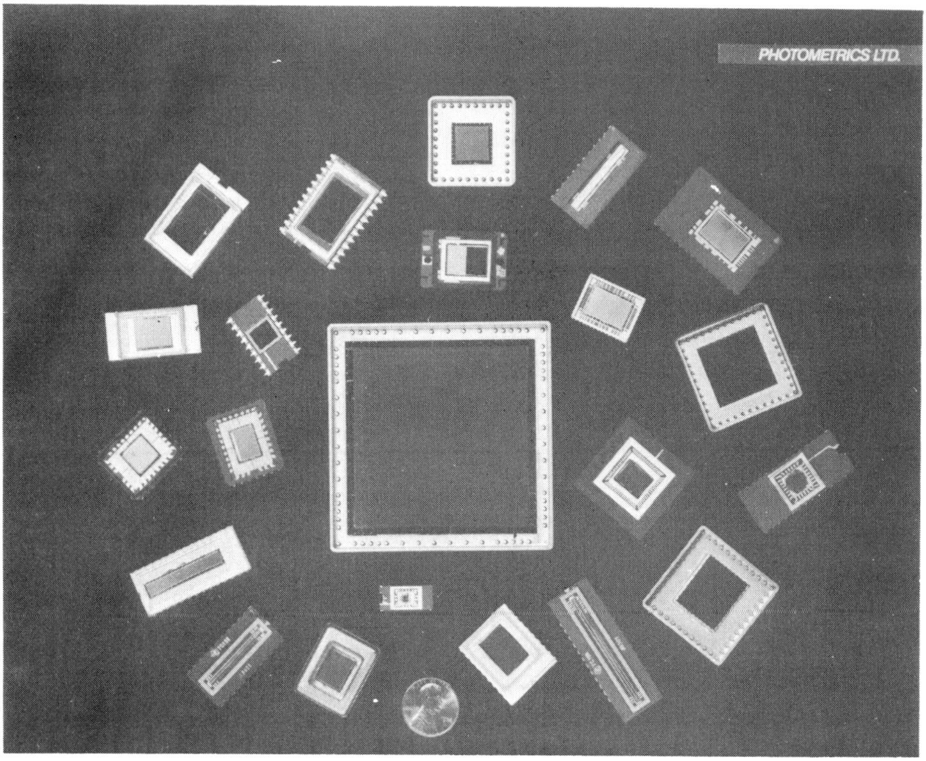

Fig. 4.1 — Charge-coupled devices and other solid-state imagers come in all shapes and sizes. This collection was gathered over the years 1970 to 1987 by Richard Aikens. Photo courtesy Photometrics Limited.

by Richard Aikens of Photometrics Limited. Several suggestions have been advanced to create much larger arrays by forming a mosaic of several CCDs using optics, fibre-optics or simple physical abutment.

4.1.2 The astronomical push

Many astronomy and space-science groups played a vital role in stimulating the development of CCDs. And indeed, stimulus was required! Initially, many manufacturers saw no 'external' market for CCD chips. Bell Labs, for instance, were regulated by law at that time against external sales. Even today there are CCD

† The sizes of CCDs, 512×512 pixels etc., may seem rather strange. Why not 500×500? The reason is 'binary arithmetic' or 'powers of two'. For example, 2 to the power 9, writen as 2^9, is 512. Computers work with binary numbers and computer memory is specified in the same way. In other cases such as 320×512, the numbers yield a size and shape consistent with standard TV scan rates in the USA.

manufacturers who construct devices for inclusion in 'systems only' and not for sale separately. Then there was the problem that many of the companies with the technology to make CCDs realised that they would be competing with other divisions of their own organisations which were manufacturing state-of-the-art TV camera tubes. Moreover, the sort of CCD required for scientific applications was harder to manufacture and not really compatible with TV and home video applications. Some American, British and European manufacturers did tackle both the scientific and the commercial markets, whereas the major Japanese firms stuck firmly to commercial products. Perhaps worst of all, astronomers wanted to use CCDs in an **integrating** mode, rather like exposing a photographic plate, and this implied that the CCD chip would have to be cooled to a low temperature to eliminate thermally generated signal or 'dark current'. Very few manufacturers had, or even wished that they had, that capability!

Fortunately, there were already many groups interested in and familiar with imaging technology — usually with vidicon-type systems — who were alert to the potential of CCDs. Gerald M. Smith, Frederick P. Landauer and James R. Janesick of the Advanced Imaging Development Group at the NASA Jet Propulsion Laboratory (JPL) operated by the California Institute of Technology (Caltech) in Pasadena, and Caltech scientist James Westphal were among the first to recognise the potential advantages of such an imaging device for astronomy and space applications.

In 1973, JPL joined with the National Aeronautics and Space Administration and with Texas Instruments Incorporated (Dallas) to initiate a programme for the development of large-area CCD imagers for space astronomy, in particular for the proposed 'Galileo' mission to Jupiter. Originally scheduled for 1981, that mission is yet to fly!

During the period 1973 to 1977 Texas Instruments (TI) developed CCD arrays of 100×160 and 400×400 photo-elements, 500×500 pixel arrays were soon to appear with the final goal being an 800×800 pixel array. Testing and evaluation of these devices was carried out at JPL by Fred Landauer and a young engineer named Jim Janesick, who just happened to be an amateur astronomer. Having already approached one astronomer at a national institute about testing a CCD on a telescope, and having been turned down, Jim luckily met and teamed up with Dr Bradford Smith, a planetary scientist at the University of Arizona's Lunar and Planetary Laboratory. In early 1976 they obtained the first astronomical imagery with a CCD. Using the superb 61-inch telescope (designed for planetary imaging) on Mt Bigelow in the Santa Catalina mountains near Tucson, they obtained CCD images of Jupiter and Saturn using a special filter to pick out methane gas in the atmosphere of those giant planets. Then they turned the telescope to Uranus (Fig. 4.2) and immediately thought something had gone wrong. It looked like a 'doughnut'! Since the CCD camera and associated tall racks of electronic equipment from JPL were designed for laboratory work, the entire experiment was hair-raising. However, after checking focus and everything else, Brad realised that it must be correct, they were observing 'limb-brightening' of Uranus in the methane band for the first time. As Brad recalls vividly, all who participated and who saw those images agreed that the potential of the CCD was superior to the other imaging equipment of the time.

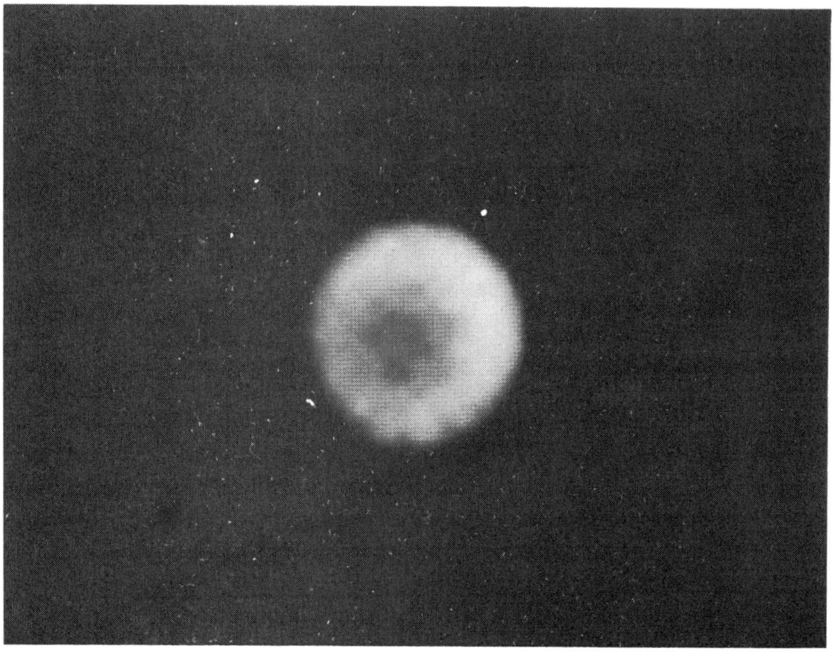

Fig. 4.2 — The first astronomical CCD image. An image of the planet Uranus obtained in 1976 through a methane-band filter with a Texas Instruments CCD. Photo courtesy of Bradford Smith.

At about the same time, NASA awarded contracts for the procurement of instruments for an earth-orbiting space telescope. One of the people awaiting NASAs decision was John Lowrance at Princeton University. John was working with SEC vidicon technology in the 1970s when CCDs were invented. SEC (or secondary electron conduction) vidicons seemed like the most appropriate imaging devices — at least in terms of their wavelength of response — for space ultraviolet/optical experiments such as the Orbiting Astronomical Observatory (OAO). While NASA and various advisory bodies deliberated, John continued to pursue the development of SEC vidicon systems. Finally, after many delays, NASA abandoned the OAO series in favour of the space telescope. By that time CCDs had been invented, and NASA JPL had begun their studies. Following a crucial meeting at which CCD results were demonstrated, the initial plan to utilise SEC vidicons for the space telescope was dropped and the concept of principal investigator instrument teams was introduced. A proposal from a team led by planetary scientist James Westphal of Caltech in collaboration with JPL was accepted for the inclusion of CCD cameras on the space telescope. Jim had heard about CCDs at a committee meeting a couple of years earlier. When he insisted on knowing more about the JPL results, the chairman (Bob O'Dell) sent him along to JPL to see for himself! There he met a no-nonsense engineer called Fred Landauer. They made some tests and Jim left satisfied. He had learned that CCDs were very low-noise devices; 100 electrons (e) noise had been observed and 30 e was predicted. On his return to Caltech he mentioned these numbers to Jim Gunn of Princeton University, who was getting ready for a major project involving vidicon technology, and whose instant reaction was 'that will revolutionise astronomy if it is true!'

John Lowrance, like Jim Westphal before him, moved away from vidicon technology and began working with CCDs. Luckily there was a key player in the game right on his doorstep — the Electro-optics Division of the Radio Corporation of America (RCA) in nearby Lancaster, Pennsylvania. At the head of the RCA group working on CCDs was Dick Savoye, and Dick was enthusiastic about the astronomical applications, and moreover he believed that the technology possessed by RCA would yield devices extremely sensitive to blue light. He was right! The thinned, backside-illuminated 512×320 RCA CCDs had superb blue response. John Lowrance at Princeton and John Geary at Harvard each established good relations with RCA and began testing these devices in the late seventies. John Geary, having first tried an unthinned device on the 1.5-m and 60-cm telescopes on Mt Hopkins in April 1980, visited the RCA facility in Lancaster Pennsylvania shortly thereafter to show them the splendid results obtained so far and urge them to provide him with a thinned CCD. He received one of the very first thinned backside-illuminated CCDs manufactured by RCA; this device was considered a reject and was lying in the desk drawer of RCA engineer Don Battson. John put it on the telescope in September 1980 and that device is still in use on Mt Hopkins to the present day! One of the first CCD camera systems actually intended for astronomy was built in the Physics Department at Princeton by Ed Loh.

Meanwhile the TI chips evolved through a programme of systematic development toward the eventual goal of an 800×800 array. One of the key figures on the 800×800 programme at TI from the outset was Morley Blouke†. Several approaches to the design and fabrication of CCDs were tried. Three-phase, buried-channel CCDs were developed first, but although suitable for Jim Westphal's Wide-field/Planetary Camera for the space telescope, further study showed that these arrays would not survive the harsh radiation environment around Jupiter and so a different construction — the **virtual**-phase CCD — was developed for the Galileo mission. These devices are also attractively simple for the mass market of home video due to the fact that the external electronic drive circuitry is much simplified by the on-chip architecture of this CCD. Tens of thousands of CCDs were being manufactured under contract to NASA (the final number was 75 000) and, following the success of the collaboration with Brad Smith, JPL realised — as Jim Janesick stated in a proposal to the Director of JPL in October 1976 — that there was '. . . a need to expose and familiarise astronomers and scientists to the capabilities of the CCD for use in planetary observation and stellar studies'. A JPL travelling camera was subsequently made available, but only those universities and institutes connected with the Galileo programme had any hope of getting a TI device. Needless to say, this caused some consternation in the astronomical community.

But around this period (1974–1977) other companies were also beginning to develop CCDs. One of the first companies to actually market a device was a division

† Today Morley is no longer with TI but he is still working on CCD technology and still trying to help the astronomical community; we will come back to his efforts shortly.

of Fairchild Semiconductor who produced a 100×100 CCD; Gil Amelio had moved from Bell Labs to Fairchild. At the Kitt Peak National Observatory in Tucson, Arizona, Richard Aikens and Roger Lynds had been working on low light level imaging systems for astronomy for many years. Soon the Kitt Peak National Observatory began a programme of development of CCDs. With Steve Marcus, they began working on the Fairchild device. The Fairchild CCD201 and CCD202 image sensors were designed for TV applications and, although capable of high performance, they had a serious impediment for astronomical work due to the **interline transfer** construction (see section 4.2) which meant that they had columns of picture elements which were alternately light-sensitive and totally insensitive due to being covered by opaque metal strips; in terms of the image falling on the CCD these devices were half blind!

Not long after this initiative, Richard Aikens left the Kitt Peak National Observatory to set up his own company, called Photometrics, which has played an important part ever since in stimulating the manufacturing of CCDs and the development of scientific camera systems. Development and testing of CCDs at Kitt Peak was taken over by Albert Fowler and Patrick Waddell.

There was a time of great concern, frustration and even anger in the late seventies about the lack of access to CCD technology by the mainstream astronomical community. Development of the Wide-Field/Planetary Camera, abbreviated WF/PC but spoken as 'wifpic', was going well and many people were now aware of the sensitivity and the scientific potential of CCDs. Industry too was embracing the new technology, but commercially available products were scarce. During this interlude people were side-tracked into trying other forms of less suitable solid-state imagers which were commercially available such as the charge injection device (or CID) from General Electric (America), or the interline transfer device from Fairchild already mentioned. When 512×320 RCA CCDs appeared in the late seventies it was a welcome relief.

The first RCA CCDs were frontside-illuminated which meant they had a poor response to blue light. Soon however, the thinned backside-illuminated CCDs appeared. Clearly, RCA had 'the secret' for treating or passivating the thinned backside surface, and these CCDs displayed outstanding sensitivity over a huge spectral range — better even than the TI chips. Unfortunately there was one weakness. The design of the on-chip output amplifier was poor and so the CCD was 5–10 times 'noisier' in electrical terms than the TI CCD. Later RCA CCDs were much better, and they even brought out a 'double-density' CCD with a very large number of pixels (1024×640). Sadly, in 1985, RCA withdrew from the CCD market for commercial reasons.

In early 1980 a somewhat unexpected source of astronomical CCDs appeared. Craig Mackay of the Institute of Astronomy in Cambridge, England had been working on silicon vidicons. Progress was slow due to lack of funds. He had met Jim Westphal on Palomar Mountain in 1975 and was aware of the TI work on CCDs, but he learned of a British source of CCDs by a curious coincidence. Silicon vidicons had good spectral response, but they were 'noisy'. Craig needed a very low-noise amplifier design. He contacted an eminent designer called Ken Kandiah at the British Atomic Energy Authority at Harwell and asked him to visit Cambridge. Kandiah offered a design based on a junction field effect transistor (JFET) but recommended Craig to a David Burt at the GEC Hirst Research Centre for a design based on the more readily available metal-oxide semiconductor (MOS) transistors. When Craig met David Burt he learned that GEC had a very advanced CCD programme. Although initially sceptical of astronomical uses, David succumbed to Craig's persuasive approach in 1978. The following year Craig started a postgraduate student, Jonathan Wright, who put together a CCD drive system based on Craig's existing vidicon camera. Their prototype GEC CCD camera was tested on a small telescope in Arizona in March 1980 with spectacular success. I remember it well, because I was fortunate enough to be present for a few hours of one night of that run!

By March 1982 the noise associated with typical GEC CCDs was reported as seven electrons while selected devices gave a mere three electrons, that is, the amount of charge was uncertain by only three times the charge on a single electron — that was a truly amazing result. The GEC CCDs were illuminated from the front rather than the rear and so their blue response was poor. Very recently, chemical coatings have been applied to these devices to overcome this deficiency to a large extent. Both Craig and Jonathan subsequently formed independent commercial companies to make CCD cameras. Craig's company — called Astromed — is involved with medical and astronomical applications. Jonathan's company — called Wright Instruments — has developed very compact low-noise CCD cameras and CCD software based on IBM compatible personal computers (PCs). In a similar way, John Lowrance of Princeton also branched out into making customised, very high-quality camera systems for other institutes.

My own first encounter with CCDs goes back to 1977. During the late seventies I was lucky enough to be working at the Steward Observatory and Lunar and Planetary Laboratory (LPL) of the University of Arizona — right across the street from the offices and labs of the Kitt Peak National Observatory. Under the astute guidance of Peter Stritmatter, the Steward Observatory had an active instrumentation programme led by Peter Stockman involving imaging devices of all sorts. At LPL, Bradford Smith and colleagues were co-investigators with Jim Westphal on the WF/PC and had many other involvements with NASA programmes.

Initially working on two types of long (linear) arrays of silicon photodiodes called Digicons and Reticons, I soon became involved with area arrays of the CID type. Our need was to find a very sensitive, very stable solid-state detector which could be used with a remarkable new instrument. That instrument was an **echelle**† spectrometer, connected to the telescope by 10 metres of **fibre-optic** cable, and carefully configured to make accurate velocity measurements of nearby stars. Conceived by the Polish-born astronomer, Krystorf (Kris) Serkowski, and developed by a skilled engineer called Jack Frecker, the ambitious purpose of the instrument was to detect the very small motion of a star caused by the presence of a large nearby planet, that is, the detection of other solar systems. Unfortunately, Kris died of multiple sclerosis before seeing astronomical technology finally catch up with his foresight; Bob McMillan and others at LPL have developed this concept into a working system using CCDs (see section 2.4). An auxiliary use of the Serkowski instrument was that it could measure the amount of polarisation in light, and this had become my special interest. One day a young postdoctoral colleague of Brad Smith's came into my office

† An echelle is a special kind of diffraction grating with grooves like a staircase.

and asked if I knew whether or not CCDs were 'polarisation sensitive'. His name was Harold Reitsema, and he and Brad were concerned because the WF/PC would include a simple set of polarising filters but no one knew whether or not the CCD itself was sensitive to polarised light. I was already aware of the planetary imaging of Uranus which had been carried out by Brad and coworkers from JPL in 1976 at the 61-inch telescope in the Catalina Mountains north of Tucson, where Kris and I also observed, but since the TI chips were not available, and the Fairchild device being tested at Kitt Peak was unsuitable, I had not had any opportunity to work on CCDs — so I immediately jumped at the chance.

In 1979, Harold and I used a new travelling, prototype WF/PC team camera from Jim Westphal at the 61-inch telescope on Mt Bigelow to perform one of the first ever **polarisation** imaging experiments with a CCD camera. Shortly afterwards I met Tony Tyson of Bell Labs and learned that he had developed a CCD imaging polarimeter based on the Fairchild 100×100 device; Tony has subsequently become a pioneer of ultra-deep CCD imaging. The TI CCD was a 500×500 pixel array by then. One of our targets was the galaxy NGC 1068, a Seyfert galaxy, that is, a galaxy with an unusually bright nucleus or centre (Fig. 4.3). We concluded that the thinned

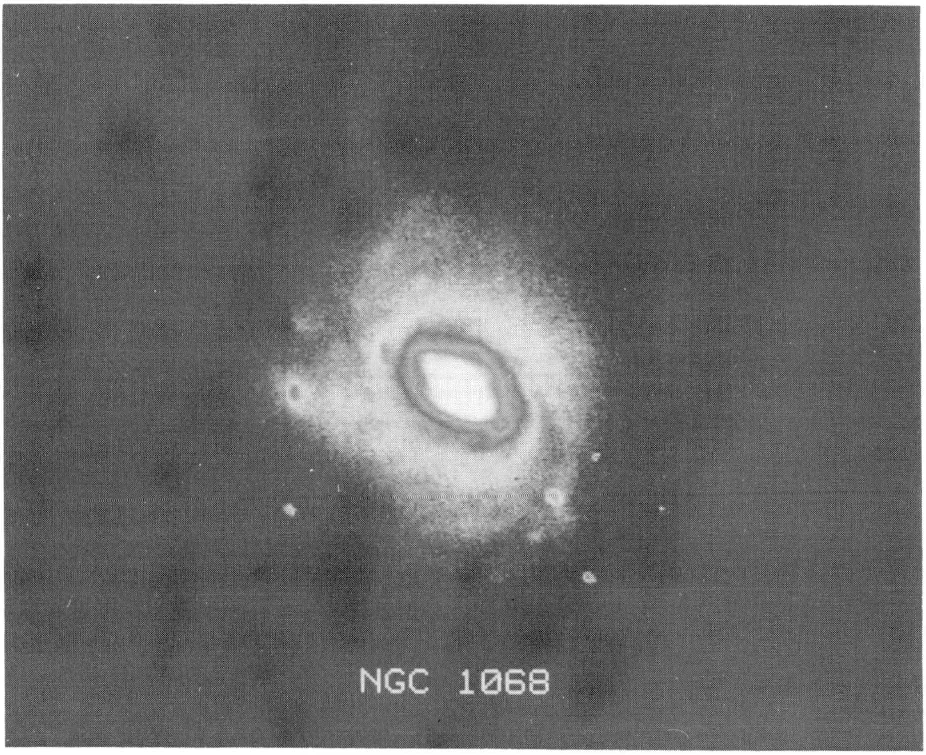

Fig. 4.3 — An early CCD image of the Seyfert galaxy NGC 1068. Note the exceptionally bright nucleus which would totally overexpose most detectors.

backside-illuminated TI CCDs were not intrinsically polarisation-sensitive and so the proposed polarimetry method would not be seriously affected by the CCD itself. It was many years before this topic was discussed again in great detail by Olivia Lupie and Pete Stockman — now at the Space Telescope Science Institute in Baltimore — in preparation for the launch of the Hubble Space Telescope.

At the end of that year I left Arizona to take up a new position at the Royal Observatory in Edinburgh, Scotland where, with the encouragement of the new Director-designate Malcolm Longair — who had also used the Westphal CCD camera for deep cosmological studies through a collaboration with Jim Gunn of Princeton and as a result of his own status as an interdisciplinary scientist on the space telescope project — I set about developing a long-term plan for the construction of CCD camera systems, and the development of software. Those plans could not have come to fruition without the total support and encouragement of two other 'new arrivals' in Edinburgh, namely Malcolm Smith, then Astronomer-in-charge of the Technology Unit and now Director of the Joint Astronomy Centre in Hawaii, and Donald Pettie, an experienced professional engineer. Donald's wealth of experience together with his friendship and patience in guiding an impetuous young scientist made a huge difference to the quality of the scientific apparatus being developed.

Our first CCD instrument was a multi-purpose, travelling system (engineered mainly by Jim Herd and Bill Cormack, with software by Colin Aspin) called an imaging spectropolarimeter or ISP (described in section 2.3). In the introduction to my proposal I alluded to the fact that, in due course, similar developments could be expected in the infrared part of the spectrum, beyond the region currently accessible to silicon CCDs. As Malcolm Smith had pointed out, this was an important issue for the Royal Observatory which was responsible for the operation of the 3.8-m UKIRT in Hawaii, and one which was fast developing into a competitive quest as the power of infrared astronomy to address major astrophysical topics, such as star formation, the nature of the Galactic Centre and the cosmological era of galaxy formation, became increasingly apparent. In November 1981, as 'first light' was recorded with the ISP — of the famous Crab Nebula (see colour plates) — I was once more in Arizona with Harold Reitsema, but little did I know that I was already 'hooked' on solid-state detectors and that a mere five years later I would once again be on a mountain top, this time commissioning a state-of-the-art **infrared** electronic camera! That story is postponed until Chapter 9. Harold Reitsema moved on too, to Ball Aerospace in Colorado where he played an important role in developing image processing software for the Giotto spacecraft CCD camera which sent back superb pictures of the close encounter with Comet Halley; Fig. 4.4 is a computer-enhanced CCD image of the rocky nucleus of Comet Halley.

By June 1981, the date of the Harvard–Smithsonian conference on solid-state imagers, the number of independent astronomy groups working on CCD systems had grown from five to 20. Devices in use came exclusively from TI, RCA and GEC (UK). Astronomers were clearly pushing the technology as hard as they could, yet with only three manufacturers, one of whom had low-noise devices (GEC), one of whom had high quantum efficiency devices (RCA) and the other of whom should have had devices with both properties but was (a) having problems with blue sensitivity and (b) not making them available for sale anyway, it was understandable that people began to worry. When production of the GEC devices moved to EEV

Fig. 4.4 — A computer-enhanced CCD image of Comet Halley from the Giotto spacecraft at close range. Courtesy of Harold Reitsema, Ball Aerospace.

there was the inevitable hiccup in supply, and when RCA withdrew from the field, it seemed like the dream had beome a nightmare. Despite this, by March 1986 at the Tucson, Arizona conference on 'Instrumentation in Astronomy VI', the range of applications involving CCDs had become very large; the proceedings for that meeting, edited by David Crawford, ran to two hefty volumes. The 'dry' period from 1982 to 1987 in the supply and improvement of science-grade CCDs was beginning to ease, or so it seemed. Exceptionally detailed studies of the TI chips by Jim Janesick at JPL had advanced the understanding of the solid-state physics of CCDs and their optimisation, TI CCDs 'excess to NASA requirements' were becoming available in the USA and new devices by companies such as Thomson-CSF (Europe) had been studied in detail by a team at the Royal Greenwich Observatory (RGO). But even so, as David Thorne of the RGO noted in his review at the Tucson meeting, 'The use of CCDs in astronomy is now commonplace; however, the point has not yet been reached where there is a perfect or even a pre-eminent type of device'.

In 1985 astronomers learned of a most exciting prospect. It was the formation of a team at Tektronix Inc. led by Morley Blouke to produce scientific-grade CCDs with large formats and outstanding performance. The initial goal would be a 512×512 array with good-sized pixels (0.027 mm) leading to an astonishing chip with 2048×2048 pixels (Fig. 4.5). Unfortunately, by mid-1986 it became clear that some

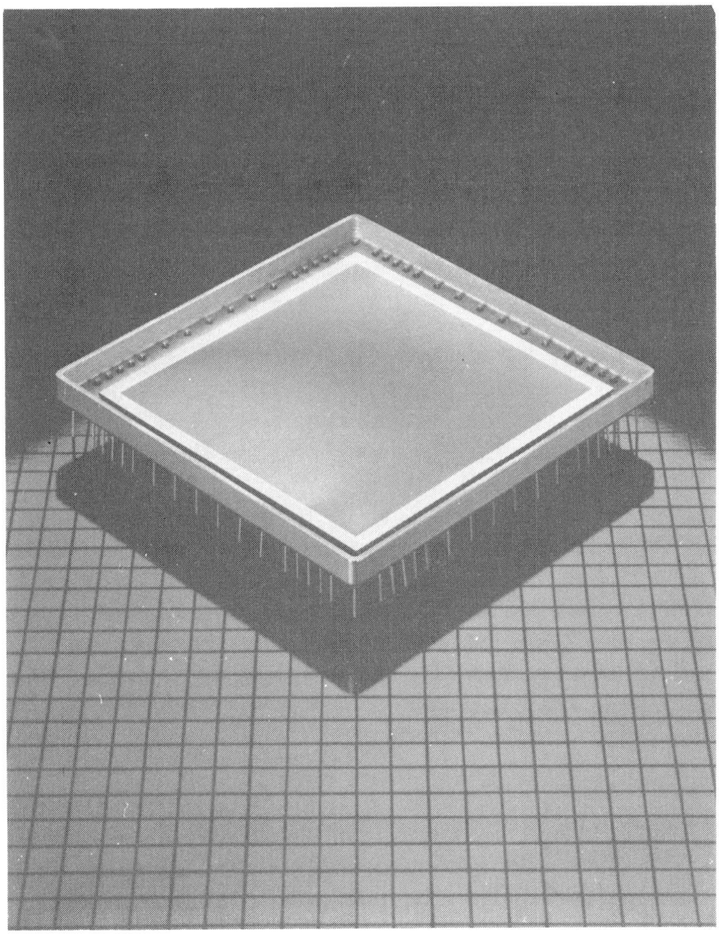

Fig. 4.5 — A close-up of the Tektronix 2048×2048 CCD detector. Courtesy of Morley Blouke, Tektronix Inc.

sort of unexpected fabrication or processing problem was resulting in large numbers of defects called 'pockets' (section 5.4) thus rendering otherwise excellent low-noise devices unusable, and hopes were again dashed. This time the situation for many astronomical groups was serious because new instrument designs and funding for instrument developments had been tied to the expected Tektronix chips.

But Morley and his team did not give up and, in collaboration with several interested parties, have valiently followed every lead in an effort to understand, model and eliminate such problems. The research at Tektronix and at JPL has led to an in-depth understanding of the solid-state physics of CCDs which, as Morley remarks, 'ought to be much easier to understand than a transistor'. Around mid-1988 Tektronix began to ship 512×512 CCDs to customers with long-standing orders.

Among the many people frustrated by the dry spell in CCD supplies was Richard Aikens, founder and president of Photometrics Ltd. In an unprecedented move he contracted with a so-called 'silicon foundry' to produce a custom CCD with 516×516 pixels — marketed as 512×512 pixels — which appears to be an outstanding success. The advantage of this approach is that the silicon foundry, having no need to consider the market value of their product, can quote for, and concentrate on, device production; a further discussion of the silicon foundry concept is contained in the final chapter. In addition, Photometrics announced the availability of a chemical phosphor coating called Metachrome II which can be applied safely to any CCD by vacuum sublimation and thereby improve its response to blue light.

In August of 1988 Lloyd Robinson reported excellent initial experimental results for another brand-new device. As the result of a National Science Foundation grant to Lick Observatory, a contract was placed with EG&G Reticon Corporation to construct a large CCD suitable for spectroscopic applications; the format chosen was 400×1200. Finally, in the same year, new initiatives at EEV (UK) and at Thomson-CSF (France) were announced. A thinning programme and a mosaic construction programme had begun at EEV. Meanwhile, funded by the European Southern Observatory and the French agency INSU, Thomson-CSF in Grenoble had developed a 'buttable' version of their excellent low-noise front-illuminated device and a 2×2 mosaic had been constructed at the European Southern Observatory near Munich. The new CCD was 400×579 pixels and Thomson-CSF had also developed new 512×512 and 1024×1024 CCDs, and a thinning programme was under way.

As of August 1988, the prospects for science-grade CCDs looks better than ever before. So, how does a CCD work?

4.2 BASIC PRINCIPLES

4.2.1 Absorption of light in silicon

The properties of any solid material depend on both the atomic structure of the atoms of the material and the way the atoms are arranged within the solid, that is, the crystal structure. Atoms are composed of a central part called the nucleus which is positively charged, and a family of tiny orbiting electrons which are negatively charged. The electrons can exist in stable orbits near the nucleus only for certain definite values of their energy. When individual atoms come close together to form a solid crystal, electrons in the outermost orbits — often called the upper **energy levels** — of adjacent atoms interact to bind the atoms together. Because of the very strong interaction between these outer or **valence** electrons, the outer orbits and therefore the upper energy levels are drastically altered. The result is that the outer electrons are shared between the different atomic nuclei. A diagram depicting the 'energy levels' of the electrons for a combination of two atoms will therefore have two permitted levels near the core of each atom. A combination of three atoms will have three levels near the core because the outer electrons of all three atoms can be shared. The higher, unoccupied orbits are also split, indicating that they too can in principle take two or three electrons. But even the tiniest sliver of a real crystal will contain many hundreds of millions of atoms, and so there is a huge number of split levels associated with each atom in the crystal because of the sharing of outer electrons. In other words, the energy levels or orbits are spread out into a 'band'. The lowest band of energies, corresponding to all the innermost orbits of the electrons, are all filled with electrons because there is one electron for each atom. This band of allowed, filled energy levels is called the **valence band**. Conversely, the upper energy band is empty of electrons because it is composed of the combined unoccupied higher energy levels or orbits of the individual atoms in the crystal. It is called the **conduction band** for reasons that will become apparent. Since the individual atoms have a gap between the permitted inner and outer orbits, that is, a gap in energy between the inner filled levels and the outer unoccupied levels, the energy region between the valence band and the conduction band in the crystal must be a **forbidden energy gap**. Fig. 4.6 summarises this description. Note that the crystal must be pure and contain

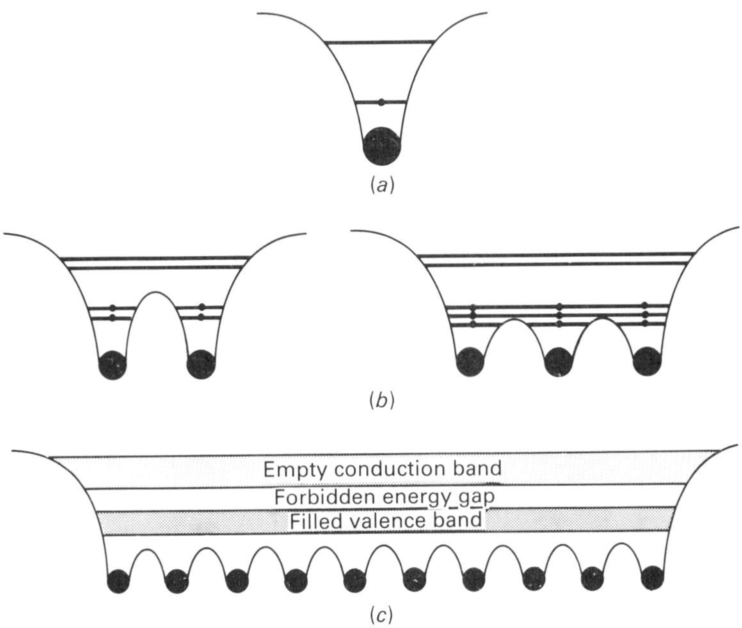

Fig. 4.6 — The development of a conduction band and a valence band of electrons in a solid crystalline material such as silicon.

atoms of only one kind, otherwise additional energy levels corresponding to the impurity atoms will be formed. More importantly, the periodic or repetitive crystalline structure must be unbroken to avoid distortions in the energy levels caused by abnormal sharing of electrons.

In metals, the valence and conduction bands overlap and so any of the many valence electrons are free to roam throughout the solid and to move in response to the force of an electric field; an electric field could be produced by attaching a battery to both ends of the piece of metal. Therefore metals are excellent conductors of electricity (and heat). An insulating material on the other hand, has a highly ordered structure and a very wide forbidden energy gap. The conduction band is totally empty of electrons and so cannot contribute to an electrical current flow. Electrons in the completely filled valence band are hemmed in, every nearby orbit is occupied, and therefore they cannot move in response to an electric field.

A semiconductor is similar to an insulator except that the forbidden energy gap is much narrower. In fact, a few electrons can be elevated from the valence band to the conduction band across the forbidden gap merely by absorbing heat energy from the random, microscopic, jostling motions of the crystal structure at normal 'room' temperature. Electrons promoted to the conduction band in this way can then conduct electricity, that is, they are free to move under the influence of an electric force-field. Interestingly, the corresponding vacancies or **holes** left in the valence band allow it to contribute to electrical conductivity as well, because there is now somewhere for the other electrons to go; descriptions of solid-state devices will often refer to **electron–hole pairs**.

Since the number of electrical charge carriers (electrons in the conduction band, holes in the valence band) is much smaller than in the case of a metal, semiconductors are poorer conductors than metals but better than insulators. The width of the forbidden energy gap in semiconductors is an important quantity. It is usually expressed in a unit called **electron volts** (see Appendix 2). One electron volt or **eV** is the energy that one electron would acquire under the influence of a voltage of one volt. Most semiconductor crystals have a band gap energy around 1 eV. This energy is roughly 30 times larger than the thermal or heat energy in the crystal atoms at room temperature. Since the number of electrons which can be promoted to the conduction band by absorbing heat will vary with the temperature of the crystal, those semiconductors with larger band-gaps are preferred since transistors and other devices made from them will be less sensitive to environmental changes in temperature. For this reason silicon is preferred to germanium. If, however, the semiconductor is cooled to a low temperature and held there, then the random elevation of a valence electron to the conduction band can be virtually eliminated. Table 4.1 is a section of the periodic table of the elements which shows that the primary semiconductors like silicon and germanium belong to the 'fourth column' elements, which include carbon. These elements have four valence electrons. Compounds of elements on either side of the fourth column can be formed and these alloys will also have semiconductor properties; gallium arsenide (GaAs) and indium antimonide (InSb) are III–IV (or 'three–four') compounds and mercury–cadmium–telluride is a II–VI (or 2–6) compound. The compound number corresponds to the number of valence electrons.

To understand how silicon is used to detect light we must think of the quantum

Table 4.1 — Part of the periodic table of elements showing the location of semiconductors

IB	IIB	IIIA	IVA	VA	VIA	VIIA
		13	14	15	16	17
		Al	Si	P	S	Cl
		Aluminium	Silicon	Phosphorus	Sulphur	Chlorine
29	30	31	32	33	34	35
Cu	Zn	Ga	Ge	As	Se	Br
Copper	Zinc	Gallium	Germanium	Arsenic	Selenium	Bromine
47	48	49	50	51	52	53
Ag	Cd	In	Sn	Sb	Te	I
Silver	Cadmium	Indium	Tin	Antimony	Tellurium	Iodine
79	80	81	82	83		
Au	Hg	Tl	Pb	Bi		
Gold	Mercury	Thallium	Lead	Bismuth		

nature of light and speak in terms of a stream of packets of electromagnetic energy called **photons**. When a photon is absorbed in the crystalline structure of silicon, its energy is transferred to a negatively charged electron — or photoelectron — which is then displaced from its normal location in the valence band of that atom into the conduction band. The vacated site is therefore positively charged and, as mentioned before, is referred to as a hole. When the electron reaches the conduction band it is no longer held in the grip of the silicon atom and can migrate through the crystal. Migration can be stimulated and controlled by applying an electric force-field to the silicon crystal by means of small metal plates called **electrodes** or **gates** connected to a battery or other voltage source. This is essentially how a **photodiode** works. The more photons absorbed each second the more electron charges released each second and the larger the measurable current.

Absorption of photons in silicon is a function of the photon energy (or wavelength). Red (or low-energy) photons pass deeper into the silicon before being absorbed. Eventually, for the reddest light, that is the lowest energy photons, there is simply not enough energy to elevate a valence electron to the conduction band. In other words, for each semiconductor there is a wavelength of light beyond (redder than) which the material is insensitive to light because the photons are not energetic enough to overcome the forbidden energy gap (E_G) in the crystal. To be precise the cut-off wavelength is given by

$$\lambda_c = hc/E_G$$

where h is Planck's constant and c is the speed of light. These wavelengths are given in Table 4.2.

When silicon atoms in the crystal structure are deliberately replaced with other atoms the semiconductor is said to be **doped** (Fig. 4.7). If the impurity atom has more valence electrons than the semiconductor then it will donate these **n**egative charges to the conduction band; such a material is called **n-type**. Conversely, if the impurity atom has fewer valence electrons than the semiconductor then a **p**ositively charged hole is left in the valence band ready to accept any available electrons; this material is a **p-type** semiconductor. In p-type material there is an excess of holes and so electrons are said to be the minority carriers of charge, whereas the opposite is true for n-type material. Junctions between p- and n-type regions are used many times in semiconductor structures to produce different devices.

When a pn junction is formed, electrons from the n region tend to diffuse into the p region at the junction and fill up some of the positively ionised states or holes in the valence band thus making that region more negative than it was. Similarly, the diffusion of holes from the p to the n side leads to an increasingly more positive electrical potential. A narrow region forms on either side of the junction in which the majority charge carriers are 'depleted' relative to their concentrations well away from the junction. Since the concentration of electrons in the n-type material is very much larger than in the p-type material, the flow of electrons would tend to be one-way were it not for the fact that the diffusion process itself begins to build up an electrostatic potential barrier which restrains the flow of electrons from the n-type region; the build-up of electrons on the p side starts to repel further diffusion. The magnitude of this potential barrier (V_0) depends on the impurity concentrations, that is, on the number of donor electrons at the junction that transfer into nearby acceptor levels, and is just equal to the required shift of the energy bands needed to ensure that the Fermi level (E_F) remains constant throughout the crystal; the Fermi level is the energy at which there is a 50/50 chance of the corresponding electron energy state or orbit being occupied by an electron. For an intrinsic semiconductor, E_F lies halfway between the valence and conduction bands whereas for an n-type doped semiconductor the Fermi level moves up toward the conduction band and conversely p-type doping lowers the Fermi level. The potential drop across the depletion layer is 0.6 V for silicon and 0.3 V for germanium about half of the forbidden energy gap. Fig. 4.8 shows how this condition is represented by an energy band model of voltage or potential versus distance from the junction. The width of the junction region increases with increasing voltage (V_0) and decreasing number density of acceptor atoms.

If a positive voltage is applied to the p side of the junction it will tend to counteract or reduce the built-in potential barrier and attract more electrons across the junction, whereas a negative voltage on the p side will enhance the internal barrier and increase the width of the depletion region; these conditions are called **forward** and **reversed bias** respectively. Therefore, on one side of a pn junction there is a region which is more negative and on the other side there is a region which is more positive than elsewhere in the crystal. When light of the correct wavelength is absorbed near the junction an electron–hole pair is created and the potential difference across the junction sweeps the pair apart before they can recombine. Electrons are drawn toward the region of greatest positive potential buried in the n-type layer which therefore behaves like a charge storage capacitor. Of course, as more electrons accumulate the positive potential is progressively weakened.

Table 4.2 — Forbidden energy gaps and long-wavelength photo-absorption limits for some semiconductors

Name	Symbol	Temp. (K)	Band gap (eV)		Cut-off wavelength (μm)
Cadmium sulphide	CdS	295	2.4		0.5
Cadmium selenide	CdSe	295	1.8		0.7
Gallium arsenide	GaAs	295	1.35		0.92
Silicon	Si	295	1.12		1.11
Germanium	Ge	295	0.67		1.85
Lead sulphide	PbS	295	0.42		2.95
Indium	InSb	295	0.18		6.9
antimonide		77	0.23		5.4
Mercury	$Hg_xCd_{1-x}Te$	77	0.10	(x=0.8)	12.4
cadmium			0.5	(x=0.554)	2.5
telluride					

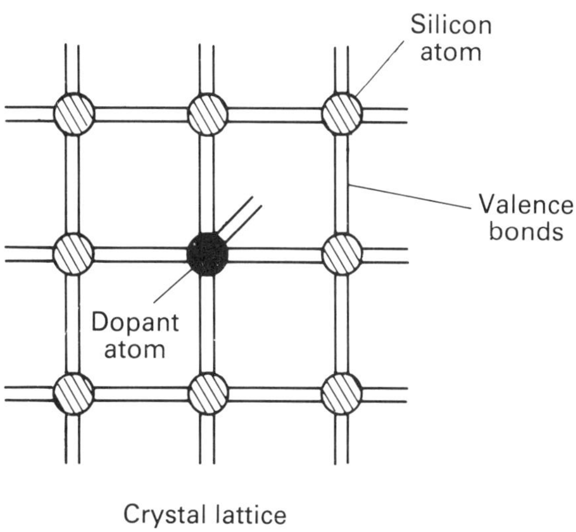

Fig. 4.7 — An impurity atom or dopant in the silicon crystal lattice.

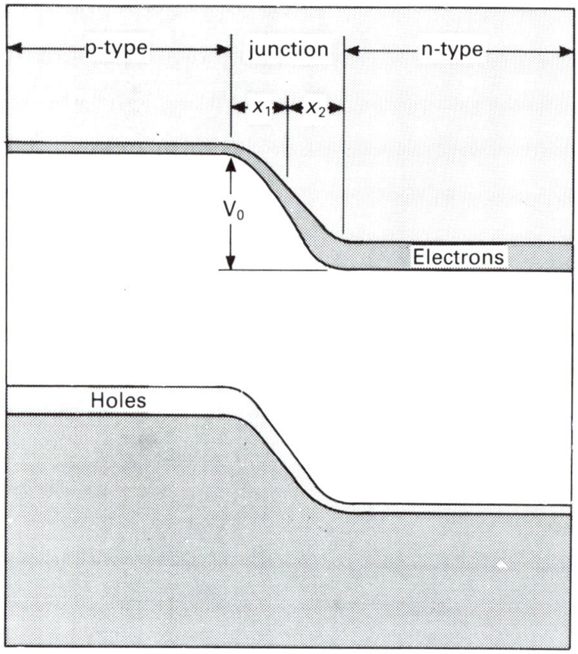

Fig. 4.8 — An energy-level diagram of a pn junction. Two differently doped layers of semiconductor silicon in contact.

4.2.2 Charge storage, charge coupling and clocking

Basically, a CCD is an array or chequerboard grid (Fig. 4.9) of numerous individual picture elements (pixels) each one of which can absorb photons of light and utilise the energy to release an electron within the semiconductor. But, if we are intent on making an imaging device then we do not want the photogenerated electrons to migrate away from the site of impact of the original photons. To trap the electron in the locality of a pixel requires a special electrostatic field to draw in the charged electron to a specific spot. What happens to the next photon? Clearly we need to create a storage region capable of holding many charges. This can be done by applying metal electrodes to the semiconductor silicon with a thin (0.1-μm) separation or buffer layer made from silicon dioxide which is an electrical insulator. The resulting structure behaves like a simple parallel plate capacitor which can therefore store electrical charge. It is called an **MOS** structure. What happens is that an electric field is generated inside the silicon slab by the voltage on the metal electrode. If the material is p-type (the usual case) then a positive voltage on the gate will repel the holes which are in the majority and sweep out a region **depleted** of charge just as in the pn junction. When a photon is absorbed it produces an electron–hole pair but the hole is driven out of the depletion region and the electron is attracted towards the positively charged electrode. The MOS capacitor is the combination of two parallel plate capacitors, namely the oxide capacitor and the silicon depletion region

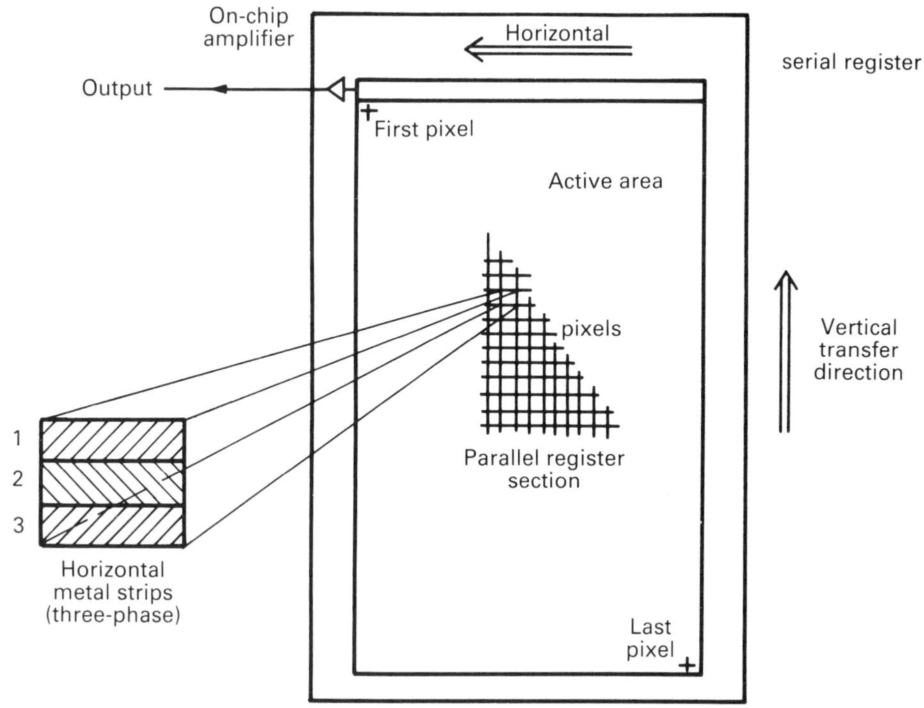

Fig. 4.9 — The layout of a CCD showing numerous square pixels laid out in a grid.

capacitor, and therefore the capacitance is proportional to the area of the plates (electrodes) and inversely proportional to their separation. Since the voltage on the plate can be controlled the depletion width can be increased or decreased, and so the capacity to store charge can also be controlled. The depletion region (Fig. 4.10) is an electrostatic 'potential well' or 'bucket' into which many photogenerated charges can be collected. As the voltage on the electrode increase, the 'depth' of the well increases; other ways are needed to create side-walls to the well. Eventually, at a certain 'threshold' voltage even the minority charge carriers due to impurities, electrons for a p-type semiconductor, will be drawn to the electrode.

It was Jerry Kristian and Morley Blouke who first pointed out that taking a picture with a CCD is a bit like measuring the rainfall over a rather large plantation! Suppose you distribute a large number of buckets in a rectangular pattern of rows and columns over the field. After it has stopped raining, measure the amount of water collected in each bucket by shifting the entire array of buckets towards a conveyor belt located at one end of the field, loading the buckets one whole row at a time onto the belt and then allowing the belt to convey each bucket from that row to a metering station where the amount of water in each is recorded. Then transfer another row onto the conveyor belt and keep repeating the sequence until all rows have been transferred to the metering station. By knowing the amount of water in each bucket the pattern of rainfall over the field can be generated. To visualise this

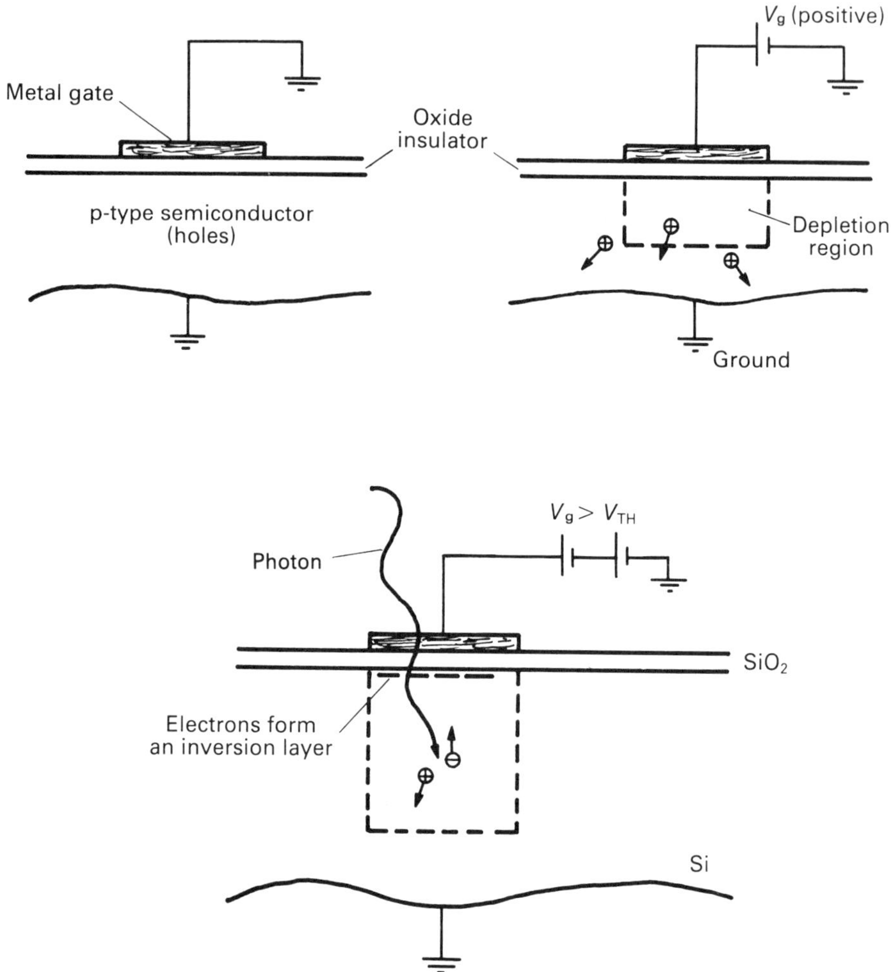

Fig. 4.10 — A single metal–oxide semiconductor (MOS) storage well, the basic element in a CCD.

pattern as a picture we convert the amount of rainfall in the bucket to a shade of grey — black for no rain, white for a full bucket — and mark this colour on our picture at the original location of the bucket. If instead of rain we think of photons of light, and our field of buckets is actually the pixels of the CCD detector in our camera, then we will have reconstructed the visual image of the scene. We will use this analogy of 'buckets of water' several times, but it is imortant to remember that in reality the photoelectrons accumulate near the surface.

A simple CCD structure is shown in Fig. 4.11; it is called a **three-phase** structure. Semiconductor silicon (in this case p-type) is covered with a thin electrical insulating layer made from silicon oxide on top of which are placed three sets of metal electrode strips. One of the three strips is set to a more positive voltage than the other two, and so it is under this one that the depletion region or bucket forms, and where the

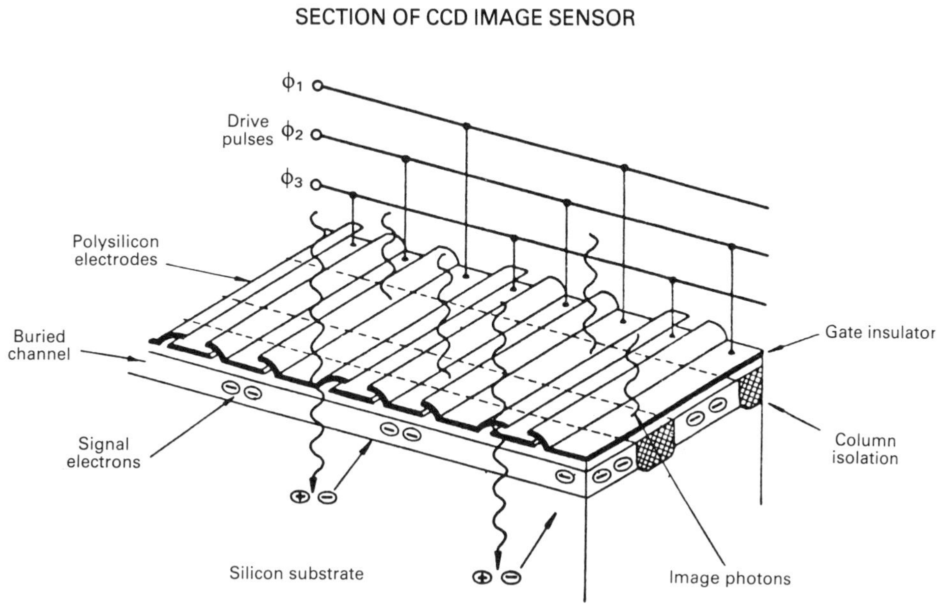

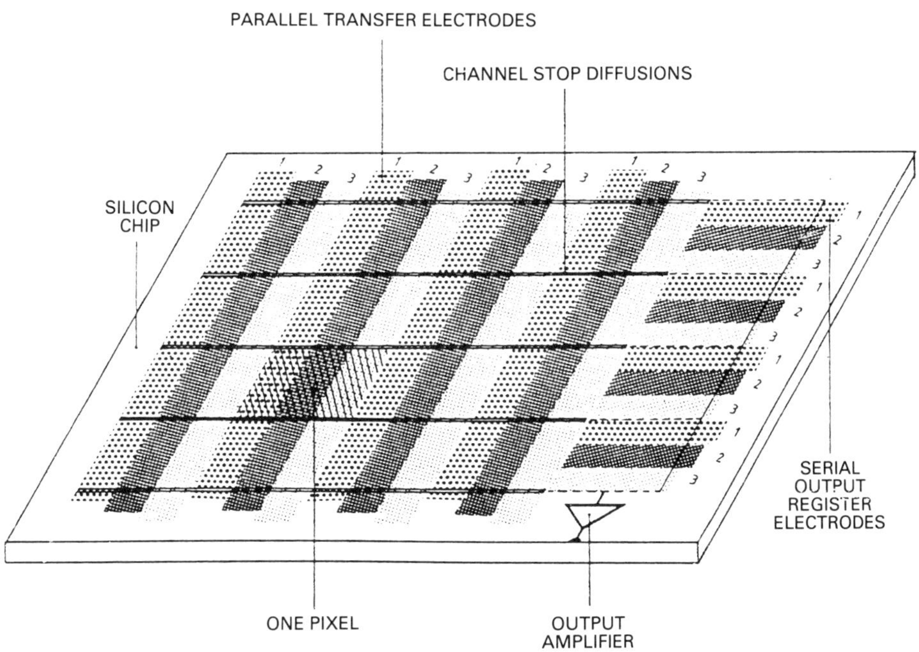

Fig. 4.11 — A schematic view from above showing the basic layout of a three-phase CCD.

electrons accumulate. We have created two walls of the well. By heavily doping the silicon crystal structure with a certain impurity it is possible to create a narrow channel which totally obstructs any movement of charge along the length of the electrode. So these **channel stops**, and the triplet of electrodes define the basic picture elements, with photogenerated charges being drawn to and collected under the middle (most positive) electrode — like water in a well (Fig. 4.12).

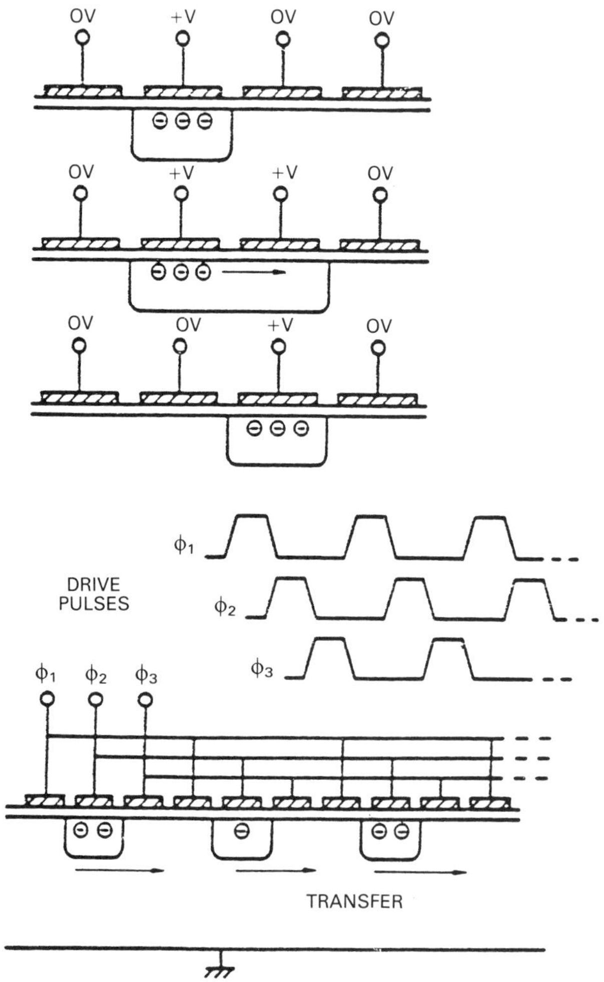

Fig. 4.12 — Charge-coupling in a three-phase CCD and timing waveforms for a three-phase CCD. In practice the degree of overlap between one electrode and the next depends on the CCD.

The novel feature of the CCD, which gives it its name, is the way in which the photogenerated charge, and hence the image of the scene, is extracted from the storage/detection site. It is called '**charge-coupling**'.

To transfer charge from under one electrode to the area below an adjacent electrode one must first raise the voltage on the adjacent electrode to the same value as the first one. This is like lowering the floor of the well. The charges will now flow like water and be shared between both buckets (refer to Fig. 4.12). Transfer can be in either direction, and by connecting sets of electrodes together the entire charge stored on the two-dimensional imaging area can be moved simultaneously in that direction. When the voltage on the original electrode is reduced to zero volts the transfer is complete because the collapse of the storage well pushes any remaining charges across to the new electrode. Since it takes three electrodes to define one pixel, then three of the above transfers are required to move the two-dimensional charge pattern by one pixel step along the direction at right angles to the electrode strips. The process of raising and lowering the voltage can be repeated over and over and is known as **clocking**. These drive or clock pulses can be described in a diagram called a **timing waveform** (as in Fig. 4.12) and are not difficult to produce electronically.

4.2.3 Interline and frame-transfer CCDs

Most CCDs are manufactured for use in video cameras and therefore the charge coupling or transfer must be done at very high speed (TV frame rates) and with light falling continuously onto the CCD. To achieve this, several manufacturers adopt an approach called **interline transfer**. In this scheme charges are moved sideways at high speed by one pixel to be relocated in a pixel which is shielded from light by an extra overlying strip of opaque metal as shown in Fig. 4.13(b). The charges are then coupled lengthways to transfer down the shielded column to the output row of the CCD.

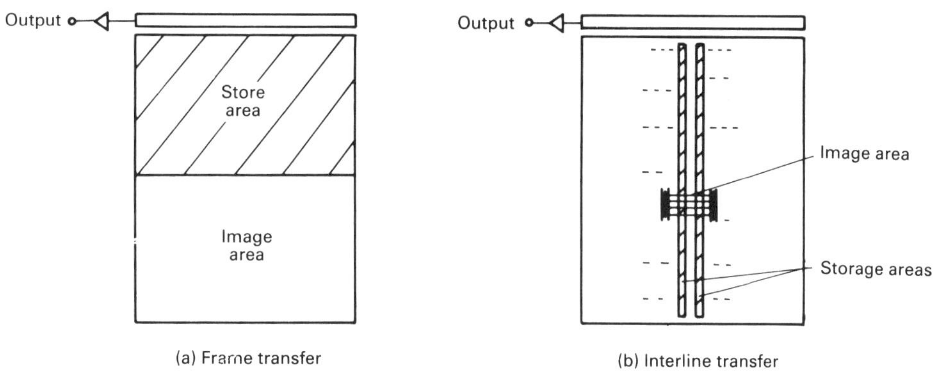

Fig. 4.13 — (a) Frame-transfer CCD. (b) Interline-transfer CCD.

Another approach is that of **frame transfer**. In a frame-transfer CCD as in Fig. 4.13(a) there are no blind spots in the imaging area, instead a duplicate imaging area, contiguous with the first, is covered by an opaque mask. The entire charge distribution representing the scene is shifted very rapidly lengthways until it disappears under the mask. It is then transferred lengthways once more, but at a slower rate, to be finally read out. The transfer rate for readout is a little faster than the permitted time which one can dwell on the image scene, whereas the image-to-storage transfer rate is extremely high to minimise blurring due to the illumination which is still present.

Frame-transfer CCDs are best suited to astronomical applications, especially since the mask covering the storage area can be removed to yield a CCD of *twice* the original size! Blurring of the optical scene during readout is not a problem in astronomical applications because a light-tight electronically controlled shutter can be used to block off the incoming illumination.

Several companies have manufactured frame-transfer CCDs which have been used for optical astronomy, including Texas Instruments, RCA, Ford/Photometrics and Tektronix in the United States, EEV/GEC in the UK and Thomson-CSF in France.

4.2.4 CCD outputs

So far we have indicated that the charge distribution representing the image scene can be coupled or transferred lengthways by indefinitely sending clocking pulses to the three electrodes. But for a CCD of finite size then all the charges are simply going to pile-up at one end!

In practice, of course, the transfer direction is terminated (as already shown in the previous diagrams) by a special row of CCD pixels called an **output register** whose electrodes are arranged at right angles to the main imaging area of the CCD — like the collection conveyor belt in the rainfall-measuring model — so that it can transfer charge horizontally rather than vertically. Since the output register is a single row it is usually called the **serial register** whereas the main area of the CCD is called the **parallel register**. At the end of the output register is a single output amplifier. Fig. 4.14 is a photomicrograph of the output corner of a typical CCD. A complete clocking sequence therefore consists of the following:

(1) A vertical shift of the entire image scene by one pixel. This delivers a row of charge to the output register.
(2) A horizontal shift through all the pixels in the output register. This delivers each charge in that row to the output amplifier, one pixel at a time, where the charge can be detected and converted to a voltage which can be measured and recorded.
(3) Another vertical transfer. This delivers the next row in the image.
(4) Another horizontal transfer.

The above process is repeated until *all* the rows in the CCD have been delivered to the output register and out to the output amplifier.

In general, the pixels (electrodes) associated with the output register are larger than those in the imaging area to ensure that they have more storage capacity and are therefore much less prone to saturation. We will discuss the effect of saturation later.

The output amplifier of the GEC P8603 is shown schematically in Fig. 4.15. This type of output has the advantage of being simple and non-critical. A packet of electrons with charge Q is allowed through the output gate onto an effective storage

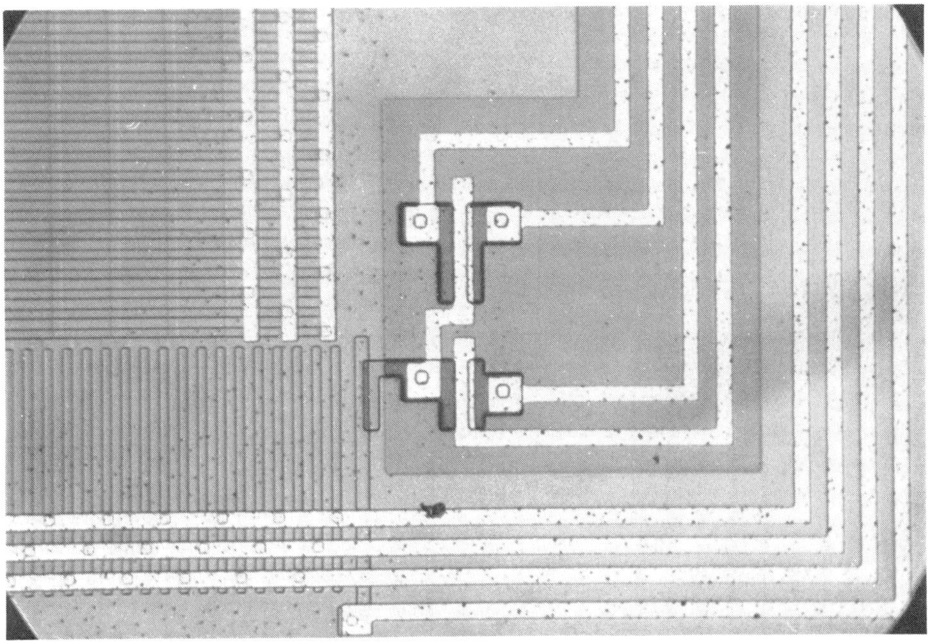

(a)

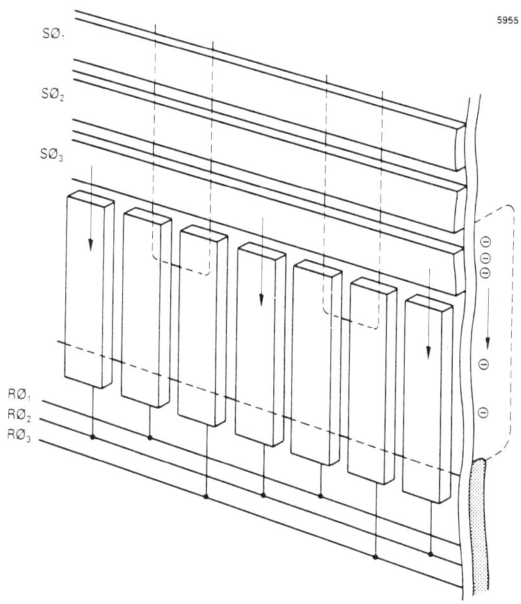

(b)

Fig. 4.14 — A micrograph (a) of the output corner of an RCA three-phase CCD. The large parallel register and the serial register at right-angles to it, as well as the output amplifier, can be seen. Photo courtesy of Pat Waddell. (b) A schematic view of the serial register interface. Courtesy EEV.

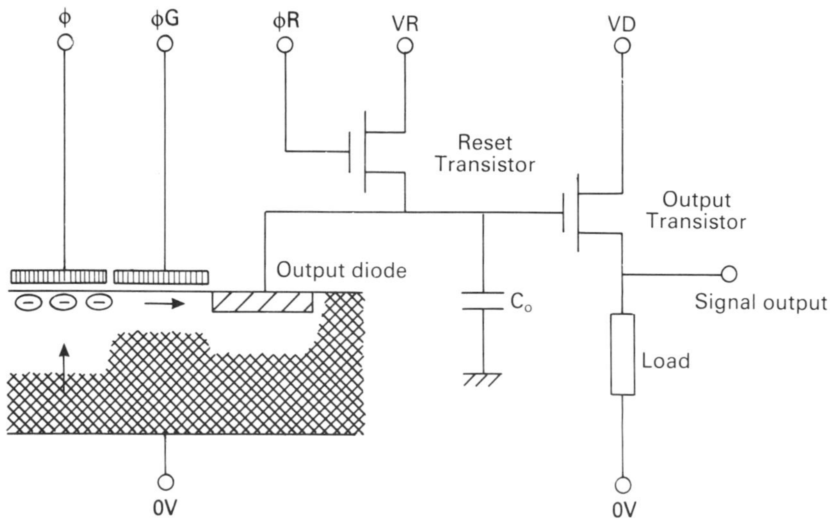

Fig. 4.15 — Output stage of the GEC P8603 CCD showing how charge is extracted in pulses by reading and resetting the output periodically. Courtesy EEV.

capacitance C (essentially a reverse-biased diode) which causes an instantaneous change $V=Q/C$ in the voltage of the input line of the on-chip transistor which in turn yields a voltage change at the output line. Before the next pulse arrives the storage capacitor is 'recharged'. The readout is destructive, that is it can only be done once, but this is of little consequence. Unwanted electrical noise due to the periodic resetting of the output diode can feed through to the output transistor and degrade the sensitivity of the CCD to faint signals, but fortunately this type of **reset noise** can be totally eliminated by subsequent electronic signal processing (section 5.6).

4.2.5 Buried-channel CCDs

Once the concept is grasped of how charges are created, stored and transferred, then a CCD becomes a remarkably simple, aesthetically beautiful idea with so many attractive properties that it is hard to resist!

While it is true that the simple CCD structures described so far can be constructed, that is metal electrode strips deposited directly onto an insulating layer on the top of a slice of uniformly doped silicon, they are far from ideal. In these CCDs the most positive potential lies at the silicon surface immediately under the insulating oxide layer; electrons are therefore stored and transferred at the surface of the silicon semiconductor and hence they are called **surface-channel** CCDs. Unfortunately, the surface layer has many crystal irregularities and defects in the crystal lattice which can readily **trap** charge but not so readily release it! The result is very poor charge-coupling and severe image smear.

To avoid the surface trapping phenomenon it is advantageous to 'grow' another layer of silicon onto the existing p-type substrate to separate it from the insulating layer. If this is a more highly doped n-type layer a more complex depletion region is created with a potential minimum (or collecting layer) which is substantially buried inside the bulk silicon. This is the **buried-channel** CCD shown schematically in Fig. 4.16. All modern CCDs are of this construction. Recall the discussion earlier of

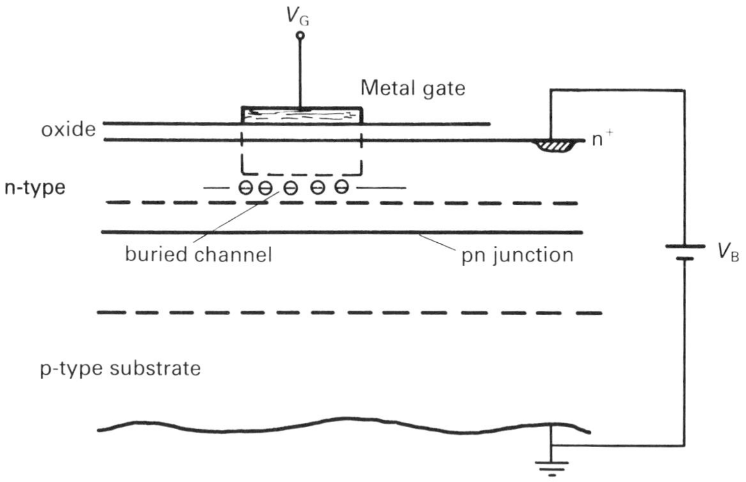

Fig. 4.16 — A single storage site in the buried-channel CCD. As the gate voltage increases the depleted zones (dashed lines) finally meet. The collection layer lies well below the surface at the overlap between the gate depletion and the depletion of the pn junction.

the pn junction in which the most positive potential occurred on the n side but within the depletion region. All that has been added on the n-type layer is an insulating oxide and a metal control gate. The most positive potential still lies well away from any surface. If a voltage more positive than any of the gate voltages is applied to this n-type layer, a depletion region will form at the pn junction. An additional depletion under the gate is generated by a difference in gate potentials and the exact depth of the buried channel can therefore be controlled by the electrode potentials as before. Charge transfer within the bulk silicon is very efficient because the number of trapping sites is considerably fewer and much less noisy. The analogy with buckets of water still holds, and charges stored at the embedded collection layer can still be charge-coupled from one pixel to the next by pulsing or clocking the metal electrodes. A penalty is paid in terms of a somewhat reduced storage capacity, i.e. a buried-channel CCD will saturate before an equivalent surface-channel CCD. Incidently, similar benefits are accrued if the output amplifier is also constructed in buried-channel form.

As we shall see in later chapters, CCDs are no longer operated with positive voltages on the storage electrode and near zero volts (the substrate value) on the adjacent 'barrier' electrodes (see section 5.4).

In summary, therefore, to construct a practical high-performance CCD one starts with the buried channel, a region where electrons are confined in depth (z) below the gate but still free to move from side to side (in x and y); this is usually n-type on a p-type substrate. Next, narrow columns of heavily doped n-type material are diffused into the normal n-type region to produce channel stops which subdivide the x-direction and constrain electrons to move only in the y-direction along columns. A serial register is formed by leaving a gap at one end between the endpoints of all the channel stops and the edge of the CCD. Electrons emerging into this narrow strip could move in x once more. Finally a silicon dioxide insulation layer is grown on top and the basic three-phase electrode gate structure is added as rows or strips across the entire parallel area to define subdivisions (pixels) in the y-direction; the serial register receives similar treatment but the gates are at right-angles to the parallel section (refer once more to Fig. 4.14).

4.2.6 Two-phase, four-phase and virtual-phase CCDs

There are alternative structures to the simple and compelling three-phase CCD outlined above.

If bidirectional charge motion is not required then the CCD can be doped along the direction of vertical transfer in such a way as to provide a barrier to backward-flowing charges. This is a **two-phase** CCD. The principle is illustrated in Fig. 4.17(a). An implant is diffused into one half of the substrate below each of the two electrodes. The presence of this layer affects the depth of the depletion region immediately beneath it in such a way that the depletion is always greater under the implant. With one electrode low and the other at some fixed voltage level, the result is a staircase-shaped well. When the low electrode is raised to a higher voltage than the fixed electrode, charges are forced to move left-to-right as shown in Fig. 4.17(a).

Texas Instruments have manufactured a single-clock CCD known as the **virtual-phase** CCD (Janesick *et al.*, 1981). The principle of this device is very similar to a two-phase CCD as shown in Fig. 4.17(b). Consider again the two-phase CCD. One electrode was left at a constant voltage to produce an intermediate depletion region. There is really no necessity to do this with a semi-opaque overlying metal gate since the same effect can be achieved by a diffusion directly into the silicon substrate. Only one electrode need be physically present and the other half of the pixel is left clear and uncovered except for its implant — this is the 'virtual' electrode or virtual phase. As in the two-phase device, the roles of other electrodes are taken by a series of potential steps created by different levels of surface doping as shown in Fig. 4.17(b). These potential steps direct the flow of charge as the voltage on the single electrode is driven above and below the potentials of the uncovered or virtual electrode. Apart from a considerable simplification in the number of clocks, this kind of CCD structure was chosen to permit an improved blue response by minimising the amount of absorption due to the polysilicon (conducting) electrodes when it is frontside-illuminated. It is intended that these CCDs be used in the Galileo Mission to Jupiter because with implants instead of polysilicon electrodes there is less likelihood of gate shorts caused by ionising radiation. Texas Instruments market a version of this device architecture with pixels which can be colour-sensitised by addition of a grid of tiny colour filters.

A few CCDs have been made with four physical electrodes per pixel which would correspond to the four potential steps apparent in the two-phase and virtual-phase

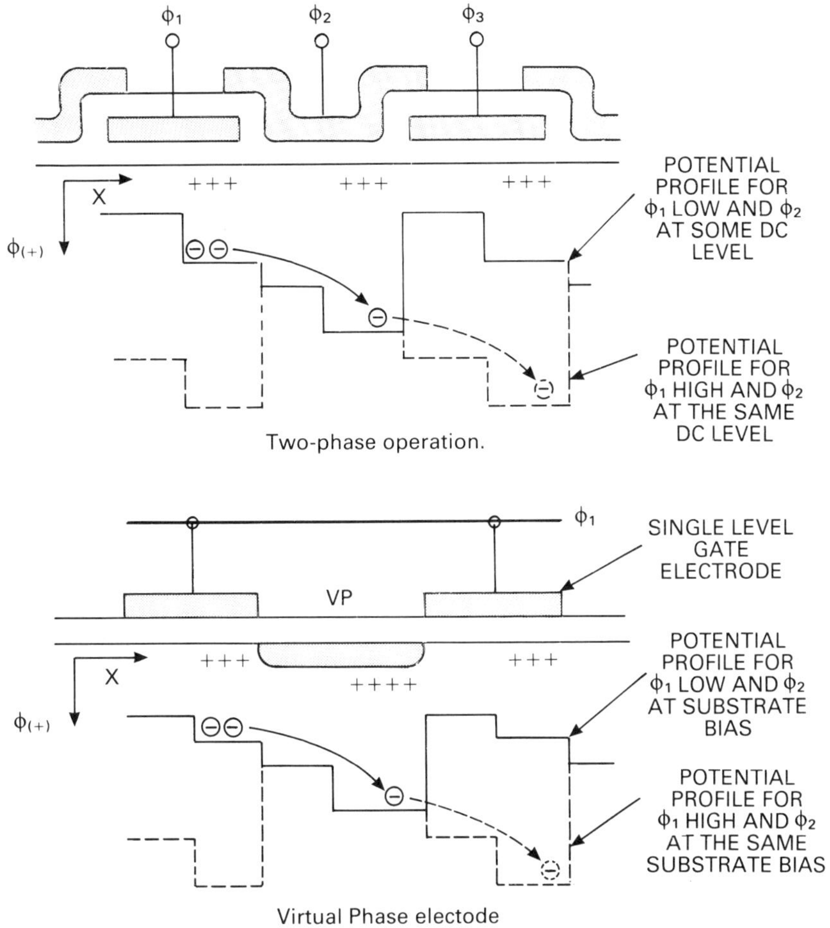

Fig. 4.17 — A two-phase CCD and the Texas Instruments virtual(or single)-phase CCD. Selective doping eliminates the need for the other electrodes by creating different 'floor levels' within a pixel.

devices. This complexity gives complete control over all combinations of phases and can be used for special applications involving two alternating image scenes — so-called 'chopped' signals — or to increase charge storage capacity.

4.2.7 Backside-illuminated CCDs

So far we have been describing CCDs in which the illumination or photon flux was compelled to pass through the overlying electrode structure in order to reach the depletion (or storage) region in the silicon. In practice this simple approach results in severe absorption of blue light — the term 'blue photons' is also used — in the electrodes before they can be detected by the silicon; the device has almost no blue response at all!

To get around this problem the CCD can be turned over and illuminated from the backside (Figs 4.18, 4.19). Before this becomes effective however, the thick silicon

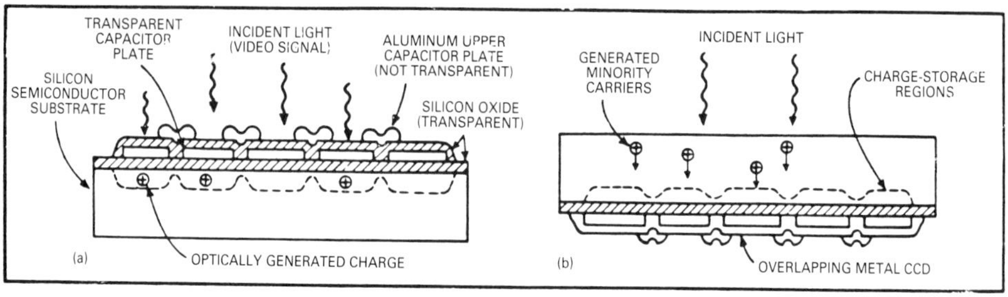

Fig. 4.18 — Backside illumination of a CCD.

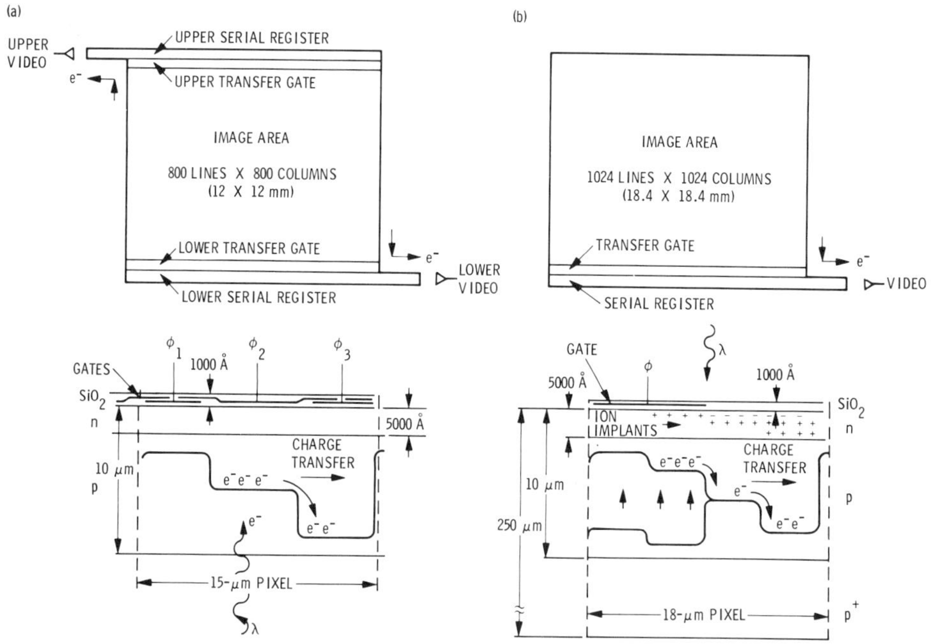

Fig. 4.19 — Detailed comparison of the structure and dimensions of the thinned backside-illuminated three-phase and thick front-illuminated virtual phase devices from Texas Instruments. Courtesy Jim Janesick.

substrate must be reduced in thickness (either mechanically or chemically) to only 10 μm or so. Such thinned, backside-illuminated CCDs have excellent response in the blue and violet. If thinned too much they lose their red response because the red photons need more absorption length, and if this is not there, they will pass right through the silicon!

One disadvantage of the thinned CCDs is that they are more mechanically fragile and prone to warping. Some thinned CCDs, in particular the RCA chips, are mounted to a supporting substrate. Thinned CCDs also permit **interference fringing** to occur due to multiple reflections internal to the CCD substrate or between the silicon and the supporting substrate in a manner similar to fringes which can be seen when two flat glass plates are placed almost in contact but at a tiny angle. If the plates were truly flat and parallel the fringes would become the same as for a Fabry–Perot Interferometer described in Chapter 2. At least for imaging, this problem can be eliminated by computer processing and calibration techniques discussed in more detail later.

4.3 NON-ASTRONOMY APPLICATIONS OF CCDs

To conclude this chapter it is appropriate to examine other driving forces and non-astronomical applications of CCDs. Currently, CCDs are making their greatest contribution as imaging devices. Over 20 companies, worldwide, manufacture CCDs in some form. Very few are mentioned here because many of these CCDs are designed for colour TV operation at frame rates (exposures) of 1/60th of a second; many have microscopic colour filters on the CCD pixels and are unsuitable for astronomical imaging applications. Several companies, most notably the Japanese and the Kodak company in the USA, have striven to produce very large arrays of very tiny pixels with each pixel 'colour-sensitive' by the application of some kind of microscopic colour filter. (Bell Labs announced a colour camera in 1972!)

Although CCD cameras were used on the Vega and Giotto spacecraft for the encounter with Comet Halley, and they will be used extensively on future space missions, including the Hubble Space Telescope and the Galileo mission to Jupiter, and although CCDs are available at every major observatory in the world, the astronomy market remains small. Incidentally, the CCD on the Vega spacecraft was manufactured in the Soviet Union.

CCDs have numerous non-astronomical applications however, which are beginning to flourish. For instance, CCDs are now used in many 35-mm cameras as light sensors and the profusion of hand-held home video 'camcorders' which sell for around \$1000 to \$1500 all employ CCDs. Almost any and every application in which a picture of a scene would help is open to CCDs, but they are particularly valuable in circumstances of low light levels, or low contrast levels, or situations in which the light emission is very red, or when a comparison of images at very different wavelengths is significant. Many uses in the medical field have been found. CCDs are useful in remote surveillance of inaccessible objects and as a tool in optical testing. Because a CCD image is readily digitised and available to a computer for digital image processing and enhancement, and because of its stability, CCDs can replace almost all existing forms of imaging device.

Richard Aikens of Photometrics Ltd lists 23 representative applications including light microscopy, electron microscopy, fluorescence microscopy, opthalmology, pathology, radiology, hydrology, geology, oceanography, fluorescence spectroscopy, atmospheric sciences, high-energy physics, Raman spectroscopy, deep-water imaging, electrophoresis and fusion research.

Craig Mackay of Astromed gives several detailed examples of ways in which CCDs can be used in medical applications. For example, modern microscopes are capable of such high image quality that only a sensitive CCD detector is capable of recording the image with sufficient quality. In transmission or reflection mode, microscopic structures with too low a contrast to be measured with conventional imaging systems can be easily visualised with a cooled, slow-scanned CCD camera. For fluorescent studies the extreme sensitivity enables experiments to be done with much lower levels of ultraviolet excitation thus minimising damage to delicate structures, and with smaller quantities of fluorescent dyes which reduces the effects of the dye on the sample under the microscope. Another remarkable example is two-dimensional gel electrophoresis in proteins. On a conventionally stained two-dimensional gel the faintest dark spots must be detected against a bright transparent background. Much greater sensitivity may be obtained by tagging the proteins with a small amount of fluorescent marker dye and using a cooled, slow-scan CCD to detect exceptionally weak light emission, now against a dark background, from spots containing only a few thousand protein molecules. Amazingly, as Craig points out, astronomical image processing software for handling crowded star or galaxy fields is of immediate use for the automatic extraction of spot parameters from gel images. This is perhaps a nice reminder of the subtle relationship between pure and applied science.

These and many other uses of CCDs should ensure that they endure for a long time to come. Indeed, as Bill Boyle remarked to me, the 'best may be yet to come'.

REFERENCES AND SUGGESTIONS FOR FURTHER READING

Amelio, G. F. (1974) Charge-coupled devices, *Scientific American*, February.

Benyon, J. D. E., and Lamb, D. R. (1980) *Charge-coupled devices and their applications*, McGraw-Hill.

Boyle, W. S. and Smith, G. E. (1971) Charge-coupled devices — a new approach to MIS device structures, *IEEE Spectrum* Vol. 8, No. 7, 18–27.

Janesick, J. and Blouke, M. (1987) Sky on a chip: the fabulous CCD, *Sky and Telescope*, September, Vol. 74, 238–242.

Janesick, J. R., Hynecek, J. and Blouke, M. M. (1981) A virtual phase imager for Galileo. In Solid state imagers for astronomy, *Proc. Soc. Photo-Opt. Instr. Eng.*, Vol. 290, 165–173.

Kristian, J. and Blouke, M. (1982) Charge-coupled devices in astronomy, *Scientific American*, October, Vol. 247, 66–74.

Mackay, C. D. (1986) Charge-coupled devices in astronomy, *Annual Reviews of Astronomy and Astrophysics*, Vol. 24, 255–283.

McLean, I. S. (1988) Infrared astronomy's new image, *Sky and Telescope*, March, Vol. 75, 254–258.

Sequin, C. H. and Tompsett, M. F. (1975) *Charge transfer devices*, Academic Press, New York.

5

Practicalities of operation

The characteristics of commercially available CCDs suitable for low light level astronomical imaging applications vary widely. Moreover, when faced with optimising the performance of any given CCD for properties such as high quantum efficiency and exceptionally low noise, it appears that each device is in some way unique. There are many practicalities to be faced, both for the designers and builders of CCD-camera-based instruments and for the astronomer, and these range from the peculiarities of individual chips themselves to the methods of computer-control and data analysis. These problems and how researchers tackle them are treated in an illustrative way in the next four chapters; there is no intention here to provide a 'constructor's manual'.

5.1 CLOCK VOLTAGES AND BASIC ELECTRICAL FACTORS

CCD manufacturers provide a **data sheet** which gives the electrical pin connection diagram for the device, the names and symbols for each pin, the voltages or range of voltages to be applied to each pin and the **timing diagram**, this is, a diagram showing the time sequence of and relationship between the CCD drive signals. Voltages applied to CCDs are of two types. Fixed voltages, referred to as **dc bias** levels, which remain unchanged after switch-on, and pulsed or **clock** voltages which can be switched back and forth between two voltage levels known as the **high** and **low** levels. As described in Chapter 4, clock voltages are applied in a precise order and time sequence to **charge-couple** the electrons in one storage well to the next. Inevitably, the type of mounting package used, its size and the number of pins and their names, differ from device to device and from one maker to another. For instance, the early Fairchild 100×100 CCD came in a 24-pin dual-in-line package whereas the Tektronix 2048×2048 CCD comes in a 52-pin four-sided flat-pack. Fortunately, the situation is not as confusing as might at first appear. Certain functions are required in all CCDs, and in particular in the **frame-transfer** image sensors used in astronomy (Fig. 5.1), namely:

(1) **Serial (horizontal) register clocks**. One pin for each phase or electrode used to

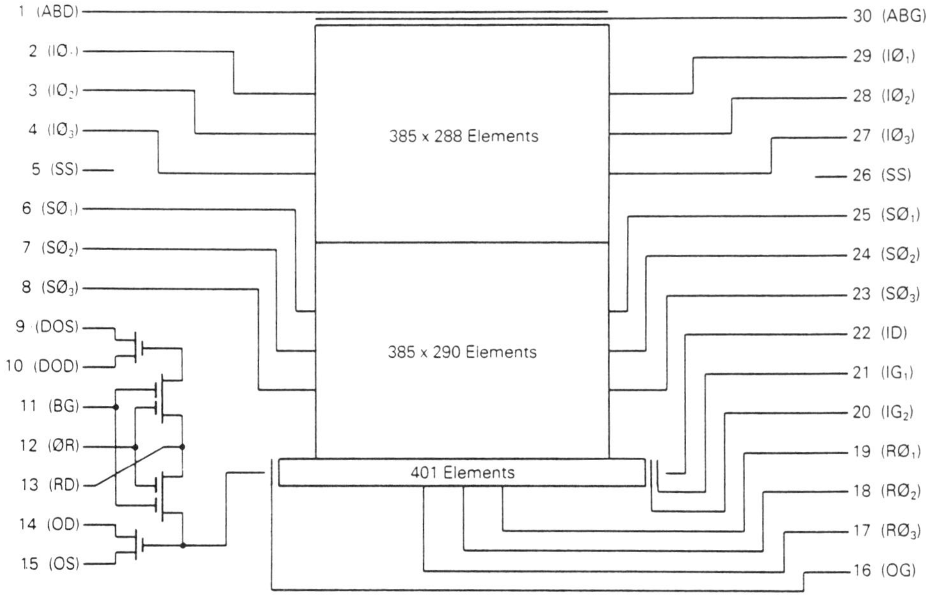

Fig. 5.1 — Device schematic of the EEV/GEC P8603 frame transfer CCD to illustrate pin connections and terminology. Each manufacturer will provide information of this kind on a data sheet. Courtesy of EEV.

define the pixels in the horizontal register is required, i.e. three for a three-phase CCD. In larger CCDs such as the Tektronix devices, two completely separate serial output registers are provided.

(2) **Parallel (vertical) register clocks.** Again one voltage line is needed for each phase. In frame-transfer CCDs the two-dimensional vertical or parallel register is split into two identical sections — the **image section** and the **store section**— which can be controlled separately and therefore two sets of pins are needed.

(3) **Reset transistor clock.** A single, periodically recurring voltage pulse is required to reset the CCD output amplifier or, more accurately, the output charge collecting capacitor during the readout process.

Each of these clock voltages will have a specified high and low level and, since the voltage difference or 'swing' — usually in the range 5 to 10 V — can have important consequences on performance, it is often arranged to select the levels on demand from the CCD controller. The total range of voltage required to operate a CCD is generally less than 20 V.

The most important dc bias voltages are as follows:

(1) **Substrate voltage** (V_{sub}). The reference for all other voltages. This voltage is usually, but not always, kept at ground or zero volts.

(2) **Reset drain voltage** (V_{RD}). This voltage is applied to the 'drain' terminal of the on-chip 'reset' field-effect transistor (FET) at the CCD output to establish the

level to which the output node (capacitor) must return after each charge packet is read out.

(3) **Output drain voltage** (V_{OD}). This voltage applied to the drain terminal of the on-chip output amplifier determines the operating point of that transistor.

(4) **Output gate voltage** (V_{OG}). The output gate is essentially an extra (last) electrode in the serial output register.

Pulsed voltage signals, corresponding to individual charge packets, emerge from the **source** of the output transistor (OS) — also called the video output — which is connected to ground via another transistor external to the CCD to provide a constant current load. The output transistor source current (IOS) is important for most CCDs.

On the Tektronix CCDs (Fig. 5.2), being of more recent design, a few additional options have been added. In particular, the vertical register (the B register in the Tektronix data sheet) can be clocked up or down to two quite separate serial registers (called the A and C registers). The A register terminates with an on-chip amplifier designed for low-noise slow-scan operation whereas the C register has an output amplifier optimised for TV video rates. In addition, the Tektronix chips have a separately clocked gate known as a 'summing well' which has the storage capacity of two serial pixels.

Almost all CCDs have electrical input connections and test points; the manufacturer will specify whether these pins should be fixed at a high or low voltage. Remember also that voltages are relative to the substrate voltage which may not always be zero volts.

For a three-phase CCD with one electrode set high (numerically) and the other two low, and with each of the recommended dc bias voltages applied, a CCD will store a two-dimensional charge pattern which is the precise analogue of the optical image. Actually, in practice it is preferable to integrate with two electrodes set high and only one set low to form the inter-pixel barrier. This mode of operation results in sharper discrimination between pixels and reduces pixel-to-pixel variations. To remove the charge pattern at the end of the exposure a shutter is closed to avoid image smear and a regular periodic stream of pulses applied to each of the three phases or electrodes. Electronic circuitry controls the sequence of pulses, their total number and type (vertical, horizontal and reset), and their high-to-low swing in voltage.

Normally, a CCD manufacturer will provide an initial set of operating voltages for a particular chip, but will not necessarily optimise these voltages for cooled, slow-scan astronomical work. Builders of commercial astronomical CCD camera systems can provide this service. As already hinted, small changes of one-tenth of a volt to clock swings or dc bias values can often yield substantial improvements in low light level behaviour for individual CCDs. Recognising that there would be differences in clock voltages and clocking schemes from CCD to CCD, but curious nonetheless to see the range of voltages which different workers had settled on, Dr Carol Christian of the Canada–France–Hawaii Telescope Corporation decided to collate as many clock voltages as possible, and has kindly made them available to add to my own collection for inclusion in this book. In Appendix 5, Tables A5.1 to A5.3 summarise the results for RCA CCDs, Table A5.4 gives clocks for the 800×800 TI chips, and

512 × 512 PIN-OUT

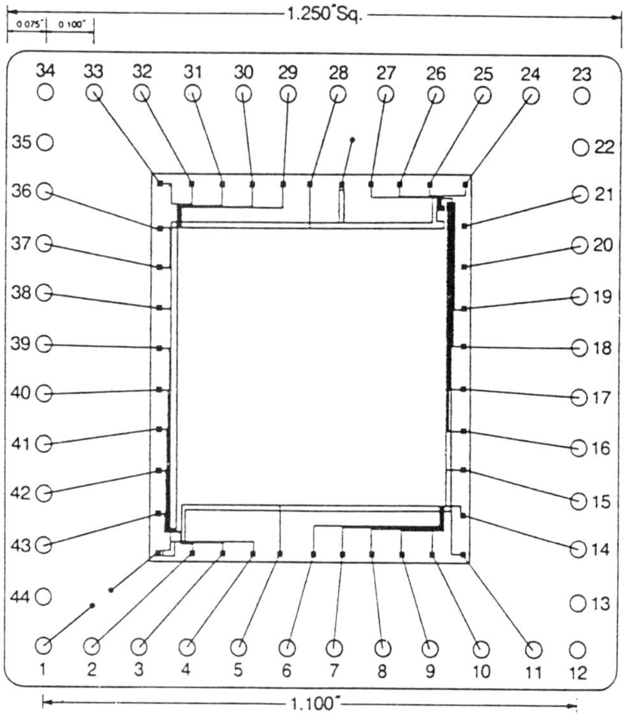

Note: Pin 1, backside contact & package body.

Functional Diagram

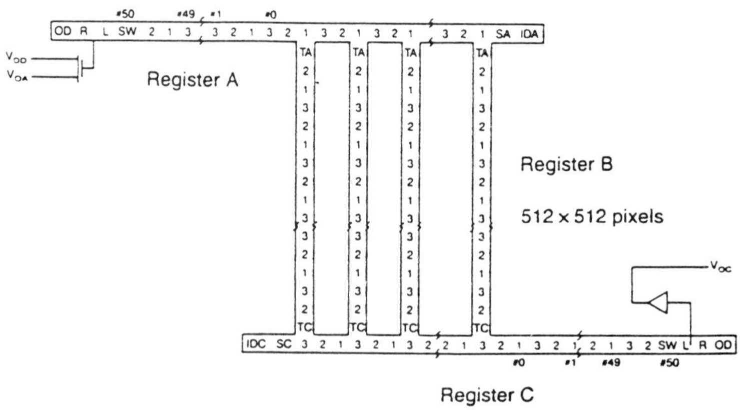

Fig. 5.2 — Schematic pin connection and functional diagram for the Tektronix 512×512 CCD showing the double horizontal (serial) readout registers, one for TV rates, the other for cooled slow-scan. Courtesy of Tektronix Inc.

Tables A5.5 and A5.6 cover the Thomson-CSF and Tektronix devices. Table A5.7 deals with the EEV/GEC devices. A terminology cross-reference is given in Table A5.8. The voltages given in these tables are not a 'recipe for success', but may be considered advisory for new investigators.

Mechanically, CCDs are quite robust unless heavily thinned — although Jim Janesick recalls a horrific moment when during the course of handling one of the 800×800 CCDs at JPL a 6/32 screw somehow rolled off a shelf above the workbench on a trajectory aimed squarely at the centre of the chip! He saw it out of the corner of his eye, and then all time seemed to stand still; an eternity passed in the blink of an eye as he saw the thinned surface bow under the impact and then give way. Thinned CCDs may not be able to survive a collision with a tiny screw but they can withstand substantial illumination overloads without permanent damage. However, they are integrated circuits of the CMOS type and their tiny gate connections can be short-circuited by static electricity discharges. Precautions must be taken when handling CCDs. For example, CCDs should be stored in electrically conducting containers, earthing-straps should be tied to the wrist during handling operations, no nylon clothing worn, and the work performed in a clean, ionised (electrically conducting) airflow. Finally, some kind of protection on the drive signals is essential if power supplies are used which have a rating even slightly above the maximum recommended voltages for the CCD. Usually this is achieved by the use of Zener diodes on the drive outputs — but they must be installed the right way round, as I learned to my cost! With these precautions, a CCD camera can last a long time.

Most users of CCDs accept that individual devices may have imperfections that must be dealt with, but many are surprised by the significant effect that the CCD electronic hardware systems used at telescopes can have on data collection. The heart of the electronic system is a signal processing unit designed to sample and filter noise, but proper grounding, power supply decoupling and optimum control timing are also very important factors in achieving a low-noise system. Undoubtedly, **ground-loops** are the most common cause of noisy CCD systems.

A ground-loop forms when two interconnecting parts of an electronic system (for example, the detector and the preamplifier) are separately connected to ground, i.e. the zero potential of the earth. Unfortunately, direct connections to true earth ground are rarely possible, since most ground connections are through the third wire of the ac line which has a small but significant resistance. As a result, a voltage difference can exist between 'grounds', and currents can flow. The practical effect of this is the introduction of electrical 'noise' into the signals going between the two units. Ground-loops between the telescope and camera body and driver electronics, within the driver electronics, or between the driver electronics and the computer system can cause interference with the readout electronics, giving clearly visible sloping, 'stripe' patterns in the CCD images; these patterns are synchronised with the mains frequency. The solution to this problem is to have only a single ground point in the system, to which all the zero reference points and shields are conencted; because many wires may radiate from such a single ground point it is often called a 'star ground' (see Fig. 5.3). This solution is harder to carry out in practice than it sounds. All earth shields on cables, such as coaxial cables, and all metal cases (or chassis) must be electrically isolated from conducting surfaces that may be connected to ground at points other than the single star ground; separate pieces of ac-powered

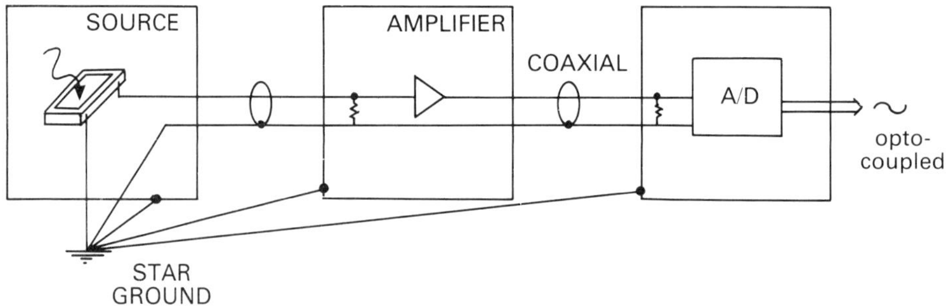

Fig. 5.3 — A simple example of the 'star' connection method for earthing (grounding) sensitive equipment with no earth loops.

equipment must have their ground connections effectively removed. All connections to the single system ground should be made as short as possible, with the lowest-resistance electrical wires available.

Quite often, low-noise CCD systems can be fabricated successfully in the laboratory, only to find that there are severe noise problems at the telescope. Invariably, the explanation will be an overlooked ground loop. Many designers carefully isolate the entire instrument from the telescope structure — even if the telescope is known to have an excellent earth ground — which means that extreme care is required when attaching any other piece of electrically-powered apparatus to the instrument in case a ground-loop is formed.

Similarly, electrical noise from motors, light dimmers, or computer parts can be 'picked up' — by capacitance coupling, inductive coupling or radiative coupling — if inadequately shielded wires or components are used in the CCD system which we should remember is capable of detecting signals of only a few millionths of one volt! **Coaxial cables** which have a surrounding braided copper shield are quite effective (90%), and **twisted-pair wire** is often used in preference to a single wire for carrying critical signals to the CCD. Most astronomical CCD systems have a low-noise preamplifier inside or very close to the cryostat to boost the CCD signal and most systems convert from the weak analogue output to healthy digital signals before transmitting the data over long cables. The physical environment at cold, high-altitude mountain-top observatories can cause 'drifts' in the operating points of many electronic components. All of these things can and must be overcome by careful design and attention to detail.

5.2 DARK CURRENT AND COOLING

Random mechanical motions of atoms, caused by the heat of normal room temperatures, within the silicon lattice will release sufficient energy to give rise to a continuous stream of electron–hole pairs in the absence of light. This process for electron–hole pair production is called 'thermal' because the energy source is heat. Depending on the type of silicon used, this **dark current** can be very substantial. At room temperature the dark current of a typical CCD is about 100 000 electrons per second per pixel (or equivalently, 1.8 nA/cm^2 for 30-micron pixels; a nanoamp (nA) is 10^{-9}A) which means that the CCD storage wells will fill up, saturate and spill over on dark current alone in just a few seconds. If the CCD is read out very rapidly and continuously at a high rate (frequency) such as 5 MHz (TV rates) then the dark current is cleared after only a brief accumulation time (about 1/60th of a second) and so the CCD can be used for ordinary TV applications. Most astronomical sources are much too faint to yield a good image in such short snapshots, especially since there is a small but definite 'noise penalty' paid for every readout. Fortunately, the solution is straightforward — it is 'cooling'. Or at least that solution seems straightforward now, but when it was first suggested to Bell Labs engineers by Jim Westphal back in 1974 it was considered radical!

Since the thermal dark current undergoes a very rapid decay with decreasing temperature, a drastic reduction in dark current can be achieved by cooling the CCD to a temperature somewhere between those of **dry-ice** (solid carbon dioxide (CO_2)) and **liquid nitrogen** (LN_2). In silicon the dark current falls to about one-third of its previous value each time the temperature is reduced by about 10°C. Mathematically, this relation can be expressed as

$$N_D = \frac{I}{e} d^2 \exp\{-(T_0 - T)/9.96\} \text{ electron/s}$$

where 'exp{}' means 2.718 to the power of the number within the brackets, I_0 is the dark current in amps per cm^2 at room temperature (T_0), d is the pixel size in cm, T is the operating temperature in Kelvin and e is the charge on the electron. A typical operating temperature for an astronomical CCD is about 135 Kelvin (135 K), that is 135 centigrade degrees above **absolute zero**, or −138°C. It is possible to end up with a dark current of only 30 electrons per pixel per *hour* instead of 100 000 electrons per second!

There is one apparent disadvantage to cooling. In order to cool the tiny CCD to very low temperatures it must be encapsulated in a fairly bulky vacuum chamber, hundreds of times bigger and heavier than the chip itself. For astronomers this is a minor and perfectly acceptable inconvenience to achieve a thousand-fold increase in integration time!

To get the CCD cold, the chip is brought into mechanical contact with a small block of copper (a good heat conductor) which is itself very well anchored (thermally) to a large reservoir at a low temperature — usually a bath of LN_2 (77 K). One arrangement is shown in Fig. 5.4. To prevent frosting, the entire arrangement is enclosed in a vacuum chamber with an optically transparent entrance window to allow light to reach the CCD. A vacuum chamber with a built-in reservoir of LN_2 is called a **cryostat**. In practice the CCD cannot be allowed to cool all the way down to the temperature of LN_2 (77 K) because, at these temperatures, the behaviour of the p- and n-type doped semiconductor silicon is itself affected and charge-coupling becomes harder to accomplish. Therefore some mechanism to control the temperature must be installed. For infrared cameras, even LN_2 is not cold enough and **liquid helium** (4 K) must be used. The design and development of cryogenic vessels for use at all orientations of a telescope is an art in itself. Many different forms and

Fig. 5.4 — An EEV/GEC CCD mounted in a socket and firmly pressed down against a block of copper; the copper has been gold-coated to prevent tarnishing. Two large heater resistors are embedded in the copper block. Courtesy Royal Observatory Edinburgh.

geometries are commercially available. Astronomical cryostats and CCD mounting arrangements are considered again from a system-design point of view in Chapter 6.

For CCD imagers there are three main sources of dark current. These are

(1) thermal generation and diffusion in the neutral bulk silicon,
(2) thermal generation in the depletion region,
(3) thermal generation due to regions called 'surface states' at the junction between the silicon and the silicon dioxide insulation layer.

Of these sources, the contribution from surface states is the dominant contributor for multi-phase CCDs. Dark current at the Si–SiO_2 interface depends on two factors: the density of interface states and the density of free carriers (electrons or holes) that populate the interface. Electrons can more easily thermally 'hop' from the valence band to an interface state (also called a mid-band state because the energy level associated with this disturbance in the crystal lattice lies in the forbidden energy gap of the normal semiconductor crystal) and then to the conduction band, thus producing an electron–hole pair. The presence of free carriers will fill interface states and, if the states are completely filled, will suppress thermal hopping (valence band electrons no longer have a stepping stone to the conduction band) and therefore substantially reduce dark current. Normal CCD operation as described earlier, however, actually depletes the potential well and the interface of free carriers, thus maximising dark-current generation. Under depleted conditions, dark current is determined by the quality of the Si–SiO_2 interface. Dark current can be significantly reduced by operating the CCD in **inversion** mode. In inverted operation the CCD well or bucket is formed by an electrode voltage near zero volts (or close to the substrate) rather than +10 V, while the 'barrier' clock level applied to the other electrodes is set to a negative voltage of −10 V. With this arrangement, holes are attracted up to the Si–SiO_2 interface at these electrodes thus populating the troublesome states with holes. Not only does inverted operation improve dark current, but the neutralisation of the interface states eliminates 'residual' images associated with saturation and problems of charge capture (see section 5.4). The inversion phenomenon was first noticed by Tony Tyson of Bell Labs while investigating RCA CCDs but its significance was not appreciated until a similar effect was uncovered later by Jim Janesick and colleagues at JPL for the TI devices.

Virtually all CCDs which are capable of inverted operation are now run in this manner and new CCDs are being manufactured with a technology called multi-pinned phase (MPP) which allows the CCD to operate totally inverted at all times while maintaining other performance characteristics. MPP CCDs fabricated at Ford Aerospace and EG&G Reticon have achieved dark levels of 0.025 nA/cm^2 at room temperature, and thinned backside-illuminated devices with a substantial part of the epitaxial layer etched away to reduce bulk dark-current contributions have attained 0.010 nA/cm^2. The new MPP technology also aids charge transfer efficiency by allowing higher operating temperatures to be used.

5.3 COSMIC RAYS AND LUMINESCENCE

With the elimination of dark current, cooled CCDs become capable of extremely long exposures. However there is now another effect which comes into play. High-energy subatomic particles (protons) entering the Earth's atmosphere from outer space will generate a shower of secondary particles called **muons** which can be 'stopped' by a sufficiently thick layer of silicon. The energy released can generate around 80 electrons per pixel per micron of thickness in the silicon. With a collection depth of 20 μm, a muon event is seen on a CCD image as a concentrated bright spot a few thousand electrons strong (Fig. 5.5). Thinned backside-illuminated CCDs and those CCDs with **epitaxial** (i.e. thin) collection layers (like the EEV/GEC CCDs) are therefore much less prone to cosmic rays than the thicker frontside-illuminated arrays. High-altitude observatories may also encounter more problems than sea-level laboratories. Because these events are so conspicuous and concentrated to a few pixels it is generally straightfoward to remove them from images of non-crowded fields and non-extended objects. Otherwise, the only approach is to take multiple frames of the same region and compare them, since there is very little chance of a cosmic ray event hitting the same pixel two times in quick succession.

Experiments conducted by Dr Craig Mackay of the Institute of Astronomy in Cambridge (England) revealed that the events attributed to cosmic ray hits were actually a combination of cosmic rays and low-level X-ray emission from several forms of UV- and blue-transmitting glass (e.g. Schott UBK7 and GG385). In fact, the X-ray emission dominated by 5:1. This discovery alleviated the concern over the

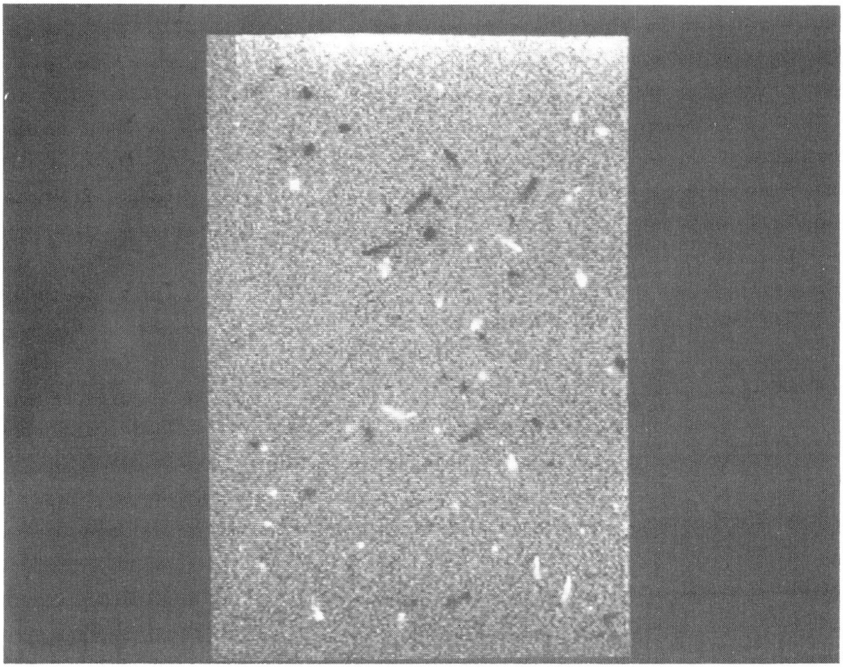

Fig. 5.5 — The difference of two 90-minute dark exposures with an RCA CCD reveal numerous events even though there has been no illumination; these are ionising particles or high-energy photons. The difference-image shows the events in the second exposure in black. Courtesy Patrick Waddell and Albert Fowler, NOAO.

fact that the muon event rate was predicted by cosmic-ray physicists to be several times lower than the observed rate of events being recorded by CCDs. With fused-silica windows the actual event rate was only 1.5–2 events cm^2/min, or 90–120 events/CCD frame/h exposure.

Another source of unwanted signal in all CCDs and similar devices — such as solid-state infrared imagers — is **luminescence**. An electrical path with lower than normal resistance in the semiconductor caused by, for example, partial shorts between electrodes in the parallel or serial registers, can act like a **light-emitting diode (LED)**. Incorrect voltages applied to certain parts of the device, in particular the output transistor amplifier, can cause similar problems (Fig. 5.6). This effect can be a serious limitation in some CCDs becuase it will limit the integration time to only a few minutes. Device selection, or alternatively, active (computerised) electronic control of the applied voltages can circumvent this limitation. The simplest example of the latter approach is to electronically switch off the output transistor during a long integration and to reduce the clock level (or voltage potential) on the pixels. Just prior to the end of the exposure the transistor is switched on again, the clock swing increased to improve charge transfer efficiency and the horizontal register 'purged' several times to clear it before the readout of the CCD frame is commenced.

One final concern expressed by some workers striving for the very lowest light

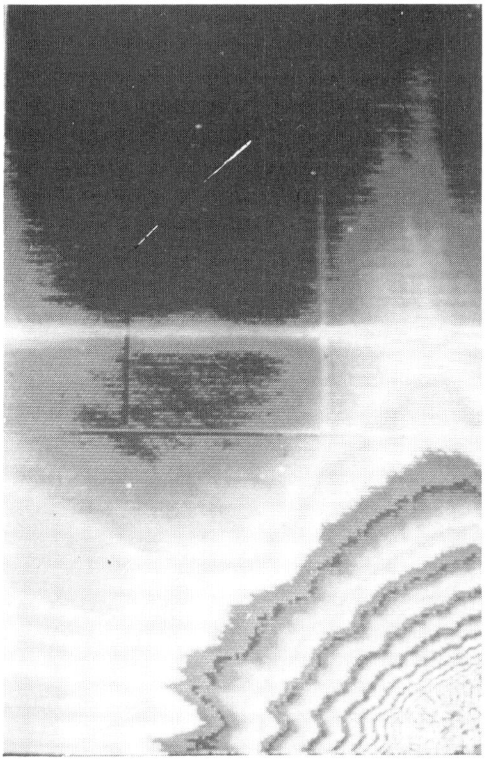

Fig. 5.6 — Intense luminescence from the output amplifier of an RCA CCD during an exposure. This effect can be avoided by turning the output amplifier off when it is not needed. Courtesy Patrick Waddell, CFHT.

levels is radioactivity from the heavy metals of which cryostats are made, such as copper, located near the CCD itself. Rumour has it that some astronomers have actually discussed the feasibility of harvesting pre-nuclear-age metals from sunken ships to fabricate 'clean' CCD cryostats for astronomy!

5.4 BAD PIXELS AND CHARGE TRANSFER EFFICIENCY

Already described in section 4.2.2 are how charges, produced by incident illumination, are accumulated under the CCD electrodes until the end of the timed exposure and then moved out by charge-coupling of one pixel to the next through a repetitive clocking (or pulsing) voltage waveform applied to the electrodes. Unfortunately, as recognised immediately by the inventors Bill Boyle and George Smith, there are several possible impediments to the success of the transfer process. Problems occur when the wells are (a) virtually empty or (b) almost full or (c) when a defective pixel gets in the way!

As the CCD wells fill up, the total stored charge can distort the electric potential of the applied clock voltage and thus charge transfer or coupling is degraded. Moreover, when the wells are completely full — a condition called **saturation** — there is nothing to stop the charge spilling over into the adjacent pair of pixels in that column; in general there is no serious sideways (row) spread because of the heavily doped channel stops. If the source of light causing the saturation is very strong then the adjacent pixels may also fill up and so the spreading continues and a streak appears up and down that column; this is also referred to as a **charge bleeding**. An example of charge bleeding is shown in the CCD picture in Fig. 5.7.

Global **charge transfer efficiency** (CTE) drops when a fixed amount of charge is 'trapped' for *each* pixel transfer. Visually, the result is the appearance of a 'tail' on point-like images; the tail will be parallel to the transfer direction. Trapped charge is not normally 'lost' but more often it is 'skimmed' to be released later, and is therefore related to a general phenomenon called **deferred charge**, as illustrated in Fig. 5.8. Deferred charge can have several origins including spurious potential pockets, design faults leading to specific potential pockets at certain locations and **charge traps** associated with impurities. For most CCDs poor CTE occurs at a level of 10 to 50 electrons/pixel and is mainly of concern only in very short exposures or in spectroscopic applications in which the dispersed background in the exposure is below 100 electrons. One way around deferred charge effects is to **preflash** the CCD with a short flood of light from an LED positioned so as to uniformly illuminate the array before *each* exposure. This action adds a constant amount of charge, sometimes called a pedestal, to each pixel which raises the base level to above 100 electrons (say), back into the acceptable region of good transfer efficiency. The technique of introducing a substantial charge pedestal was termed (by someone with a sense of humour) as adding a **fat zero**; inevitably someone nicknamed the smaller charge pedestal needed to offset the deferred charge problem the **skinny zero**! In astronomical circles the most common description is preflash. Although this method can be quite useful, it has the disadvantage of adding additional noise to the frame and it covers up an inherent performance problem with the particular type of CCD which may in fact persist despite the flashing treatment.

Apart from these general problems of charge transfer inefficiency there are other difficulties related to specific faults on a given CCD. A break in one of the polysilicon electrode strips due to a fault in the manufacturing process can effectively stop or seriously delay charge from being transferred past that point in that column; this defect is called a 'blocked column'. An example is shown in Fig. 5.9. Crystal defects in the bulk substrate silicon can sometimes spread into the depletion region during manufacture. Such defects disrupt the semiconductor nature of the silicon and, in general, behave like sponges which can soak up any charges in their area. Delayed release of charge can have a wide range of time constants (from milliseconds to hours) and columns with this sort of behaviour are usually irretrievable by image processing. Needless to say, a charge trap near the beginning of the horizontal (serial) register is a disaster! Even the design of the overlapping electrode structure can create pockets of resistance in the channel potential which trap charge. Trapping sites are particularly serious for surface-channel CCDs. Using our water-in-a-well

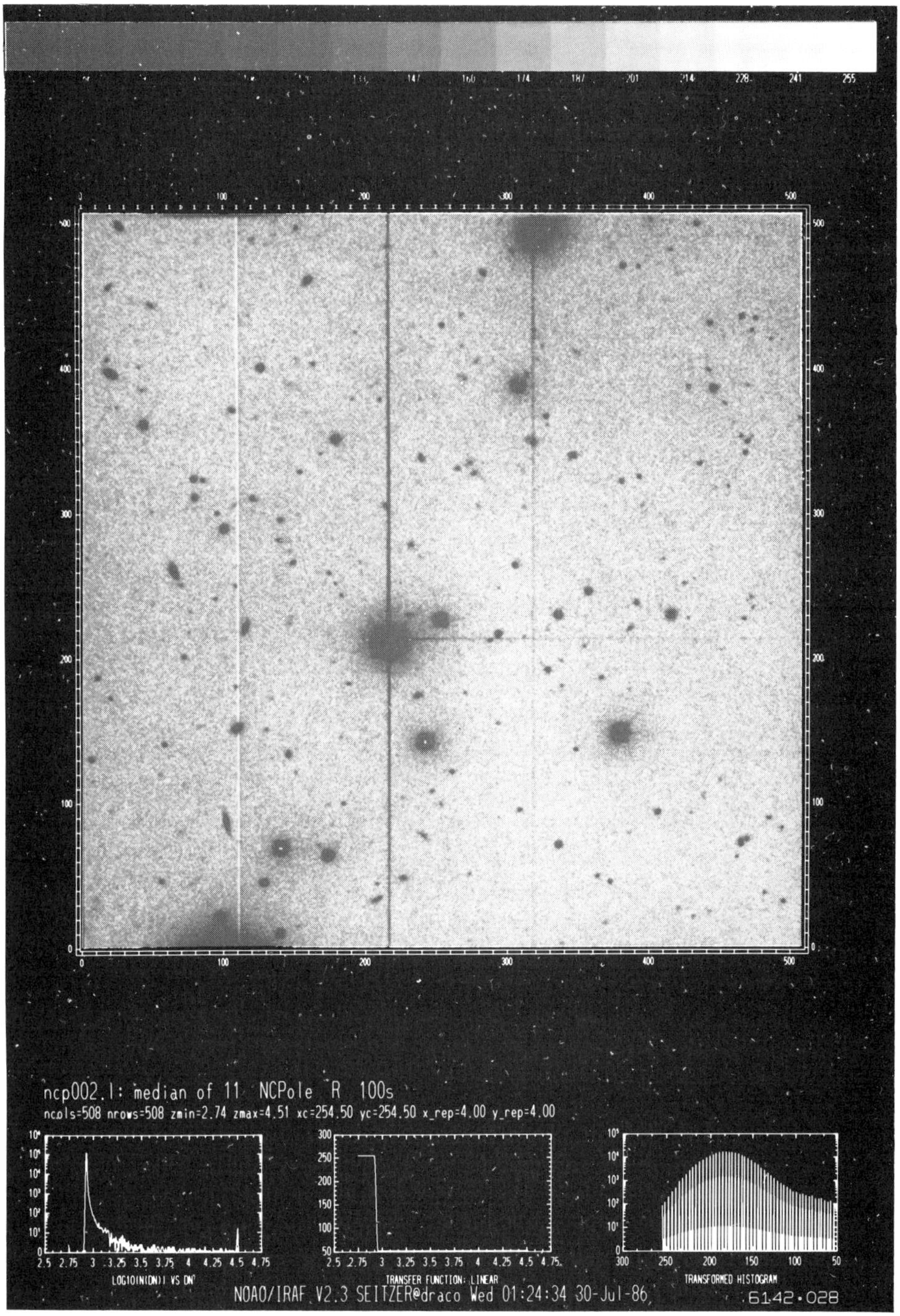

Fig. 5.7 — An astronomical CCD image (displayed as a 'negative') in which a bright star has saturated in the exposure time needed to record the fainter objects. When the CCD pixel saturates it spills over and 'bleeds' up and down the column. The horizontal trail is evidence of poor CTE in the serial register. Courtesy NOAO.

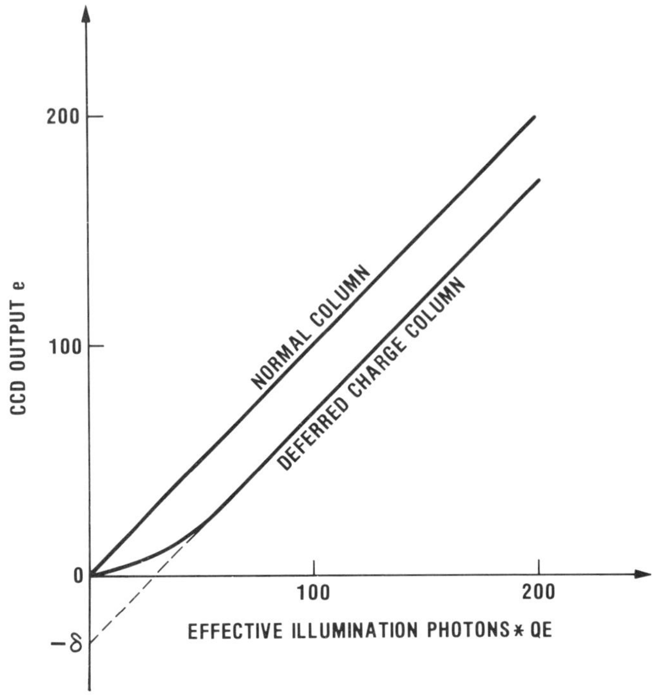

Fig. 5.8 — (a) A plot of the number of electrons detected at the output of a CCD versus the number of electrons present in the original pixel. If no charge is trapped anywhere during the charge-coupling readout then the graph should follow that for a normal column. In practice it does not, indicating trapped charge whose readout has been deferred. Courtesy Tony Tyson, Bell Labs.

analogy, traps would behave like small pot-holes at the base of the well; such pockets fill up easily but do not empty their contents merely by raising and lowering the floor of the well.

CTE is quoted in the form 0.99999 *per transfer* and, in the CCD jargon, people speak of CCDs with 'five nines of CTE'. What this means is that after one transfer, 99.999% of the original charge will have made it successfully, after the second transfer 99.999% of *that* charge will be moved on so that 99.998% of the original charge is still intact and finally, after 1000 transfers, 99.00% (that is 0.99999^{1000}) of the original charge on the initial pixel will have been transferred to its new location (the output, say), while 1% of the original signal will be spread over the intervening pixels. In practice most of that 1% is in the immediately adjacent (following) pixel. We have already mentioned that CTE is a function of temperature. It also depends on the **clock frequency** or 'readout rate' and the rise and fall times of the clock voltage pulses and their degree of overlap. The CTE in the parallel (column) direction is usually different from that in the serial (horizontal) register and there can be large variations from column to column.

Measuring the level of CTE is not as simple as it might seem at first sight. A

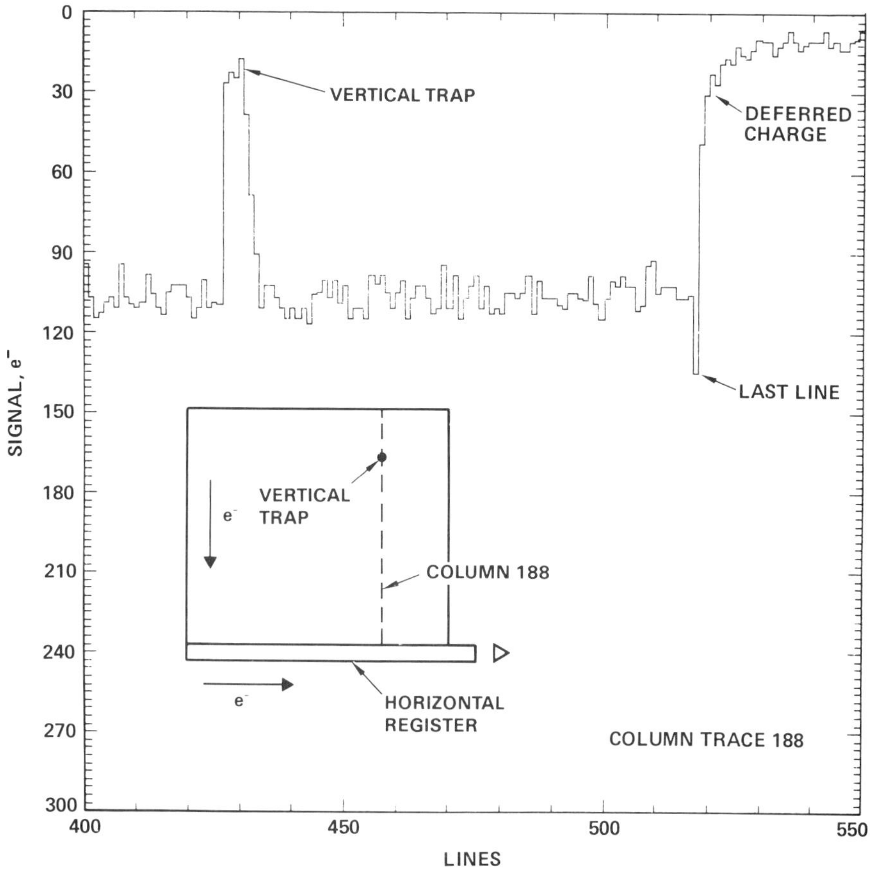

Fig. 5.8 — (b) A plot of the signal (in electrons) along a particular column of a CCD showing the loss of charge in a 'trap' and the curved response following that trap and the last pixel indicating deferred charge. Courtesy Jim Janesick, JPL.

relatively straightforward method, at least for CCDs which permit it, which is both qualitatively and quantitatively useful, is to flood the CCD with light, read out the device and **overscan** both registers to produce a resulting image which is several pixels larger, along both axes, than the real CCD. Most CCD controllers permit overscanning but some CCDs — such as the Thomson chip — have special additional pixels in the serial register which confuse the overscan experiment. This technique is also called 'extended-pixel-edge response' or EPER by some workers. When the signals occurring in the imaginary pixels following the last row and column of the real CCD are compared to the dark level and to the illumination level then any charge transfer problems leading to deferred charge will be obvious. Some investigators lump all rows or columns together to give a mean measure of CTE, but careful examination of an overscanned uniformly illuminated image reveals big variations from column to column. Some averaging will be needed to reduce the effect of readout noise. The CTE is then given by

Fig. 5.9 — Two groups of blocked columns on a TI CCD. The pixels causing the block are different in the two groups. Other poor columns can also be seen, but most of the device is flat. Courtesy of Lick Observatory.

$$\mathrm{CTE}(e^-) = 1-(Q_d/Q_0N)$$

where Q_d is the net deferred charge in the overscan, Q_0 is the charge on the last real pixel and N is the number of pixels transferred. An example of EPER at extremely low light levels is shown in Fig. 5.10.

In practice, the best appreciations of the levels of CTE come from astronomical observations themselves and especially when CCDs are used as the detector in a spectrograph.

It is also possible to utilise the sensitivity of CCDs to X-rays to assess CTE, as demonstrated by Janesick *et al.* (1986). X-ray absorption produces a large known amount of charge in a single pixel (see Chapter 9).

The spurious potential pocket problem has plagued most manufacturers and it is only recently that it has become well-enough understood. Early TI and Thomson-CSF CCDs exhibited design-induced pockets in the transfer-gate region where the parallel (imaging) area mates with the horizontal (serial) register. By design, the charge-carrying or signal channel beneath the transfer-gate was constricted by channel-stops to properly herd signal charge from the CCD columns into the horizontal register. As charge moved from a wider to a narrower region there was a

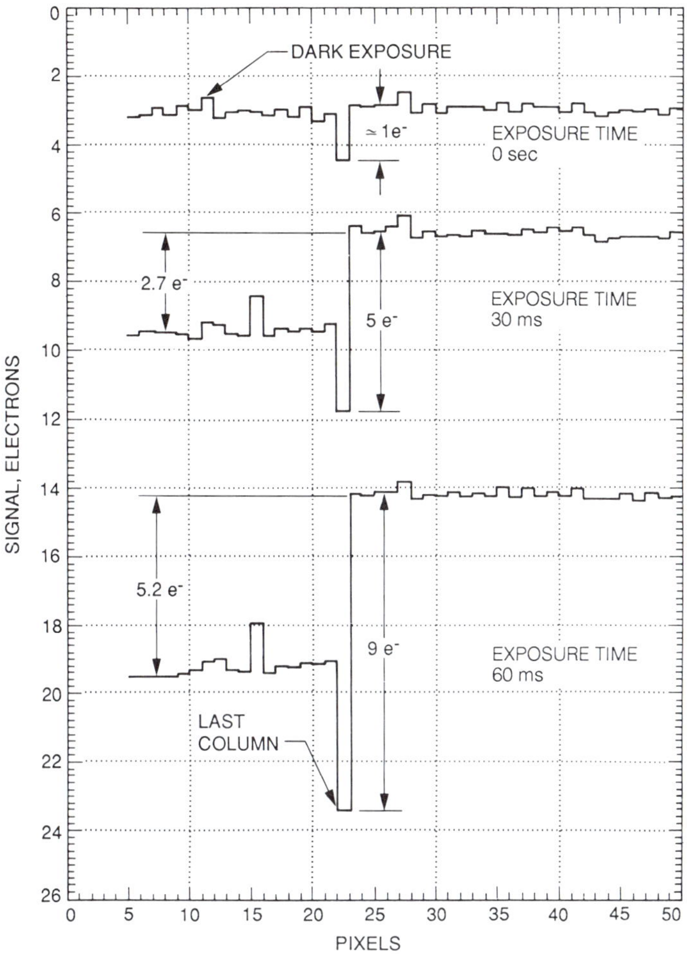

Fig. 5.10 — An 'extended pixel edge response' or overscan experiment at JPL on a Ford CCD at the five- and nine-electron levels. Linearity is excellent at the one-electron level and there is no deferred charge. Courtesy Jim Janesick, JPL.

local decrease in potential (Fig. 5.11) effectively creating a step or pocket which trapped temporarily small quantities of charge; the effect can be overcome by a preflash or by adjusting the potential on phase one to produce a 'fringing' field. Thomson-CSF corrected their problem with the design of a new set of manufacturing masks, but TI withdrew from the three-phase development before a correction could be made. Design-induced spurious potential pockets appear to be a thing of the past due to stringent scrutiny of the masks prior to fabrication.

The most serious problem that the current generation of scientific CCDs has been facing is the isolated single-pixel trap or 'pocket' which affects CTE randomly on a local rather than global level. Some new CCDs were literally riddled with pockets. In the strive for the perfect CCD the pocket phenomenon has taken its toll on CCD

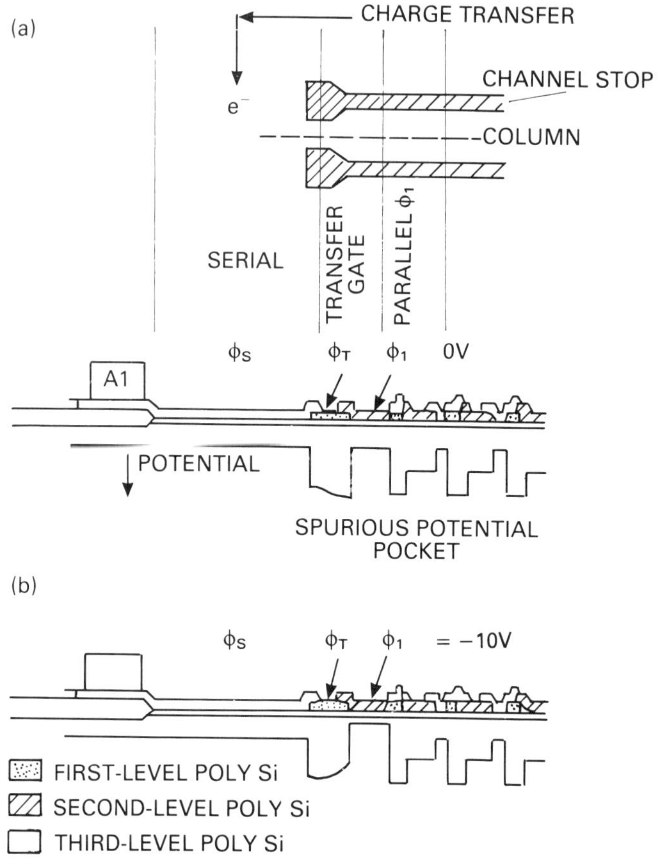

Fig. 5.11 — Schematic illustration of the electrostatic potential pocket which occurs in the 800×800 TI CCD at the junction of the parallel and serial registers. The occurrence of this spurious potential pocket or trap is caused by the narrowing of the CCD column by the channel stops at the interface to the serial register. A 'fix' is to drive the voltage on the adjacent gate to −10 V to 'pull-up' the electrostatic potential at the pocket. Courtesy JPL.

yields at several different manufacturers (Tektronix, Ford/Photometrics, Lincoln Laboratory) and has been studied with incredible intensity. Detailed tests have shown that these traps are localised to a single level of the polysilicon electrode within a pixel. When the effect first appeared — with the Tektronix devices — it threw everyone into a quandary. Why had this phenomenon not been observed before? It seemed that past RCA, TI, Fairchild, Reticon and Thomson CCDs never exhibited these single-pixel traps to such an appreciable extent — even allowing for the improvements in other performance factors such as readout noise. This was the phenomenon which put the Tektronix devices 'on hold'.

Two clues finally emerged. The TI family of 800×800 CCDs showed no large single-pixel traps and it was discovered that the impurity boron, used to dope the p-type substrate, had diffused through the thin (epitaxial) n-type layer of the buried channel construction during a high-temperature phase of manufacture — the effect was called the 'p^+ epitaxial diffusion tail' — whereas the Tek CCDs exhibited very little diffusion of boron. The second clue was the realisation that the change from the older practice of using silicon dioxide as the insulating layer on the semiconductor surface to an oxide–nitride double-layer could be relevant; the new material has significantly reduced short-circuits between the polysilicon electrodes and therefore improved the manufacturing yield. Pin-holes created by surface contamination during the oxide growing stage could allow nitride to diffuse through the oxide to the n-type semiconductor. At that tiny location the different insulator material would cause a slight variation in electric potential within the semiconductor.

Ultimately, with all other problems solved, it will be traps in the bulk silicon crystal which limit CTE. Fortunately the quality of manufactured semiconductor silicon has greatly improved over the years and will probably continue to do so. Global CTE performance of selected Ford, Lincoln and Thomson-CSF CCDs are believed to be limited by 'bulk traps' at the level of 0.9999995 (or 'six nines five') per pixel transfer. In other words, only three electrons out of a total of 10 000 electrons are deferred for 512 pixel transfers.

5.5 READOUT NOISE AND OTHER NOISE SOURCES

A CCD is by its very nature a 'digital' camera. The clocking procedure described earlier delivers a stream of charge 'packets' from pixels in the image area all the way to the output amplifier; the charge (Q) in each packet is proportional to the amount of light in that part of the original image scene. As each charge packet arrives at the output FET it causes a change in voltage to occur (of amount $V=Q/C$, where C is the capacitance at the output node); the smaller the node capacitance, the larger the voltage change for the same size of charge packet. For the RCA 512×320 CCD, $C\approx0.6$ picofarads (pF) which yields about 0.25 microvolts (μV) per electron (in the charge packet) whereas for the EEV/GEC P8600 CCD the node capacitance is only 0.1 pF which gives a healthy 1.6 μV per electron. Larger values are possible, though mainly for video applications; Kodak CCDs using an extremely small output MOSFET give 15 μV/electron yet the overall noise under slow-scan conditions is greater than 10 electrons because other noise sources become larger as the MOSFET gets smaller. Recall that CCDs can detect light levels of just a few tens of electrons, which corresponds to voltage changes of a few tens of microvolts! This is a very small electrical signal and, unless great care is taken in circuit design, such signals will be swamped by other causes of voltage variations (noise).

It is desirable that the noise performance of a CCD camera system be limited only by the output transistor of the CCD and not by any other part of the electronic system. To achieve this one must understand the noise sources associated with the CCD and take steps to get them to an irreducible minimum; this minimum is the ultimate **readout noise**. There are several potential sources of unwanted electronic noise in CCDs. These include,

(1) background charge associated with **fat zero** offsets,
(2) transfer loss fluctuations,
(3) reset or 'kTC' noise,

(4) MOSFET noise,
(5) fast interface state noise.

When a preflash is used to introduce a fat zero charge to aid transfer efficiency or eliminate charge skimming, then the consequence is a noise equal to the square root of the total number of charges in a pixel.

During charge transfer a fraction of the charges are left behind. However, this fraction is not constant but may fluctuate and so an additional noise component is added to the signal noise. This noise is given by the square root of twice (once for trapping, once for release) the number of charges left behind multiplied by the number of transfers and can be of order 70 electrons for surface channel CCDs but is typically 10 times smaller or better for buried-channel CCDs and astronomical light levels. For very large CCDs this effect implies that exceptionally good charge transfer efficiency must be achieved.

Noise associated with the re-charging of the output node is given by $(kTC)^{\frac{1}{2}}/e$ or about $284(C)^{\frac{1}{2}}$ at a CCD temperature of 150 K, where the capacitance, C, is in picofarads. This **reset noise** is effectively eliminated by an electronic signal-processing technique called **correlated double sampling** which is described in detail in section 6.3.1.

Other noise sources associated with the output MOSFET, such as **1/*f* noise**, can generally be made quite small by good manufacture, typically about six electrons or less.

Traps which absorb and release charges on very short timescales, causing a fluctuation in the charge in any pixel, are called **fast interface states**. This effect is very serious for surface-channel CCDs but is normally quite small (of the order five electrons) for buried-channel devices.

It is a remarkable tribute to the foresight of Boyle and Smith, and the skills of every scientist and engineer who has worked hard on CCD technology, to realise that devices are now routinely made with a readout noise of less than 10 electrons (Fig. 5.12). The lowest noises that have been consistently measured so far are in the range two to three electrons — with EEV devices by Craig Mackay and Lincoln Lab. CCDs by Jim Janesick.

5.6 UNIFORMITY OF RESPONSE

As craig Mackay has aptly put it ‘the only uniform CCD is a dead CCD’! When a CCD is exposed to a scene in which the brightness is absolutely constant or uniform across the entire chip — this is called a **flat-field** (see also section 8.3) — the actual output image from the CCD is generally far from flat (Fig. 5.13(a)). The image will be non-uniform in several ways. Some pixels will be more sensitive and some will be less sensitive than the average, and these pixels may be grouped in odd-shaped patterns or there may be a gross gradient in sensitivity from one corner of the CCD to the other. Bad pixels, blocked columns, and even hairs or other debris will stand out dramatically. Some effects will even change with different light levels; this kind of behaviour is said to be non-linear. In addition, the flat-field response of a CCD is a strong function of colour (Fig. 5.13(b)), that is, of the wavelength and passband being

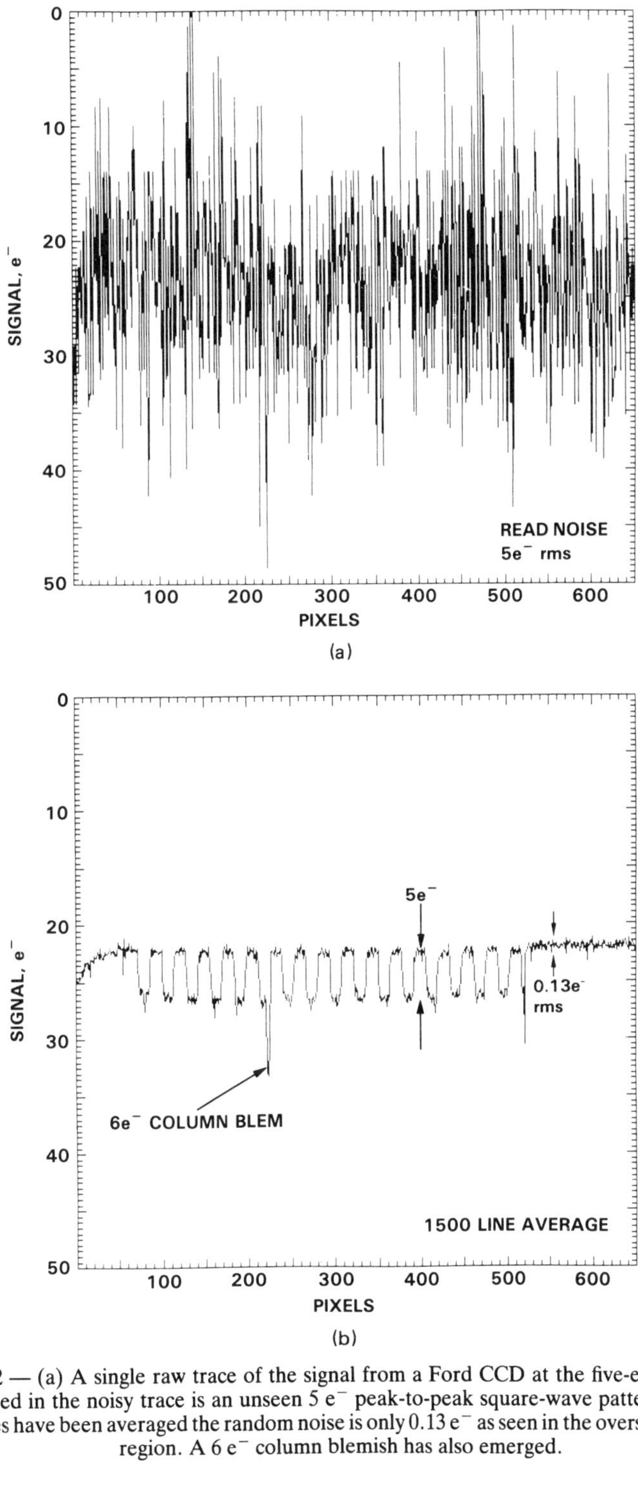

Fig. 5.12 — (a) A single raw trace of the signal from a Ford CCD at the five-electron level. Embedded in the noisy trace is an unseen 5 e^- peak-to-peak square-wave pattern. (b) After 1500 lines have been averaged the random noise is only 0.13 e^- as seen in the overscan or EPER region. A 6 e^- column blemish has also emerged.

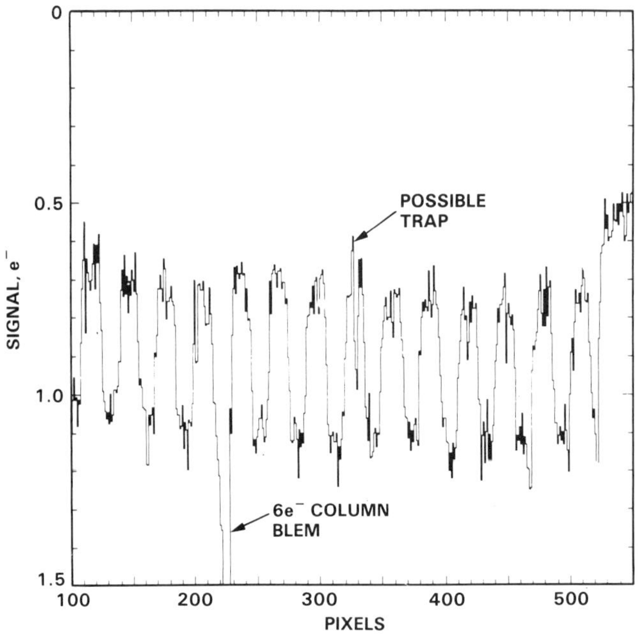

Fig. 5.12 — (c) A close-up of the trace after 20 000 lines have been averaged. Horizontal charge transfer efficiency is excellent; a very small trap ($<0.1\ e^-$) may be present at pixel location 325. Courtesy Jim Janesick, JPL.

used to illuminate the detector — mainly because photons of different wavelengths are absorbed at different depths in the silicon and the material is not homogeneous.

At first sight, a newcomer to CCD technology might reasonably wonder if CCDs are all they are advertised to be! How is it possible that astronomers can tolerate such non-uniform detectors? The reason is simply that the non-uniformities can be systematically eliminated by a division of the observed region of sky by the flat-field frame and this arithmetic task, although tedious for humans, is trivially accomplished by a computer. Pixel-to-adjacent-pixel variations in detector sensitivity of perhaps a few per cent of the average response is normal, and across the entire CCD variations as large as 20% can occur. Reducing such non-uniformities of response is a major pursuit, but their occurrence is not a serious limitation provided they are *stable*. One very big advantage of CCDs is that the pixels themselves are absolutely fixed in position, unlike in a vidicon-type TV tube where the pixel is defined by where the electron readout beam happens to land, and so stability in the uniformity of the CCDs response to light is a natural consequence — it is built-in.

CCDs are manufactured in layers from a 'mask set' produced by a CAD (or computer-aided design) system. Small errors in the mask give rise to small but significant geometric variations in pixel size or in column widths. Stresses introduced by mechanical thinning can also cause variations in response. The silicon itself is not 'pure', rather it is 'doped' by impurities to form the required type of semiconductor, i.e. p- or n-type, but there are always patterns on a microscopic scale to this doping, often with an appearance which resembles 'tree-rings', which show up in flat-fields

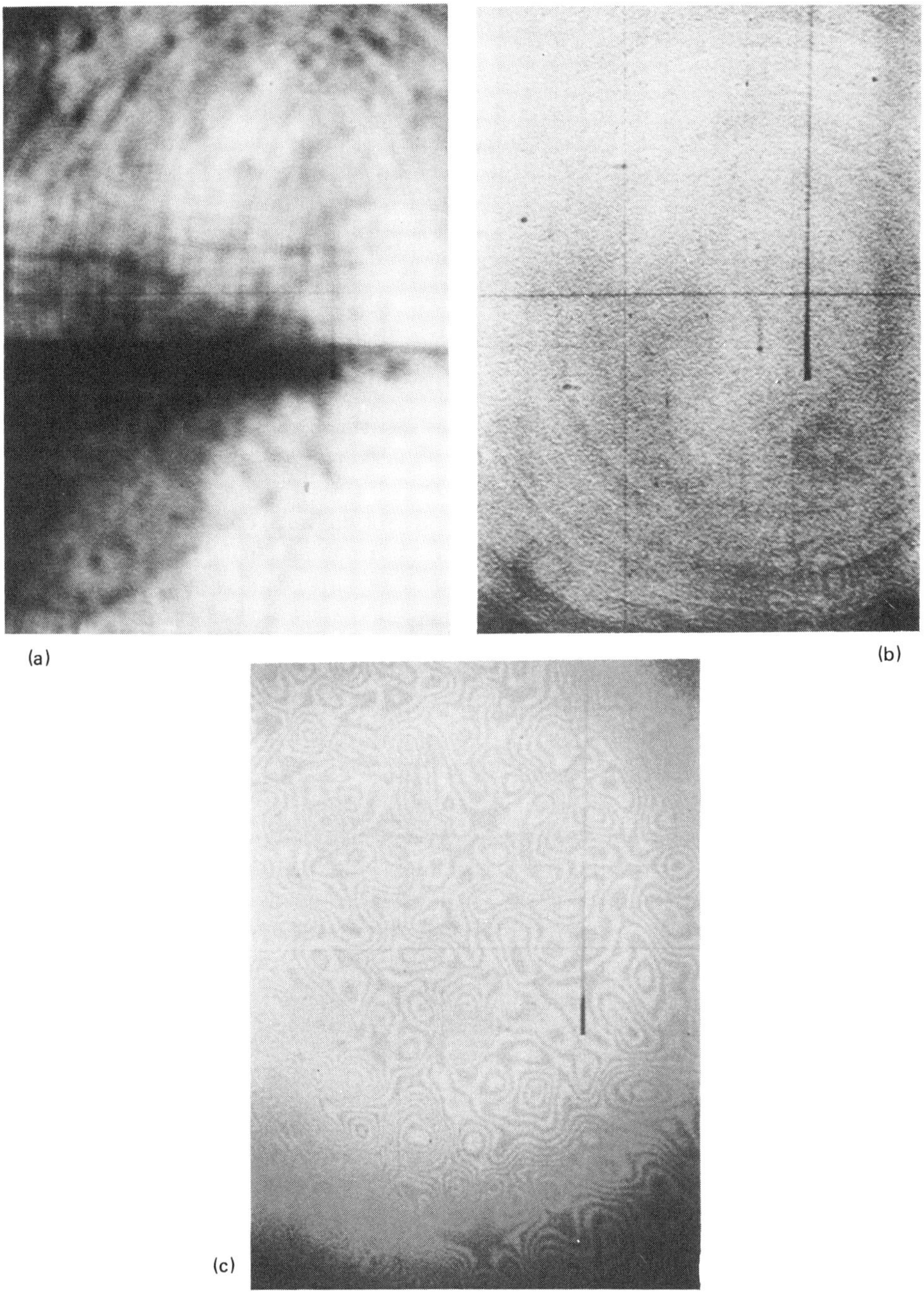

(a) (b) (c)

Fig. 5.13 — (a)–(c) A uniformly illuminated RCA CCD at two different wavelengths. The apparently non-uniform image is the response to the flat-field. In (b), 'tree-ring' structure due to doping variations is seen. In the third frame, interference 'fringes' appear due to internal reflections of a night-sky emission line in the thinned CCD. A blocked column group can be seen and the fine horizontal line is the join between the image and store regions of this frame transfer CCD. Courtesy Carol Christian and Pat Waddell, CFHT.

(see Fig. 5.13(b)). Sometimes there is also clear evidence of a cyclic or 'step and repeat' pattern.

With a CCD selected to have a minimal number of bad pixels or columns one might expect to be able to correct the remaining non-uniformity in response by simply taking a ratio of images of astronomical sources to those of a uniformly illuminated white screen. But it is not quite so simple at the lowest light levels and the highest measurement accuracies!

One problem is that the non-uniform response of a CCD is colour(or wavelength)-dependent. Photons of different wavelengths are absorbed at different depths in the CCD on average. Frontside-illuminated CCDs, in which the electrode thickness may vary across the array, will therefore cause a variation in response to light which, because of the nature of the absorbing electrodes, will vary with wavelength. In thinned backside-illuminated devices, on the other hand, surface effects and multiple reflections from the back surface of the CCD cause colour-dependent non-uniformity. This phenomenon is known as **fringing** (refer again to Fig. 5.13(c)).

Obtaining an acceptable and appropriate flat-field is 90% of the solution to high-quality, very deep imaging with CCDs. We will return to practical techniques for flat-fielding in Chapter 8 on characterisation and calibration of CCD instruments.

5.7 ULTRAVIOLET FLASHING AND QUANTUM-EFFICIENCY PINNING

One of the earliest concerns was that the 800×800 CCDs to be used for the WF/PC instrument on the Hubble Space Telescope showed very poor response to blue and ultraviolet (UV) light, which was very bad news for a telescope expected to have great UV performance. Since tens of thousands of chips had been made, and since the WF/PC was a key instrument on this very expensive satellite, a solution *had* to be found. Of all the CCDs produced for astronomy only the TI and the RCA CCDs were thinned backside-illuminated devices. The RCA chip showed superior blue sensitivity and had useful sensitivity at UV wavelengths well below 0.40 μm (or 4000 Å), but had a higher readout noise than was desirable. The thinned TI detectors had little or no response to blue light even though the exposed silicon surface was free of any electronic circuitry or other covering. Later, a second problem became apparent with the thinned TI CCDs. The effective quantum efficiency actually *increased* after exposure to light, so that the use of these CCDs as accurate, linear, brightness-measuring detectors was badly comprised. This effect was called **quantum efficiency hysteresis** or **QEH** by the JPL workers.

Initially, chemical coatings were deposited on the TI CCDs to act as UV-to-optical converters. The idea for such a coating was due to Jim Westphal who, in an effort to learn more about UV behaviour, went to the rather sparsely stocked CalTech library and was surprised to find a 1968 volume by Samson which was highly relevant. He was even more surprised to learn that not only had knowledge of UV **fluorescence** been around for over 50 years and that a salicylate, like aspirin, was a phosphor which if smeared on a photocell would give it UV sensitivity, but the book also listed two little-used materials which fluoresced under UV illumination at the relatively long wavelength of 0.5 μm — ideal for CCDs; these substances were **lumogen** and **coronene**. Jim chose coronene because it could be obtained easily and he and Fred Landauer coated one half of a JPL/TI CCD and tested it at Palomar during an hour or so of twilight. (This use of a CCD on a telescope was shortly after the successful run by Brad Smith and Jim Janesick.) Incoming UV photons caused the coronene to glow in visible light, where the CCD was sensitive, and produced an effective efficiency to blue and UV light of perhaps 14%, i.e. one in seven photons at these wavelengths were actually detected. The coronene also seemed to act as an **anti-reflection** coating and so helped to improve the normal visible sensitivity.

Much later, Martin Cullum and colleagues at the European Southern Observatory headquarters in Munich, West Germany, reported the use of a fluorescent laser dye coating painted onto *front*-illuminated EEV (GEC) CCDs to obtain quantum efficiencies as high as 25% at a wavelength of 0.35 μm (3500 Å) in the UV. The laser dye is reported as being more uniform in response in the blue (0.40 μm or 4000 Å) where coronene actually has a sharp dip in performance. The importance of achieving good UV response, as well as good optical/red response, in CCDs can be judged by the international efforts to devise better **down-converters** like coronene. At Photometrics in Tucson, Richard Aikens felt it sufficiently important to retain a chemist (Gary Simms) full-time on the team! Thanks to Gary's efforts, Photometrics now offer a new phosphor coating which they call Metachrome II which can be applied to any CCD by vacuum sublimation (Fig. 5.14).

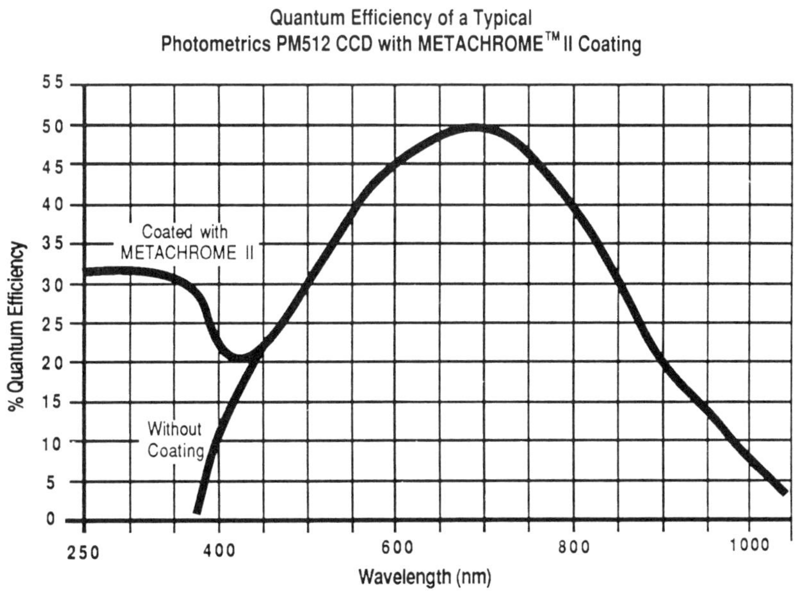

Fig. 5.14 — The enhancement in UV response obtained by 'coating' a front-illuminated CCD with a phosphor. The CCD shown here is the Photometrics/Ford 512×512 chip with Photometrics' own coating called Metachrome II. Courtesy of Photometrics Ltd.

In 1981 a totally unexpected discovery was made. A high-quality, thinned CCD which had been kept in a dry, sealed container at JPL was found to have a much higher dark current than it had a few months earlier. Checking if it might be the coronene which had somehow affected the CCD, the JPL workers illuminated the device with a UV lamp to look for fluorescence. Soon after, the dark current was found to have fallen to normal! This clue led to a fix for both problems, namely poor UV response and QEH.

What was the explanation? It seems that imperfections in the thin **native-oxide**, i.e. silicon dioxide (SiO_2), layer grown in air on the exposed backside surface of the CCD had created traps for electrons and holes. A net positive charge at the surface can develop, causing a shallow bucket or potential well which can trap electrons. This unwanted potential well must prevent photo-induced electrons from moving to the depletion region where the charge should have been collected. Since blue and UV photons are absorbed closest to the surface the result is a loss of sensitivity in the blue and UV. Traps can also absorb electrons from the silicon itself, electrons produced perhaps by photons; the electrons trapped in the unwanted potential well at the surface will affect the depth of that well and so the UV quantum efficiency will appear to change with illumination level. Time constants for the release of trapped charge can vary from seconds to days depending on the energy level and the CCD temperature. If the energy level of surface traps is midway between the valence and conduction levels, thermally induced dark current could be quite large.

The 'fix' — exposure to UV light — works because energetic UV photons in the wavelength range 2000–2500 Å (note: these wavelengths are not transmitted by the Earth's atmosphere) preferentially fill the most energetic or active trapping sites, creating a *negative* charge on the CCD surface. Since like charges repel, the negatively charged surface generated by the 'UV flood' eliminates the unwanted potential well and encourages electrons formed near the surface to move away from the surface toward the depletion region. Thus there is no loss of quantum efficiency and no QEH. The energy level of surface defects is forced away from the middle of the forbidden band and so the dark current is also reduced.

Fortunately, the negative charging procedure can be done near room temperature, and when the CCD is cooled to below $-100°C$ or 173 K, the charge is 'immobilised', and provides a permanent field until the CCD is next warmed up. The quantum efficiency of the CCD is driven to its theoretical maximum and held or 'pinned' there; this is the **QE-pinned** condition.

UV flooding requires wavelengths between 2000 Å and 2550 Å and about 10^9 photons per pixel (at about 2139 Å) which can be provided by a zinc lamp. It also appears to require the presence of oxygen during the flood. At room temperature, the improvement disappears in a few hours, especially if humid air is present. In a constant vacuum at $-120°C$ the improved UV response can last for at least six months.

The decay of the UV sensitivity of the coronene-coated TI chip was noticed in an interesting series of communications between Jim Janesick at JPL and Bev Oke at the Palomar Observatory. Bev had UV sensitised his precious TI CCD from JPL and looked after it by keeping it dry and under vacuum in the CCD camera cryostat, whereas Jim had so many devices to test at JPL that cryogenic storage was impossible — so he kept them in a drawer! Initially each suspected that the other had made some error of measurement during testing. Eventually devices were exchanged, findings corroborated and the explanation found!

UV flooding is not a perfect solution. Apparently it works better for some CCDs, in particular for CCDs which are insufficiently thinned than for others; thinning should be such as to reach all the way to the active layer where depletion occurs. For some CCDs the process will only work if oxygen (O_2) gas is present. The best results occur if the flooding is done warm and then the CCD cooled immediately afterward. Further it appears that if the UV radiation is too hard, i.e. too short a wavelength (e.g. 1850 Å) the process is reversed!

Negative surface charging can also be chemically induced. Exposure of a thinned CCD to nitric oxide gas (which is poisonous) for a few minutes dramatically improved the performance at 4000 Å in the blue. The effect was stronger than for UV flooding and somewhat more permanent. An alternative method for obtaining good blue response is the use of an arc or corona discharge near the CCD. It is hypothesised that the corona discharge or arc produces ionised oxygen, which charges the silicon surface by oxidation. Nitric oxide is also a strong oxidising agent. Very little use has been made of corona discharge methods because of the possible danger to the CCD from the high-voltage arc. However, as pointed out by Lloyd Robinson of Lick Observatory, high-voltage ion pumps have been used regularly on some CCD vacuum cryostats without destroying the CCDs. Lloyd even noticed an improved response in the TI 500×500 CCD in one of his cameras when the ion pump was switched on! Early users of the RCA CCD were concerned however, that ion bombardment of the silicon surface was causing some damage and most people try to avoid the use of ion pumps. The original success of the RCA chips in achieving superlative blue response was, in retrospect, not given sufficient attention. Having produced scanned electron-beam silicon vidicon targets for several years, RCA knew the importance of surface charging versus electron–hole pair recombination. Their 1976 patent application describes doping of the thinned backside layer to create a potential whose effect is to send photogenerated electrons inward toward the charge collection layer! In 1986 the JPL workers once again came up with an even better method of obtaining control of the backside CCD surface of the thinned TI CCDs. Attempts were made to deposit a thin electrically conducting coating on the CCD that would allow an adjustable electrical voltage at the surface. Problems with good insulation between the conductor and the CCD limited the success of the proposed electrically controlled layer, but in the attempt it was discovered that simply the act of depositing an ultra-thin layer — just 4 Å thick — of the right kind of metal, namely gold or platinum, produced an astonishing and permanent improvement in the UV, the extreme UV (XUV) and even the soft X-ray region. Terming the new surface coating the **flash gate**, Jim Janesick described both theory (Fig. 5.15) and experimental results in a major paper at the sixth 'Instrumentation in Astronomy' conference held in Tucson, Arizona in 1986. Over coffee afterwards, we remarked to each other just how far the concept of CCDs had come since the classic paper of Boyle and Smith in 1970.

Even more recently, while investigating oxide growth on the CCD surface, it was found that if the oxide was removed by an acid, the resulting quantum efficiency was very low initially, and moreover not one of the above surface treatments had any effect. Gradually, by exposure to air, the oxide grew back onto the surface and lo and behold the QE and the response to surface treatment returned to normal. Why? The explanation lies in the reduction in the number of surface traps as Nature smoothly and uniformly grows back the oxide layer. The most recent result has been the

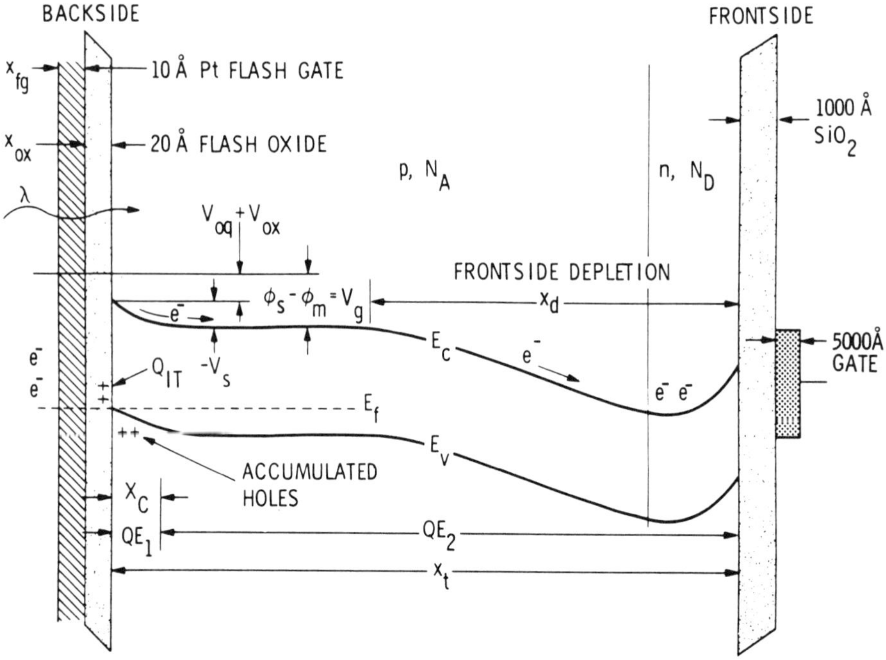

Fig. 5.15 — A cross-sectional view of a thinned, buried-channel backside-illuminated CCD with an ultra-thin coating of platinum called a 'flash-gate'. The sketch is not to scale. The thickness of the flash gate is only 0.001 μm. Courtesy Jim Janesick, JPL.

discovery that if the CCD is chemically etched (eaten away with an acid), then oxidised by several days' contact with deionised water or by a one-hour exposure to steam at 100°C (Fig. 5.16), the CCD response *with no other surface treatment* is essentially perfect! Even devices that would not have responded well to UV flood are retrieved. Apparently, this new 'good' oxide layer must give an overall negative charge to the CCD surface whereas the original 'bad' oxide layer had so many traps that the surface was positively charged.

I cannot help but wonder if the astronomical/space community have rediscovered what somebody somewhere in the semiconductor industry must have known already, although perhaps through a more empirical 'tried and tested' approach. SiO_2 has been used as a critical part of device fabrication for 25 years or more, and its study is still actively pursued. Undoubtedly, the recipes and techniques used are a carefully guarded secret by the various semiconductor companies! Recent research has indicated that chemical bonding of the oxide with hydrogen can lead to critical effects; the formation SiOH acts to 'accept' electrons while SiH behaves as a trap for holes, i.e. a repellent for electrons.

Because of acid corrosion and the awkwardness of cleaning off all the acid, it is unlikely that the flash oxide or the ultra-thin flash gate can be retro-fitted to existing CCDs. Treating CCDs with UV floods and nitric oxide gas are therefore likely to remain with us until the older devices outgrow their usefulness. Finally, there is an

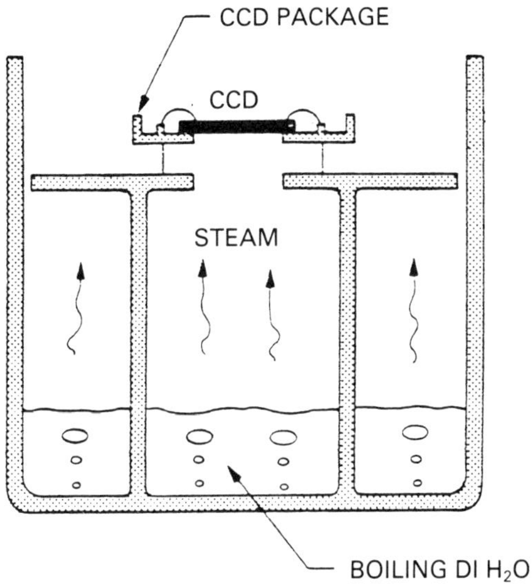

Fig. 5.16 — A steam bath for a CCD! Deionised (DI) boiling water is used to assist in the preparation of the backside surface of thinned CCDs. Courtesy Jim Janesick, JPL.

interesting story from Lloyd Robinson concerning the 'care and treatment' of CCDs. Having carried out both UV flood and nitric oxide treatments successfully on a TI 800×800 device, and after several months of use at the telescope, an accident occurred which resulted in the vacuum vessel containing the CCD being opened to air while the CCD was still cold. 'Visually, the CCD looked like it had been out in the rain', said Lloyd (Fig. 5.17(a)–(c)). After some thought and much discussion the CCD was washed in warm water and then in ethanol (an organic solvent) and dried in a vacuum chamber. It looked clean and respectable again, and once more was subjected to the nitric oxide treatment. This time the response was more uniform than before (Fig. 5.18), to within 10% in fact, and with at least 50% quantum efficiency at 4000 Å in the blue and the same 6500 Å in the red. In addition, the CCD had useful response down to at least 2500 Å in the ultraviolet. Lucky!

REFERENCES

Janesick, J. R., Elliott, T., Daud, T., and Campbell, D. (1986) The CCD flash gate, Instrumentation in astronomy VI, David L. Crawford, *Proc. SPIE*, Vol. 627, 543–582.

Robinson, L. (ed.) (1988) Instrumentation for ground-based optical astronomy: present and future, *Santa Cruz Summer Workshop in Astronomy and Astrophysics* (9th; 1987), Springer-Verlag, New York.

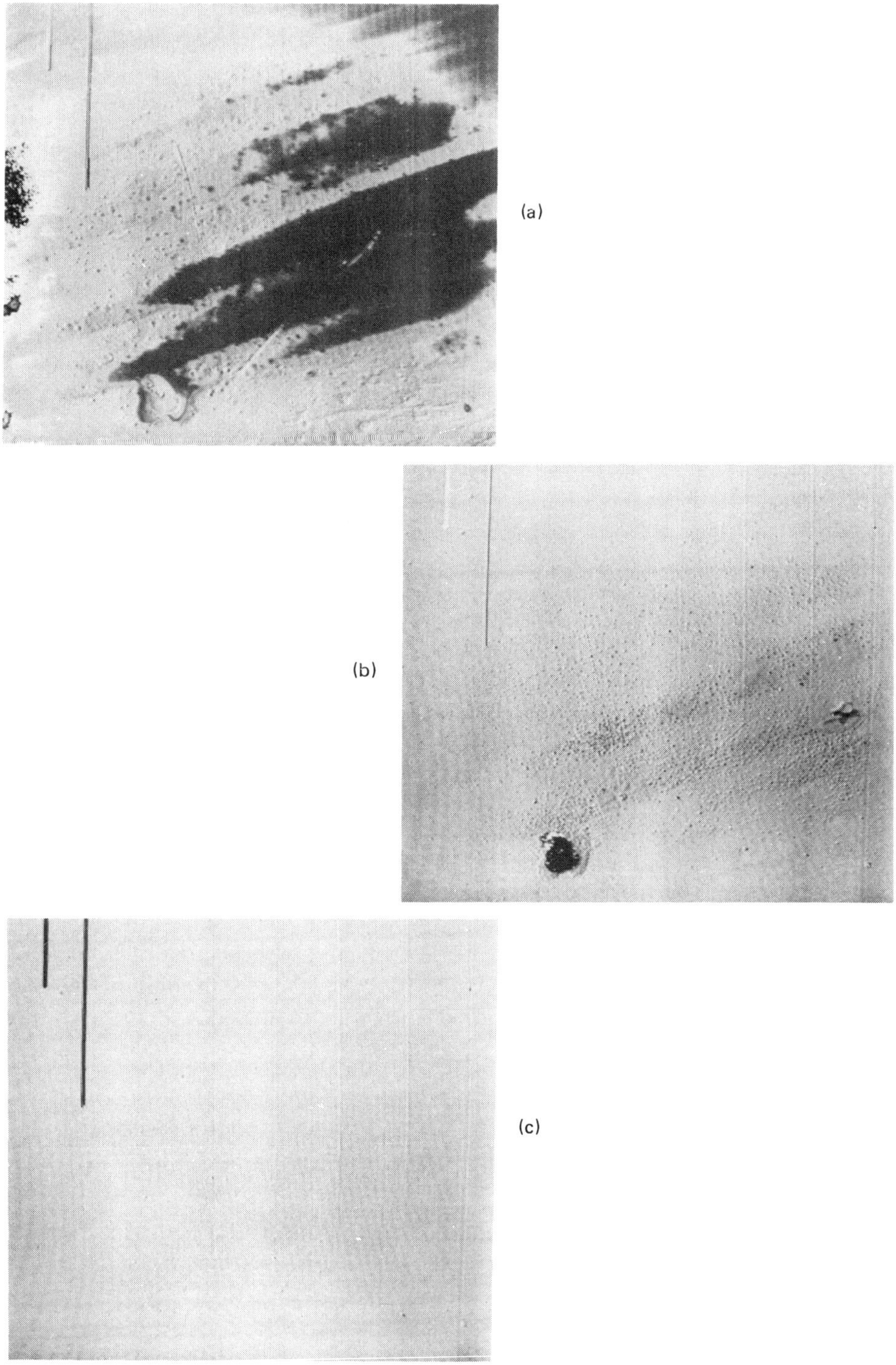

Fig. 5.17 — (a) The TI CCD shown earlier after it got wet from a cryogenic mishap. (b) After cleaning. (c) After nitric oxide treatment. Courtesy Lloyd Robinson, Lick Observatory.

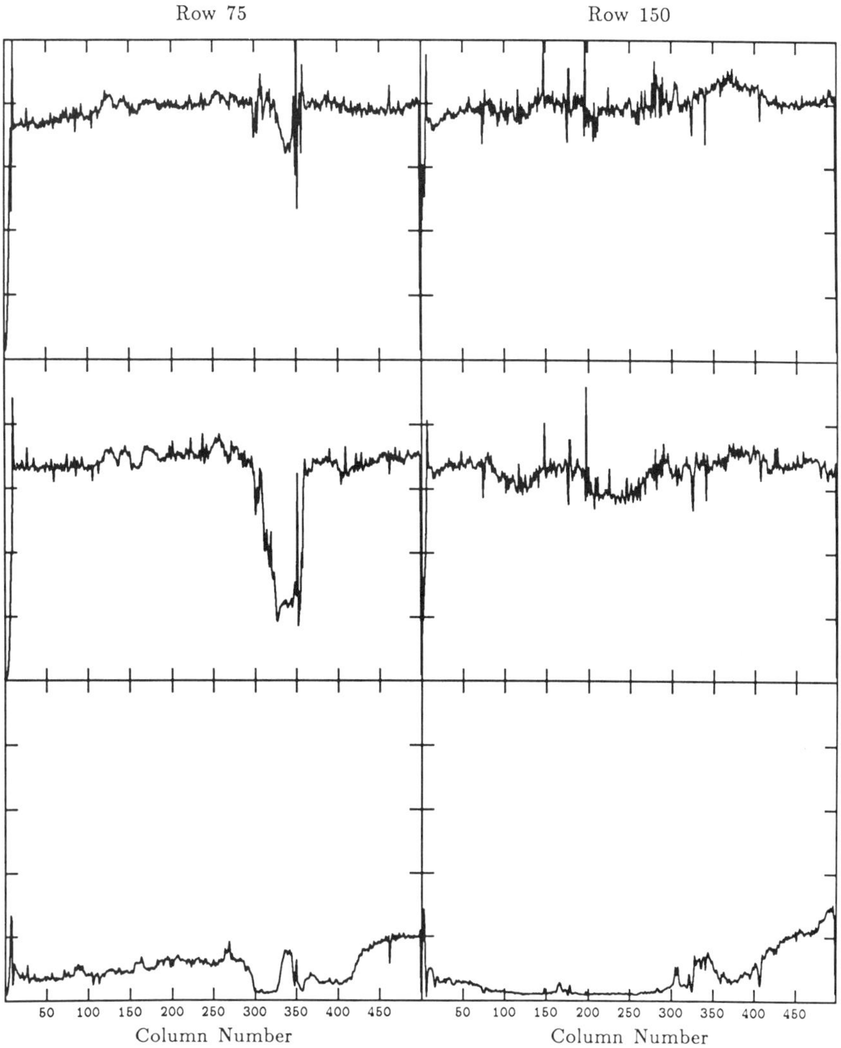

Fig. 5.18 — Relative response of rows 75 and 150 of the unhappy CCD shown in Fig. 5.17. Bottom panel was before the accident. Middle panel after cleaning. Top panel after nitric oxide treatment. The response is better and more uniform than ever before. Courtesy Lloyd Robinson, Lick Observatory.

6

Electronic and computer control systems

6.1 BASIC REQUIREMENTS

There are as many types of CCD-based instruments and control systems as there are engineers and scientists who build them! Commercially available CCD camera systems suitable for astronomical applications can now be obtained from several companies (see Appendix 6), but during the 'CCD boom' period of 1975–1982 these companies either did not exist or were not long established, the demand for cooled slow-scan CCD cameras was considerable, and many astronomical institutes with detector development experience felt the need to compete for 'who could build the best system'.

Although each CCD control system must provide appropriate clocking pulses, signal processing electronics, digitisation, synchronisation with other pieces of apparatus — such as a spectrograph — data-handling and data display, there are many ways in which to implement these requirements in hardware and software. Moreover, since speed, power consumption and versatility are sought in CCD control systems then they too have evolved rapidly as new electronic products have appeared on the market. Optical and mechanical considerations differ from telescope to telescope, and the science content of the project may also define a certain construction path. In astronomy, the net result is that almost every major observatory does a 'home-job', and no two systems actually look alike! To some extent this can lead to confusion and non-standardisation, and it certainly leads to the impression that it is hard to implement a CCD control system. The latter is not true. It is not very difficult to get a CCD system going.

Achieving low-noise performance, reliability, flexibility and ease-of-use, well, that is a different matter!

In this chapter we explore in a general way some of the more important issues in the design of astronomical electronic imaging instruments and explain the technical details of techniques such as correlated double sampling, without which cooled, slow-scan CCD cameras could not achieve their ultimate sensitivity.

6.1.1 Overall system layout

When starting to design any system, the first step is to lay out, in pictorial form, all the essential building blocks and their interconnections. Invariably, a modular approach results in the best design. As an example, the block diagram from the IRCAM infrared imaging system developed by the author and his colleagues is shown in Fig. 6.1. This infrared camera is basically similar to an optical CCD camera. When examined at a level more detailed than the block diagram, it is seen that each subsystem could use one of several different methods for detailed implementation, yet the overall concept is essentially the same.

Whatever your choice of **host computer** and style of electronic hardware — choices which may be dictated by financial resources, compatibility with existing equipment or perhaps even the specific application — certain building blocks must be present. These are:

(1) the detector,
(2) an optical system (lenses, mirrors, filters, shutter),
(3) a cooling system,
(4) signal-processing hardware (amplifiers, noise suppression),
(5) an analogue-to-digital converter (or A/D),
(6) clock-driver and dc bias circuits,
(7) timing logic and synchronisation circuits,
(8) an electronic interface to a computer,
(9) a host computer+peripherals (magnetic disks and tapes),
(10) an image-display system.

These 10 items form the basis of many a CCD system layout. In fact, the above list applies to almost any form of detector system used in astronomy if items (3)–(7) are taken in their most general sense.

The optical configuration depends on the application. To some extent this is also true of the cooling system. For instance, applications at TV rates may require only modest cooling systems. One can also select different host computers and interfaces, different image-display systems and different pieces of hardware such as A/Ds. It is also feasible to put a lot of 'intelligence' into the timing logic by using a microprocessor, and even 'down-load' the timing information from a larger computer which accepts information in human-readable form. Thus, there is no single, uniquely infallible recipe for putting together a practical CCD system.

6.1.2 Optical considerations

It is beyond the scope of this book to describe and analyse the principles of optical design of astronomical instruments. However, one important issue which we will discuss is **optical matching** of stellar images to CCD pixel sizes, together with a brief discussion of some innovative optical techniques which have largely been inspired by the existence of CCDs.

In most astronomical applications, CCDs (and in fact, all solid-state imaging sensors) are placed in a vacuum chamber and cooled. Light must reach the CCD through a 'window' which must be of high-quality transparent material of low **refractive index** or with a durable **antireflection** coating on both faces to minimise

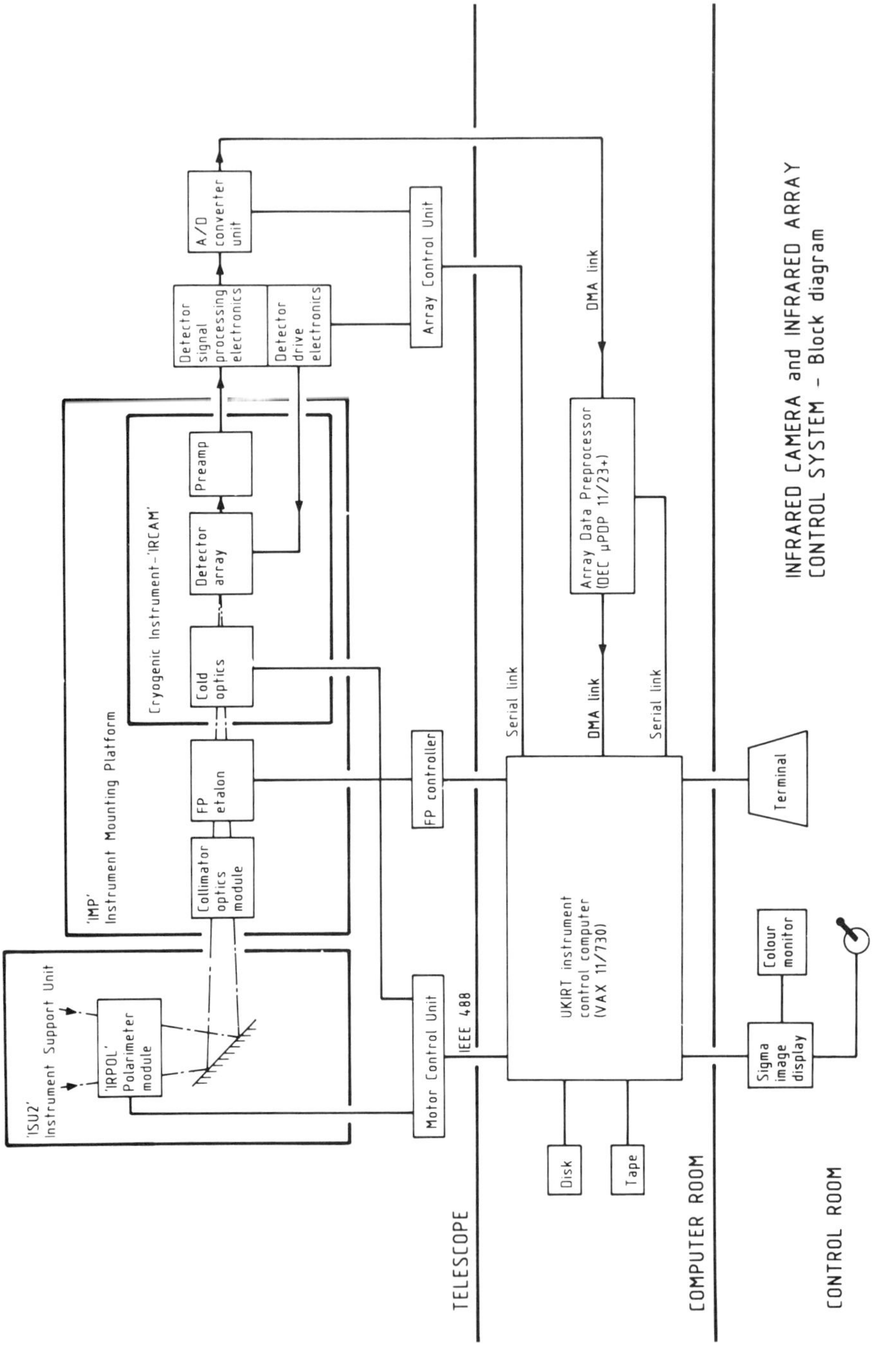

Fig. 6.1 — A block diagram layout of an entire system is an essential starting point for any design. This one is for the infrared camera IRCAM for the 3.8-m UKIRT.

unwanted reflections which would appear in the CCD picture as 'ghost' images. The surfaces of the window must be polished very flat, usually to one-quarter of the wavelength; some groups request the optical manufacturer to introduce a small 'wedge' angle between the front and back faces (of about 0.5°) to minimise interference fringing and ghost images. Fused silica — an amorphous form of crystalline quartz — is often selected since it has good transmission from the UV to the near-infrared, depending on the grade of material; crystalline quartz itself is **birefringent,** that is, it is sensitive to the polarisation of light, and so is widely used in such contexts. At wavelengths longer than 1 μm, infrared transmitting materials are required. Calcium fluoride, zinc selenide and germanium are frequently used in infrared optics. Of these only calcium fluoride transmits well in the visible region. In infrared imaging systems the *entire* optical instrument — including filters — must be cooled to a very low temperature and so optics and detector are therefore usually in the same vacuum container. At optical wavelengths this is not necessary. The housing for an optical CCD may be considered independently of the rest of the optical system, whether that system is a simple box with a rotatable wheel carrying glass filters, or a large diffraction grating spectrograph or one of the many other optical sytems used in astronomy (see Chapter 2).

The distance from the detector surface to the outermost mechanical surface or face of the camera chamber, which generally must include an electric shutter, should be less than the **back focal length** of the *last* optical component in the 'instrument' proper. The back focal length is the actual distance between the surface of the last optical component and the focal plane where the final image is formed; the back focal length of an optical component can be quite different from its effective focal length. Simply put, if this condition is not met then the image falling on the CCD cannot be brought into focus — and it woudn't be the first time that that had happened! On the other hand, if the distance is smaller than the back focal length then a mechanical piece known as a 'spacer' can be inserted.

Of greatest importance in the optical design is that the compact or point-like image of a star must be carefully 'matched' to the physical size of the CCD pixels by optical magnification so as to maximise the observing efficiency, i.e. get more light onto a pixel and therefore keep the required integration time to a minimum to achieve a good image. Preferably, this should be done without compromising the ability of the CCD camera system to obtain very accurate brightness measurements. There are at least three basic conditions:

(1) speckle mode,
(2) wide-field imaging,
(3) spectroscopic mode.

If an optical image could be obtained in the total absence of turbulence in the Earth's atmosphere it would be 10–20 times smaller (on a large telescope) than the usual blurred 'seeing' disk. This size is limited by the diffraction of light; for a given wavelength, the **diffraction-limited** image size of a perfect telescope (that is one with excellent optics) depends inversely on the diameter of the telescope mirror (or lens). These small images can be seen as tiny **speckles** which move around in a high-speed video recording; a sufficiently fast snapshot will 'freeze' the atmospheric turbulence.

The speckle mode is most applicable to very large telescopes and relatively short wavelengths. The resulting magnification is usually very high so that the 'seeing' disk would cover most of the CCD array.

Wide field imaging, a normal case, occurs when the image of the telescope diffraction pattern or the seeing disk, which ever is the larger, is about the same size as a pixel. This maximises the field of view and the detection efficiency. The magnification in this case is usually quite modest; in fact, sometimes a demagnification or **focal reducer** is required. As a general rule however, most designers favour a reduction of field of view in order to 'over-sample' the image two or three times. That is, rather than match exactly to the diameter of the seeing disk they spread the image across two or more pixels for optimum sampling. While the field of view on the sky is certainly reduced, brightness measurements are more accurate because any problems due to interpixel gaps and intrapixel sensitivity structure are minimised.

Optical matching to a CCD in **spectroscopic mode** is similar to that in the normal camera mode except that it will usually be relevant to choose the pixel size to correspond to the **spectral resolution** of the spectrometer, and this is partially determined by how narrow the entrance slit is made. Sometimes this will be smaller than the seeing disk, although designers do try to ensure that all of the light is accepted without loss of spectral resolution; the consequence of this is that the spectrograph has to be physically very large. As in the case of spatial resolution, it is best to oversample the spectral resolution with two or three pixels.

To calculate the required magnification we first of all need to know the **focal plane scale** of the telescope at the focal station being considered. For example, at the prime focus of the 3.6 m CFHT the scale is 13.70 arcseconds per millimetre (often written as "/mm), whereas, at the Cassegrain focal station the scale is 7.33 "/mm. At the **infrared** Cassegrain focus of the 3.8-m UKIRT the scale is a mere 1.52 "/mm. Next, we need to know the physical size of the CCD pixels. For example, the EEV/GEC P8603 has square pixels which are 22 μm (or 0.022 mm) in size; the RCA SID501EX has 30-μm pixels and the TI 800$\times$800 CCD, as well as the RCA SID006EX007 'double-density' device, have 0.015-mm pixels. Multiplying the telescope scale by the pixel size yields the **pixel field of view** on the sky, which can then be compared to the seeing disk (about 1 arcsecond). For the examples above we get, at the CFHT prime focus, 0.41 arcseconds for the normal RCA chip and 0.2 arcseconds per pixel for the TI or double-density RCA CCDs; at the Cassegrain focus the values are 0.22 and 0.11 arcseconds for the 30- and 15-μm pixel CCDs respectively. To perform speckle observations at the CFHT Cassegrain focus we would need to magnify the images by three, say, whereas to perform wide-field, seeing-limited imaging at the prime focus we should de-magnify the image with a $\times$2 converter to get 0.8 and 0.4 arcseconds for the particular CCDs discussed in this example.

As CCD pixels get smaller and telescope mirrors get larger, the problems become harder of inventing optical re-imaging (or matching) systems which do not have **aberrations,** i.e. distortions which result in a loss of image quality. Again, this is a field of technology which has benefitted immensely from the developments in the computer industry. Optical designers can explore the consequences of numerous options by means of computer-simulated **ray-tracing** without ever manufacturing or experimenting with real components. Commercially, most of the emphasis of computer ray-tracing goes on the design of complex, yet extremely efficient, camera lenses for the consumer market. But many observatories now have, or have access to, optical ray-tracing facilities which are invaluable in developing the complicated, off-axis lens/mirror designs used in modern, multi-purpose astronomical cameras and spectrographs.

Of considerable importance in commercial optics is the **anti-reflection** coating. Since most high-quality lenses are actually made from multiple simpler components, it becomes crucial to eliminate reflections between these components. This is done by coating the lens with a thin layer of material which has an 'effective' **refractive index** intermediate between that of air and the original lens material. These coatings are beneficial to astronomers also, and actually have been deposited on the surfaces of some CCDs; the large refractive index (n=4) of silicon (and other semiconductors) will result in considerable reflection losses (about 40%) and therefore set an upper limit of 60% on the quantum efficiency.

In recent years optical designers have endeavoured to come up with complete solutions to image quality problems by appealing to specially designed **corrector plates,** thin lens-like pieces which 'undo' some of the effects of other components in the optical stream. Undoubtedly, the most eminent proponent of this approach is the English designer Charles Wynne; most large telescopes with prime focus cameras employ a 'Wynne corrector'. It is even possible to combine the liquid nitrogen vessel with a re-imaging optical arrangement which is a hybrid between a Schmidt and a Cassegrain telescope as is done in the Palomar four-shooter (Fig. 6.2).

An alternative approach is to make the surface(s) of the primary optical components depart from their normal spherical (or conic) shapes in such a way that they are actually compensating for the misbehaviour or aberrations which the normal spherical surface would cause. Shaping a lens or mirror in this way is called making it **aspherical;** a parabolic shape, for example, would be aspherical but it is also possible to obtain completely general shapes. Manufacture of aspherical optics is only practical with the aid of modern, precision computer-controlled milling and turning machines using diamond-tipped cutters. Diamond-machined aspheric surfaces are a particularly attractive technological innovation for infrared optical components which must be cooled to low temperatures. For example, it is possible to make an aspheric mirror by direct diamond-machining of a solid block of aluminium which has *already* been machined by normal lathes to provide its own mirror-support and attachments! The differential thermal contraction between metal supports and classical glass components is completely eliminated since mirror and support are made of the same material; for better infrared performance the surface of the aluminium mirror is gold-coated.

6.1.3 The cooling system

There are basically three levels of cooling systems which are used in solid-state CCD-type cameras for ground-based optical and infrared astronomy, namely:

(1) Thermo-electric coolers: these normally operate over the range −20 to −60°C and are suitable for TV-rate operations.
(2) Liquid nitrogen cryostats: these will cool detectors to −196°C (77 K) and are routinely used for silicon CCDs.

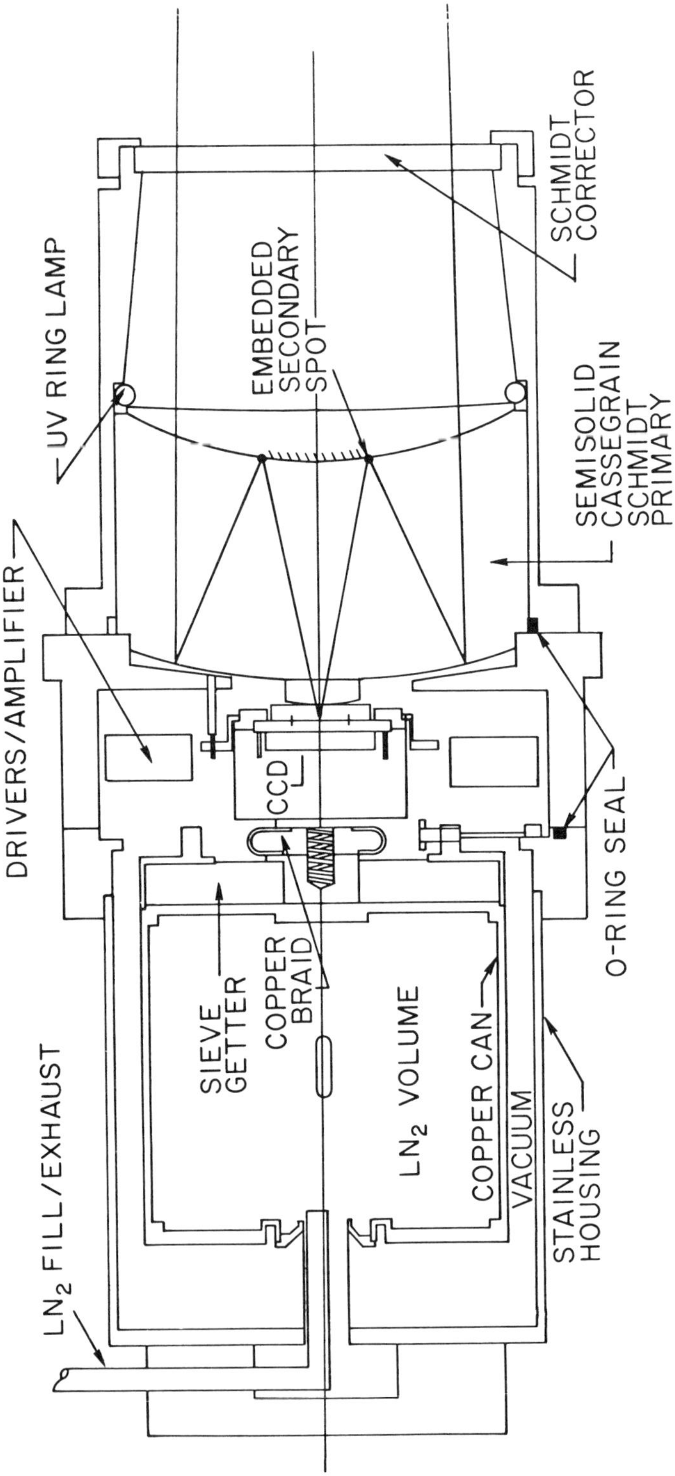

Fig. 6.2 — The combined cryostat and re-imaging optics of one of the Palomar four-shooter CCD cameras. Coutesy Jim Gunn.

(3) Liquid helium cryostats: these will cool detectors to −269°C (4 K) and are required for low band-gap semiconductor materials used in infrared arrays.

At 77 K liquid nitrogen is actually much colder than the typical CCD operating temperature of 135 K. While some CCDs do still work at this temperature, others do not because of 'silicon freeze-out' — the silicon becomes non-conducting, so we need to work at a higher temperature. To achieve an intermediate temperature between LN_2 and the warm outer body of the vacuum chamber at room temperature it is usual to set up a kind of thermal analogy to an electrical bridge circuit. One way to do this is to attach a copper block (solidly) to the outer walls of the cryostat by means of pillars of fibre-glass tube — a poor thermal conductor — and also to connect it by means of a copper wire braid to the copper face of the LN_2 reservoir. Since the rate of heat transfer conduction q_H (usually in watts) for a thermal conductor of length L and constant cross-section A is given by

$$q_H = \frac{A}{L}\,\mathrm{TCI}$$

where TCI is a tabulated property for many materials called the 'thermal conductivity integral' between room temperature and LN_2 temperature. By trimming the length and cross-section (number of wires) in the braid and the length and cross-section of the fibreglass tube, a 'balance' can be reached between the tendency of the LN_2 to cool down the copper block (and hence the CCD) and the tendency of the block to warm up to room temperature via the fibre glass pillars. In addition, a heavy-duty resistor can be embedded in the copper block to enable the block to be heated (by passing a current through the resistor) at any time. A temperature sensor — usually a diode — is also embedded in the copper block, and a circuit controls the current to the heating resistor depending on the voltage measured across the diode which has a constant current flowing through it. This allows some measure of control over the actual operating temperature.

In practice, it turns out that the cooling function is the most straightforward to accomplish and that it is the other aspects of cryostat design that cause the most headaches for designers. Among these other factors are mechanical flexure (bending), **hold-time** — how long the liquid cryogens last — **out-gassing** — the slow, natural escape of gasses trapped on interior surfaces — electrical and optical access, and the ease with which the cryostat can be attached to other instruments such as spectrographs or polarimeters. For infrared work there is an additional problem, namely that of shielding the sensitive array detector from the flux of infrared photons emitted copiously by all warm objects such as the outside walls of the cryostat. This is accomplished by a double radiation shield, with the inner shield being as cold and as light-tight as possible.

Design objectives for a good cooling system, suitable for both laboratory and telescope usage, should include the following.

(A) **Minimum detector movement,** either along the optical axis (defocus) or at right angles to the optical axis (image smear), to yield spatial stability on the CCD relative to the telescope focal plane. This means doing a careful mechanical and thermal design with detailed calculations and/or experiments to expose weaknesses.

(B) **Minimum effort to keep the system uniformly cold.** When using LN_2 as the coolant, this requirement simply means that the LN_2 should not run out too quickly, i.e. it should have a good 'hold-time'. The hold-time of a cryostat depends on its thermal and mechanical design, the amount of power generated by the device being cooled and the quality of the vacuum. Typically, the hold time in hours is about 45 times the volume of LN_2 in litres divided by the input power in watts. A hold-time of less than 12–14 hours is not acceptable for astronomy since a refill would be required during the night. For a 1-litre capacity cryostat to last 15 hours the heat input should be less than 3 W.

Moderately good vacuum techniques and methods of maintaining a good vacuum lead to excellent hold times. After initial evacuation the pressure in a cryostat may tend to rise slowly again due to 'outgassing'. A simple, widely used class of vacuum pumps which remove unwanted trace gases in the vacuum chamber by physically tieing up molecules on a surface are called 'sorption pumps'. These can be constructed as small chambers about one inch in diameter filled with a sorbent material such as activated charcoal or one of the synthetic zeolite materials known as molecular sieves. These materials have a huge effective surface area of thousands of square metres per gram. At liquid nitrogen temperatures these materials will absorb air, and at very low temperatures of about 10 K zeolite can be used to partially absorb helium.

Good, uniform physical contact between the CCD and the copper cold-block is essential.

(C) **Good accessibility.** When it is necessary to work on components inside the cryostat it is important to have enough 'finger-room', good handling arrangements and adequate safety protection for delicate electronic parts including the CCD itself. Actual handles on the cryostat or on an internal subassembly if it is large enough, are often over-looked — until the unit is dropped!

(D) **Cost and manufacturing time.** To achieve the above objectives usually means that a commercial cryostat purchased 'off-the-shelf' will require some modifications or additions, which will cost time and money. Nevertheless, this can be the best approach when a good dialogue between the company and the observatory is established. One alternative approach is to 'design your own' cryostat, stick to that design and copy it many times. This is a difficult route for most CCD camera users and, of course, unless the design really is stable then it is not clear that there are any savings of time and money!

More often, researchers find it cost-effective to construct a 'hybrid' cryostat by separating the 'cooling function' from the function of supporting the detector and associated components — often called the 'camera head'. The camera head is a simple vacuum chamber in its own right, and the cooling function can then be provided by a proven, commercially available cryostat. This is the approach adopted at the Royal Observatory, Edinburgh, where cryogenic systems are required for a whole range of optical, infrared and sub-millimetre instrumentation. Suitable LN_2 and LHe (liquid helium) vessels are available from several companies including Oxford Instruments (UK), Infrared Labs (USA) and Janis Corp. (USA).

(E) **Ease of coupling** (mechanically and optically) to existing telescopes and/or instruments.

A standard LN_2 cryostat used at Kitt Peak is shown schematically in Fig. 6.3 and again as it actually appears (Fig. 6.4(a)) complete with electronics box. With the vacuum end-plate removed a typical mounting arrangement for the CCD can be seen (Fig. 6.4(b)).

6.1.4 The electronics

In this category we find not only the amplifiers and drive electronics for the CCD, but also the other electronics needed to control the instrument itself. For instance:

(1) stepper motors, dc motors and solenoids,
(2) electric shutters,
(3) temperature controller,
(4) preamplifier,
(5) CCD driver and analogue signal processor,
(6) A/D unit.

Most CCDs require a **preamplifier** or preamp fairly close to the output pin before the signals can be sent along long lengths of cable. A typical design is shown in Fig. 6.5. In some cases the preamp is placed inside the cryostat, as close as possible to the CCD. Other workers have had good success with preamps placed immediately outside of the cooled chamber provided the lines are well shielded, by using grounded coaxial cable or twisted-pairs (section 5.1). This appraoch is obviously attractive from the point of view of accessibility. The excellent capacitance of the EEV/GEC CCD gives it a very healthy output signal in terms of microvolts per electron and Craig Mackay has found it unnecessary to have any electronics, including a preamp, closer than 1 metre from the CCD.

Good preamp performance is essential since it is at that point that small signals are being turned into large signals. Over the years, however, CCD manufacturers have endeavoured to put as much 'gain' on the CCD chip itself; for instance, the earliest CCDs could provide about a quarter of one millionth of a volt for each electron of stored charge whereas slow-scan CCDs available now are typically five times better. Recently some devices intended for colour TV operation are quoted at 15 μV per electron.

6.2 CCD SEQUENCERS AND CLOCK DRIVERS

The key feature of any control system is timing. Most of the complexity in a CCD or other advanced detector system, revolves around the **timing diagram.** Numerous events must occur in strict time order and at precise time intervals. For example, in a typical CCD system the following sequence of events must occur in time sequence:

- check instruction status from operator,
- prepare the CCD — unrecorded readouts to flush the CCD,
- open the shutter for a timed period,
- close the shutter,

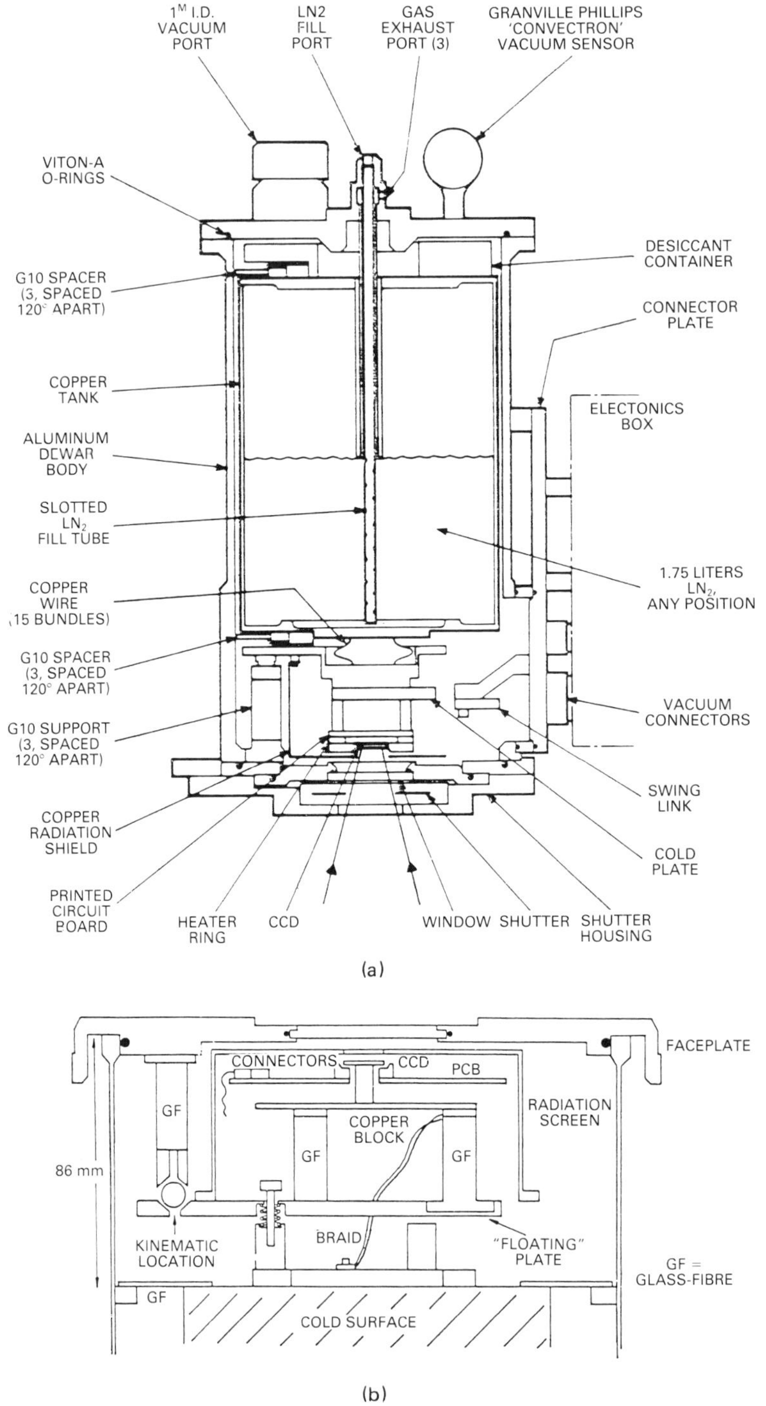

Fig. 6.3 — (a) An LN_2 cryostat of the design used at Kitt Peak (NOAO); (b) a similar design by the Royal Greenwich Observatory.

(a)

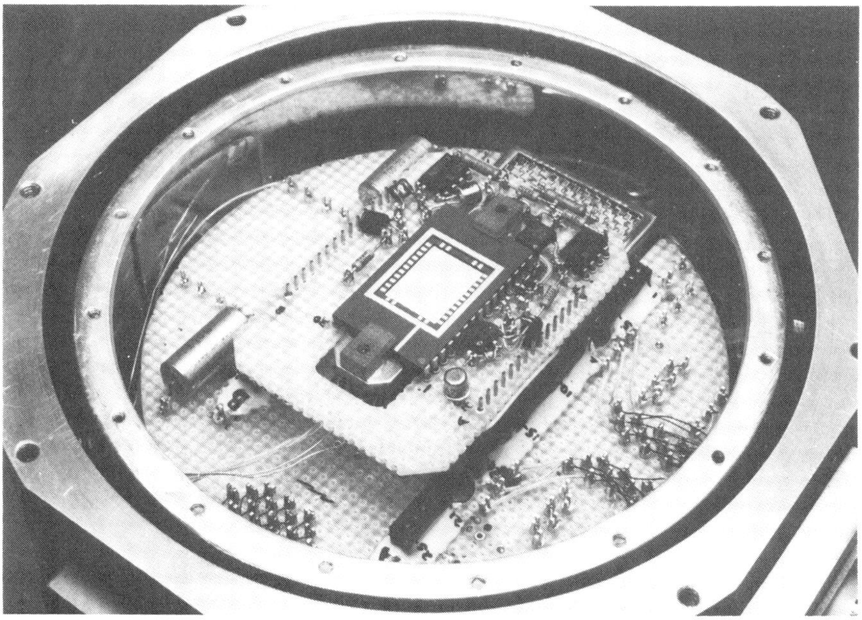

(b)

Fig. 6.4 — (a) The Kitt Peak CCD camera complete with electronics unit. (b) A view of the CCD mounted on a circuit board within the cryostat.

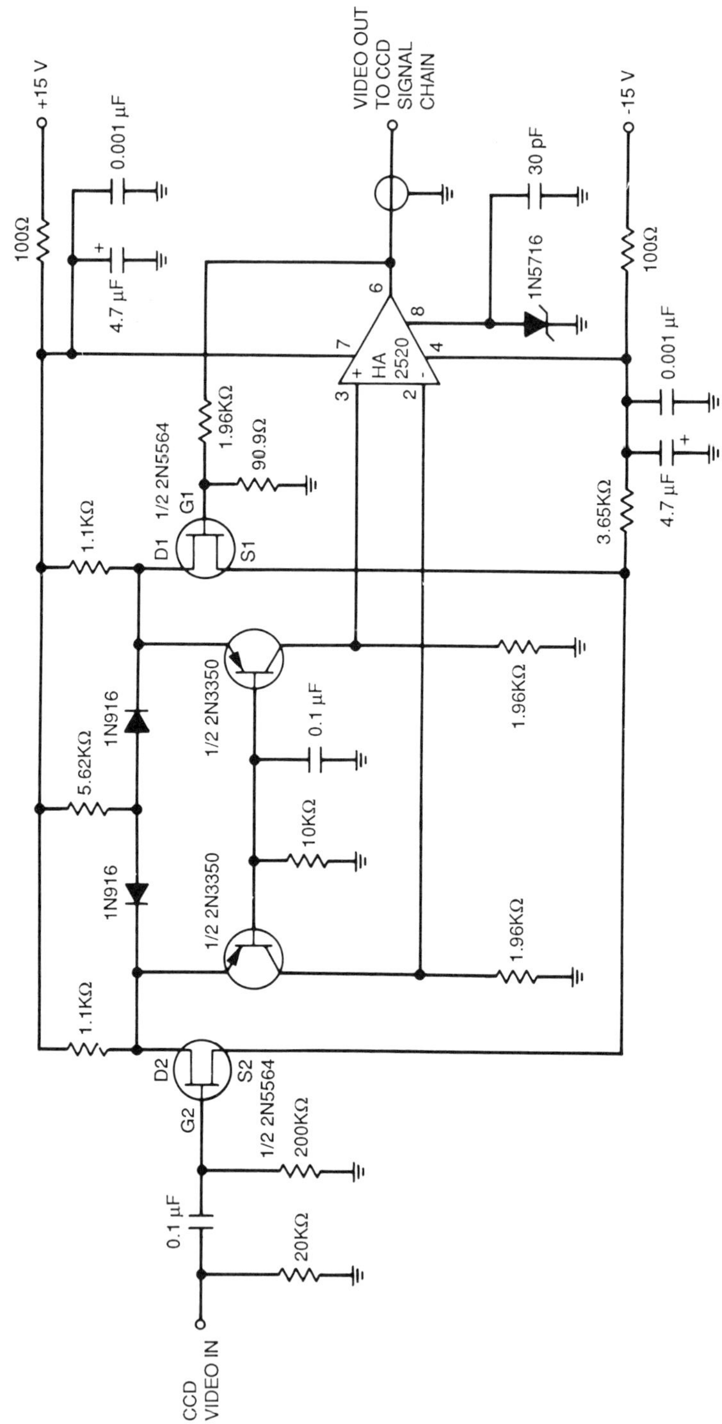

Fig. 6.5 — A tried and tested preamplifier design used by Lloyd Robinson of Lick Observatory based on an original design by Jim Janesick.

- read out the CCD according to a precise pattern,
- digitise the signal from each pixel,
- store the data in a computer
- return the CCD to a 'standby' mode, if appropriate.

That part of the system responsible for such timing activities is often called the **sequencer.**

Almost all CCD systems fall into one of the following design categories:

(1) hardwired,
(2) programmable.

6.2.1 Hardwired designs

By 'hardwired' we mean that the electronic functions are carried out by circuitry, and cannot be altered by typing instructions on a computer keyboard. This approach results in a very compact system and is favoured by some astronomers and others with travelling systems or with severe constraints on space and weight. It is restrictive in the sense that the circuitry has to be altered, perhaps by re-soldering wires, in order to effect a change, such as to another type of CCD, or to an alternative clocking sequence.

In a typical hardwired CCD camera, the digital controller for the camera is built with commercially available logic chips such as TTL or low-power high-performance CMOS. Readout of the CCD is in a fixed form, and all clocks and signal processing pulses are generated on a single board — the 'clock logic' board. This circuit board derives its time-base from a small device known as a quartz crystal oscillator. The clock logic board in turn feeds both the CCD clock 'driver' board which has more current handling capacity than logic chips for pulsing the CCD electrodes, and the 'analogue processor and digitisation' board. A fourth board, containing several low-noise well-regulated dc power supplies for setting the fixed voltage levels on the CCD, as well as providing auxiliary functions such as temperature control, completes the camera head electronics. The whole four-board controller could fit into a small box of typically 13×15×20 cm. Many of the pioneering astronomical camera systems in the USA were of this type.

In the original GEC CCD astronomical camera developed by Craig Mackay and Jonathan Wright, there were no electronic components inside the cryostat other than the CCD itself. The camera and electronics box are separated by about 1 m of cable. All electronic functions are contained on three **printed circuit** cards, usually a bit more densely packed than is possible with actual wire connections, performing (1) double correlated sampling, filtering and analogue to digital conversion, (2) three-phase clock drive and reset pulse generation, (3) dc bias supplies. On each board all critical power supplies are double-regulated (smoothed) to filter noise and high-stability components are used throughout. All CCD functions are controlled directly by the host computer connected to the CCD hardware by a single coaxial cable used as a bidirectional serial link allowing a 16-bit data transfer in 2 μs.

Another example of a hardwired system is the one developed originally by Jim Gunn of Princeton and Jim Westphal of CalTech. The clock logic is all CMOS except for the 50-ohm line drivers and recievers on the four coaxial cables providing communication between camera and data-recording system. A master clock running at 2.5 MHz is divded down to 1.25 MHz to give an 800-ns control cycle period. CCD clock waveforms were generated by a counter-divider chain feeding a set of electronic latches which allow 'wire-wrapped' programming. The clock generators are housed with the bias regulators, video signal chain, A/D, and parallel-to-serial converter and line driver circuitry in two 'saddle-bags' clamped to the LN_2 cryostat. One external signal, a line start command, generates the following sequence; (a) the serial clocks (normally running) are interrupted, (b) a vertical transfer sequence is executed, and (c) the serial clocks are started again. The first line start triggers a continuous frame signal for clock voltage control. Clocking signals go to a driver board inside the LN_2 cryostat.

In almost all CCDs the electrode voltage levels (clocks) must be within +20 V. The levels can be configured in digital form with the values being selectable by tiny switches or tunable from front-panel knobs attached to potentiometers. D/A converters — the opposite of A/Ds — are used to give an analogue output which goes to the chip through high-power **operational amplifiers**† that can cope with the rate-of-change through potentially large voltage swings.

6.2.2 Programmable designs

By 'programmable' we mean that the electronic functions can be changed by computer commands. Among the earliest proponents of this type of sequencer were the Kitt Peak National Observatory. Initially controlling CID arrays, they developed a sequencer based on the 2901 'bit-slice' microprocessor, essentially a specialised microchip capable of accepting 'instructions' in binary digit (bits) form and doing some arithmetic logic but without all of the memory management frills of a full microprocessor. The micro-instruction is 64 bits long, of which 32 bits are used to control the 2901 and the remaining 32 bits are available to control the CCD clocks, the signal processor and other units with which the camera must communicate. The key feature of the bit slice approach is that the timing is very precise and is not a function of the particular set of micro-instructions called. Timing and control signals for the CCD are generated by sequentially stepping through the microprogram instruction set, which itself is contained in eight **programmable read only memories (PROMs)** of 8 bits by 256 instructions.

If the microprogram is stored in PROM, the system can be operated without the main data-gathering computer and the program will always be available. It has the disadvantage that one must replace the PROMs with new ones to effect a timing or sequence change. This process is called 'burning' PROMs and requires some special, although compact, equipment and a computer to develop the new PROM codes on. Good quality erasable PROMs (EPROMs) are available which at least save on the purchase of these microchips. PROMs provide a simple and reliable solution when the camera is fully commissioned and performing in a scientific instrument. For development engineers, however, it can be much faster and more convenient to replace the PROMs with **random access memory (RAM)** chips. These devices must

† An operational amplifier is actually a collection of high-gain dc-coupled (transistor) amplifiers all packaged together in a single tiny unit to act as a high input–low output impedance amplifier. Depending upon how the output is 'fedback' to the input, the amplifier can perform mathematical operations such as addition, subtraction, inversion and integration — hence the term 'operational'.

have the microprogram sent to them from the main computer, which has overall control, each time the sequencer is switched on; this process is known as **down-loading.** Several different programs can be stored in the mini-computer. This is the approach Kitt Peak (NOAO) took to provide flexibility at the telescope observing environment. A block diagram of the Kitt Peak system and its physical appearance in an equipment rack together with computer and magnetic tape drive is shown in Fig. 6.6.

The CCD system controller used at the Royal Greenwich Observatory is shown in Fig. 6.7. Developed initially under contract to the Anglo-Australian Observatory, this system was later modified and further developed for the La Palma Observatory in the Canary Islands. Many different types of CCDs had to be tested in the laboratory and so these workers, David Thorne, Paul Jorden and Ian van Breda opted for the 'down-loading' technique. The waveforms to drive and read out the CCDs are defined in a computer language known as FORTH, compiled into microcode and down-loaded into the CCD controller. A similar approach is taken by commercial slow-scan CCD camera companies, although, in general, most CCD builders tend to favour FORTRAN as the high-level language.

A practical approach to CCD camera design requires room for expansion and change as new devices come along. It therefore makes sense to use a 'modular' design. For example, one might design a CCD camera system to allow the following range of configurations

(a) slow or fast readout,
(b) on-chip pixel binning (two axes),
(c) digitisation of only a subset of the image area,
(d) bidirectional charge shifting,
(e) overscanning.

These modes can be built-up from a small set of primitive waveforms. The longest of these involves the digitisation of a pixel charge value. In a typical system this will take 20–50 μs due to the integrating correlated double sampling method described in section 6.3.1. Also, the operational modes require groups of consecutive sequences of the various waveforms. For instance, in the Royal Observatory, Edinburgh, CCD system, developed by Jim Herd, Bill Cormack and the author, the waveform sequencer was based on a PROM designed to accept 16-bit commands with two parts from the microprocessor. The first part specifies which waveform is required while the second part specifies the number of waveforms to be generated, that is, a 'repeat' count. To smooth the flow of commands and eliminate timing problems a **FIFO** (first-in first-out) buffer was placed in the 16-bit data path between the microprocessor and the waveform sequencer. This method provides a clean split between the hardware and the software with the microprocessor responsible for breaking down high-level commands, e.g. 'take an image', into a stream of sequencer-compatible 16-bit commands. The PROM-based nature of the sequencer, which provides a waveform resolution of 250 ns (4 MHz clock), implies that it can be easily updated to provide any desired format of driving/readout waveform.

At the Lick Observatory in Santa Cruz, California, Lloyd Robinson has developed a CCD controller based on an 8-bit MOS 6502 microprocessor. Several

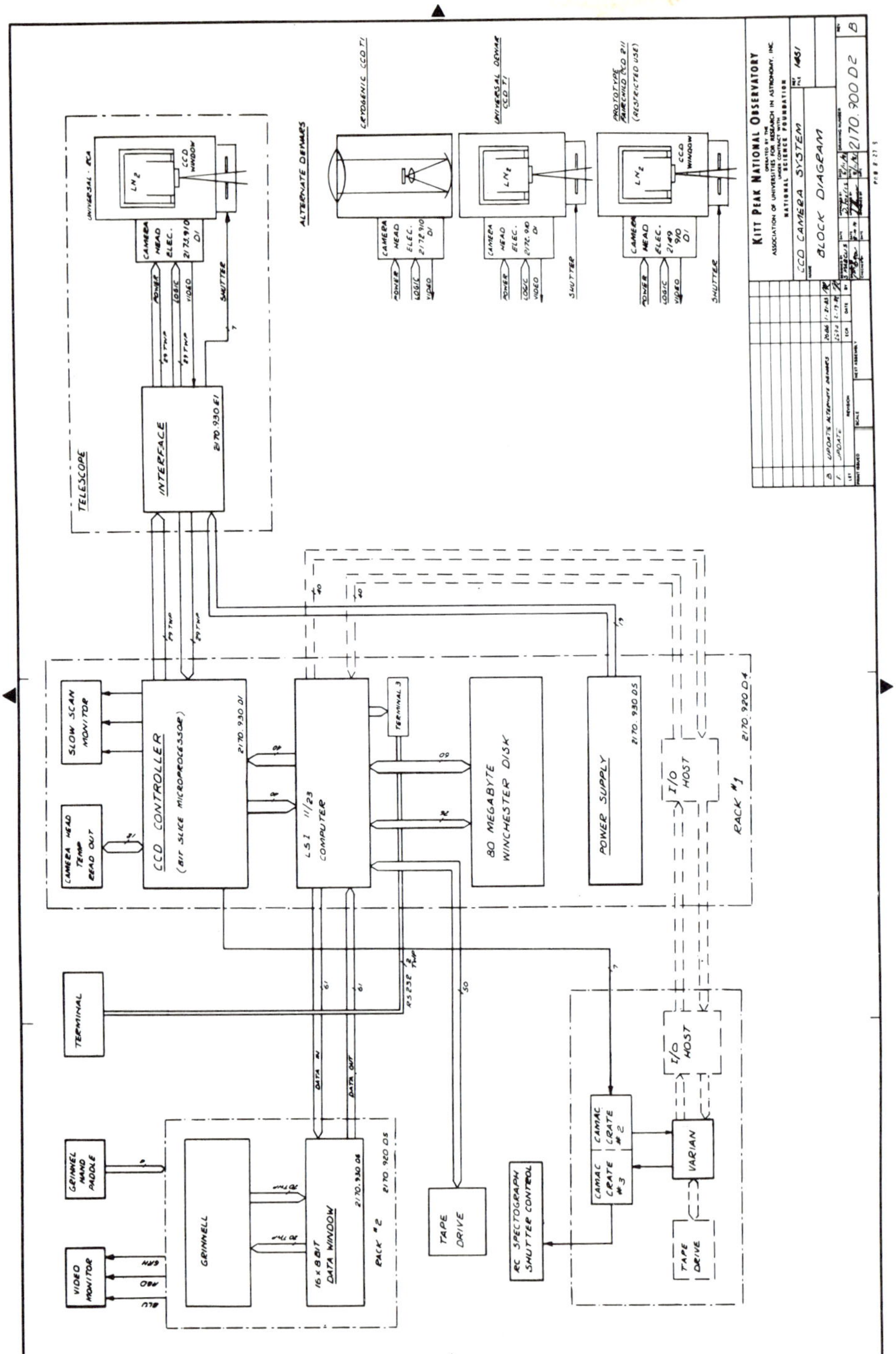

Fig. 6.6 — (a) A section of the block diagram of the Kitt Peak CCD camera system.

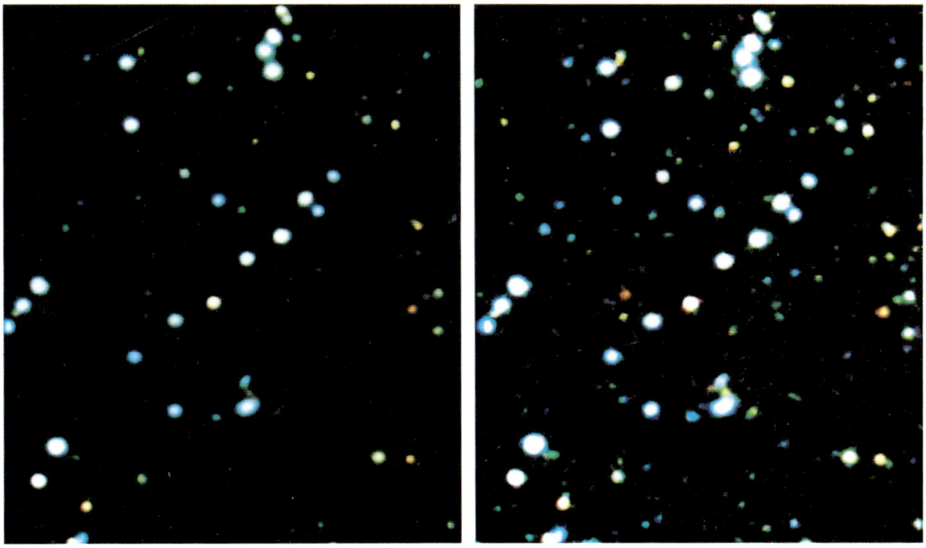

Plate 1 — A composite colour image of a field 6×8 arcminutes in size obtained with the AT&T Bell Labs CCD camera in 1986 on the Lowell Observatory 1.1-m telescope. The total exposure time was 2.25 hours and blue objects as faint as 23 magnitude are easily detected with this relatively small telescope. Two contrast levels are shown because the brightness or 'dynamic' range spanned by the CCD images is very large. At least four of the bluest objects are quasi-stellar objects. Courtesy Tony Tyson.

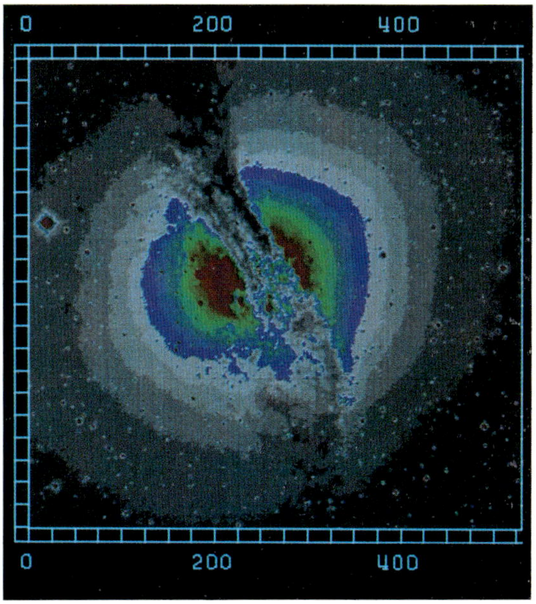

Plate 2 — A 'false-colour' CCD image of the galaxy Centaurus A showing a massive dust lane seen almost edge-on. In this picture a different colour, including grey, has been used to represent a range of brightness. This technique allows subtle changes in brightness to be highlighted by their colour.

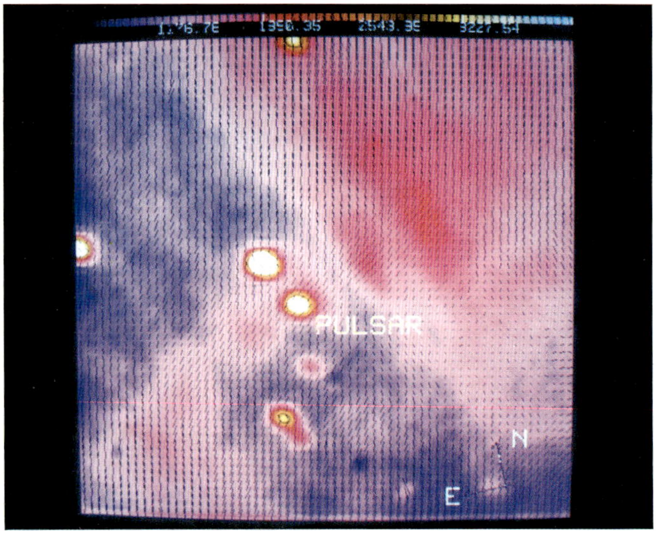

Plate 3 — A 'false-colour' CCD image at a very red wavelength of the inner regions of the famous Crab Nebula supernova remnant. The 'pulsar' or spinning neutron star at the centre is marked. Emission from the nebula is due to synchrotron radiation by relativistic electrons spiralling in magnetic fields; the emitted light is polarised. Line segments superimposed on the image trace the amount and direction of polarisation in incredible detail. The measurements were made with the Royal Observatory, Edinburgh, CCD Imaging Spectropolarimeter.

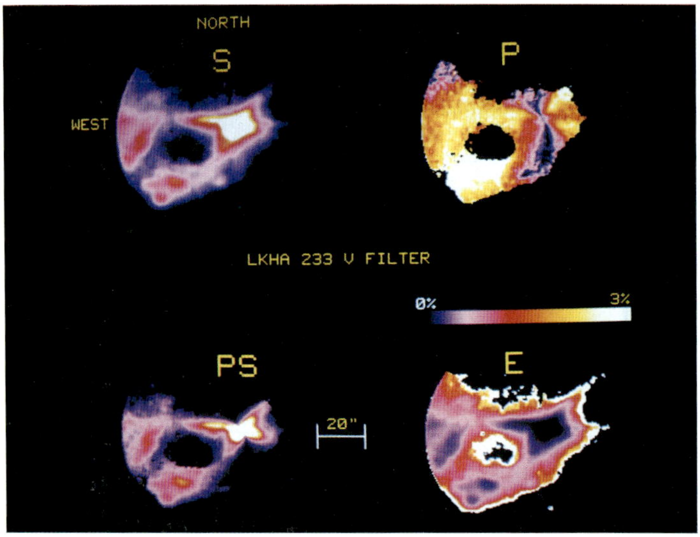

Plate 4 — A CCD image of the peculiar nebulous object known as Lick Hα 233. False colour is used to enhance subtle changes in brightness. The images marked S and P show respectively the normal image (intensity) and the fraction of light which is polarised. Light from the nebulousity is highly polarised indicating that the nebula shines by reflection. When the intensity (S) and polarisation (P) are multiplied together the image marked PS is obtained which shows a 'pinched' effect indicating that the source of light is hidden behind a narrow disk of dense material. Image E is a colour-coded map of the accuracy of the polarization measurement. Courtesy Colin Aspin, ROE.

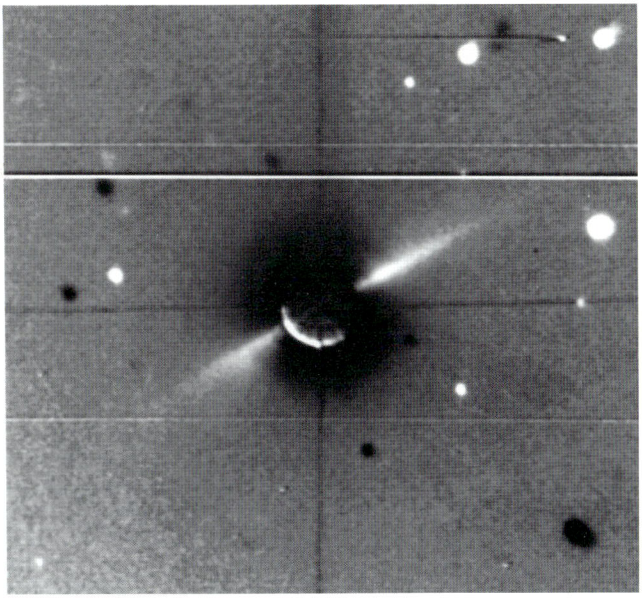

Plate 5 — A CCD image of the bright star Beta Pictoris made with the aid of a small circular 'occulting' mask. This mask blots out most of the direct starlight and, together with the sensitivity of the CCD, reveals the presence of a faint but very extensive edge-on ring or disk of illuminated dust particles in orbit about the star. Photo courtesy Brad Smith, University of Arizona.

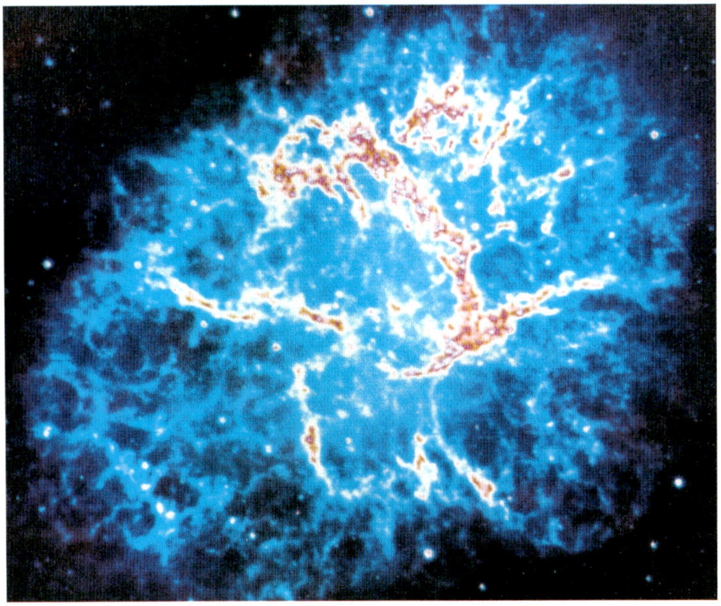

Plate 6 — A montage of three CCD images using a EEV/GEC chip on the 1-m Jacobus Kapteyn Telescope, La Palma. The object is the Crab Nebula supernova remnant and a special filter with a narrow transmission range in wavelength was used to pick out the network of radiating filaments in the nebula. The drift scan method (see text) was employed. Courtesy Royal Greenwich Observatory.

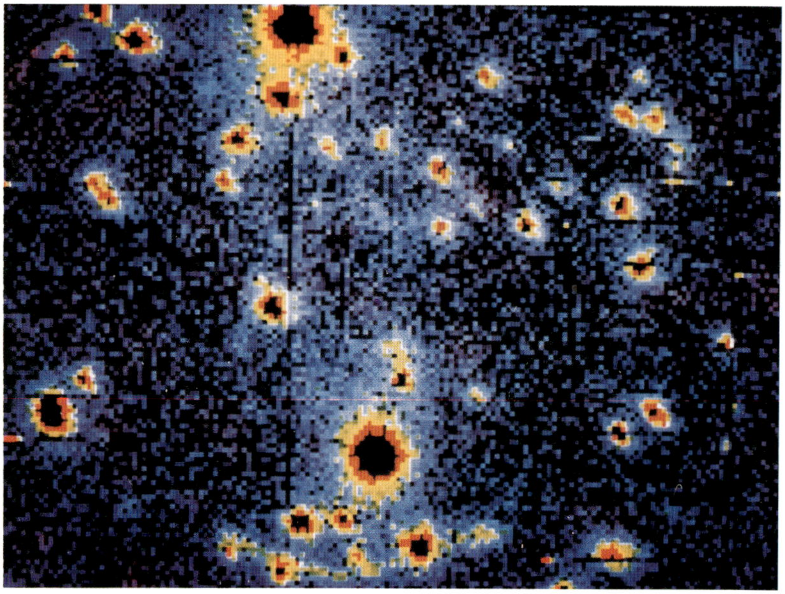

Plate 7 — A false-colour, computer processed CCD image of the cluster of galaxies called Abell 370 obtained by astronomers from the Toulouse Observatory using the 3.6-m CFH telescope on Mauna Kea, Hawaii. A 'Giant Luminous Arc' 500 000 lightyears long can be seen at the lower edge. It is believed to be a 'gravitational ring' caused, according to Einstein's General Theory of Relativity, by bending of light rays from a background galaxy in the strong gravitational field of the Abell 370 cluster. Courtesy European Southern Observatory.

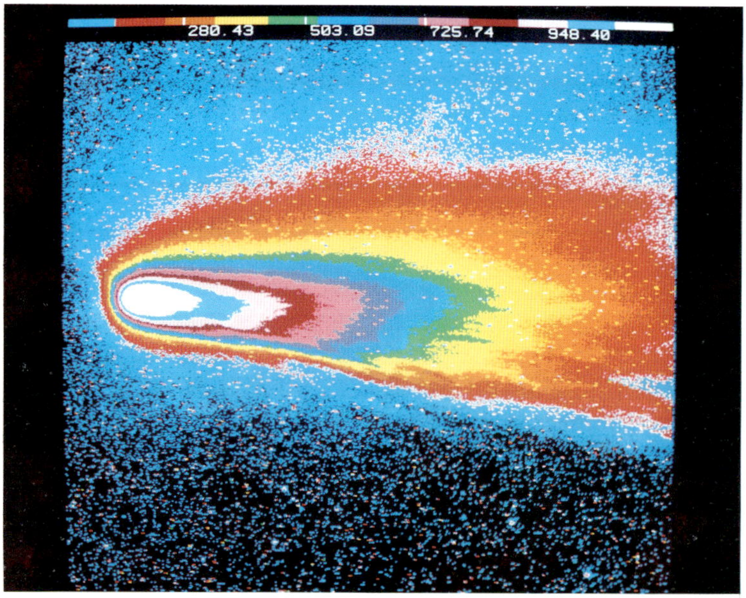

Plate 8 — A false-colour, computer-processed image of Comet Halley. As can be seen from the colour bar along the top, each colour represents a numerical range in brightness. All picture elements (pixels) with brightness above a certain value are set to white and all those below a certain value are set to black. Courtesy Anglo-Australian Schmidt Telescope and the Royal Observatory, Edinburgh.

Plate 9 — Images at very different wavelengths may be combined by computer processing of their digital representations. Plates 9 and 10 show colour-coded composite images of the supernova remnant known as the Cygnus Loop. The optical images were obtained with a CCD camera on the 1.52-m (60-inch) telescope on Mt Palomar, the X-ray data are from the Einstein satellite and the radio image was obtained with the Very Large Array (VLA) in New Mexico. Red represents Hα emission from hydrogen atoms, green represents the emission from doubly-ionized oxygen [OIII] atoms which is so rare as to be considered 'forbidden', and blue represents X-ray emission. Courtesy of Jeff Hester at IPAC.

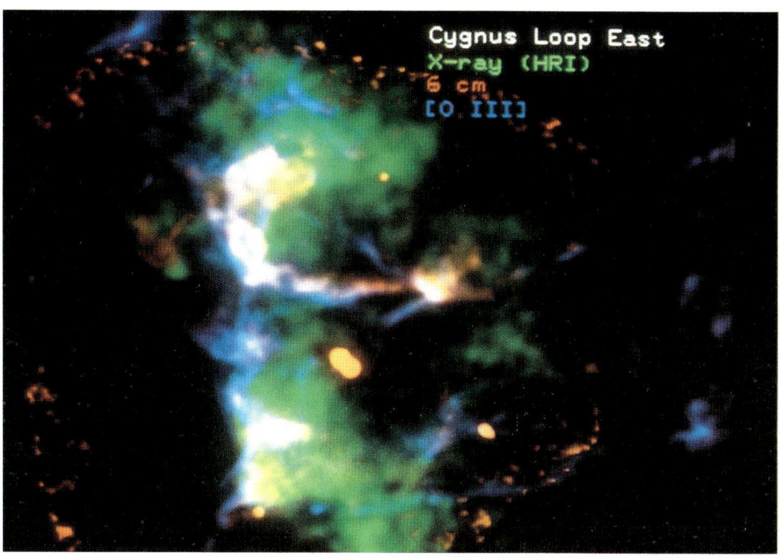

Plate 10 — Cygnus Loop East as in Plate 9 except that the colour code is now green, X-ray, blue, [OIII] optical emission; and red, radio emission at a wavelength of 6 cm. The picture has been smoothed to match the radio beam.

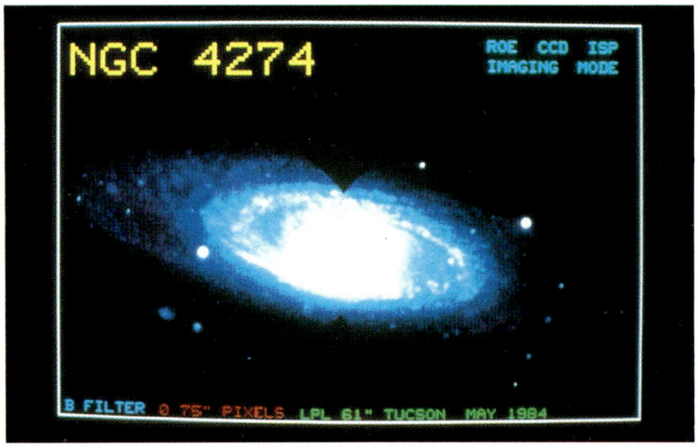

Plate 11 — A computer-generated mosaic of four overlapping CCD images of the spiral galaxy NGC 4274. Computer programs are used to manipulate the digitised images to bring them into registration and balance brightness levels. The detailed structure of the spiral arms remains difficult to see.

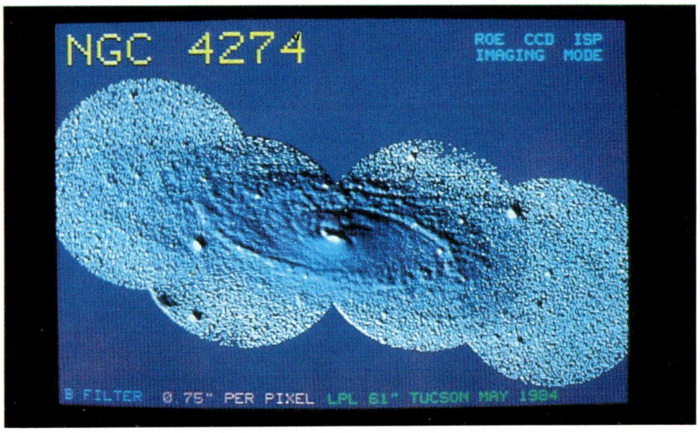

Plate 12 — The same composite image of NGC 4274 as shown in Plate 11 after the digital image data have been processed by a computer program which highlights gradients using a bass-relief effect. The images were obtained with the Royal Observatory, Edinburgh, CCD camera.

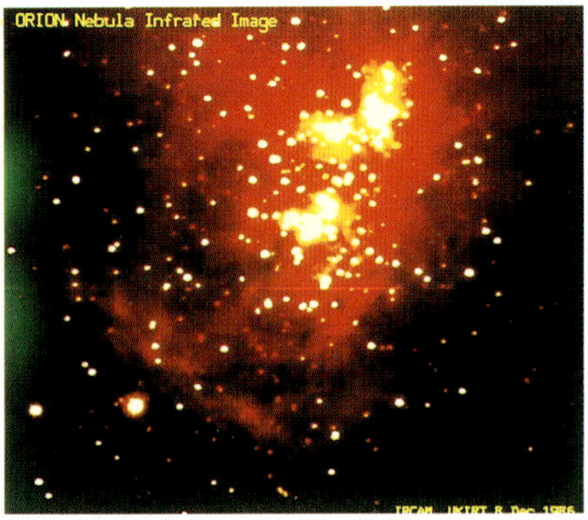

Plate 13 — An infrared image of the Trapezium star cluster in the Orion Nebula obtained with the IRCAM on the 3.8-m UK Infrared Telescope on Mauna Kea, Hawaii. The wavelength is 2.2 μm, twice the red wavelength limit of optical silicon CCDs. Since the range in brightnesses encountered in this image exceeds 2000:1 it is necessary to limit the 'window' of brightness to be displayed.

Plate 14 — The same data and same false-colour palette as in Plate 13 but the 'window' of brightness levels has been made smaller and has been shifted to much fainter values.

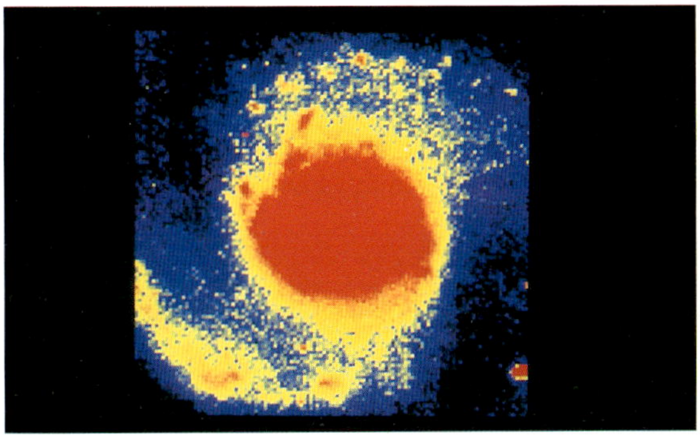

Plate 15 — A false-colour image of the nucleus of the galaxy M51 at the infrared wavelength of 2.2 μm using a simple colour palette with relatively few hues. The knots in the inner spiral arms show clearly but the nucleus itself is lost in this presentation. UK Infrared Telescope photo.

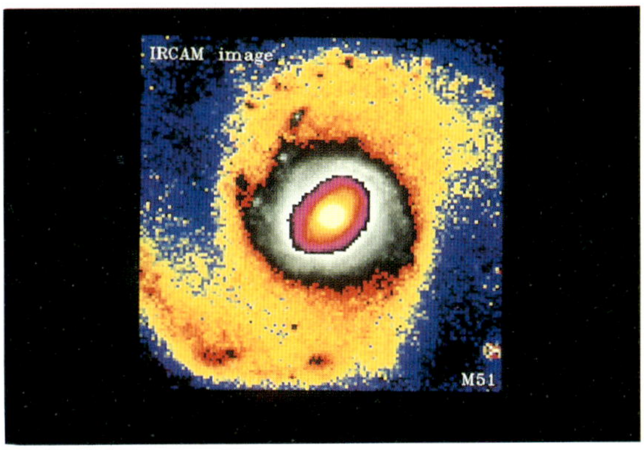

Plate 16 — The same IRCAM data as in Plate 15 but this time displayed using a colour palette with many more hues, including grey and including some repetition. Now it is possible to trace the innermost contours of brightness of M51. Since at this wavelength obscuration by dust and gas in M51 is negligible, this computer-enhanced image reveals that the nucleus is not round but 'bar-like'. UK Infrared Telescope photo.

Plate 17 — An optical colour picture of the giant gas cloud called Messier 17 and an infrared colour-composite image of the same region with blue, green and red used to represent the invisible wavelengths 1.2 μm, 1.65 μm and 2.2 μm respectively. Note that the dark bay in the optical image has been 'filled-in' with stars. Photos courtesy of Ian Gatley, NOAO.

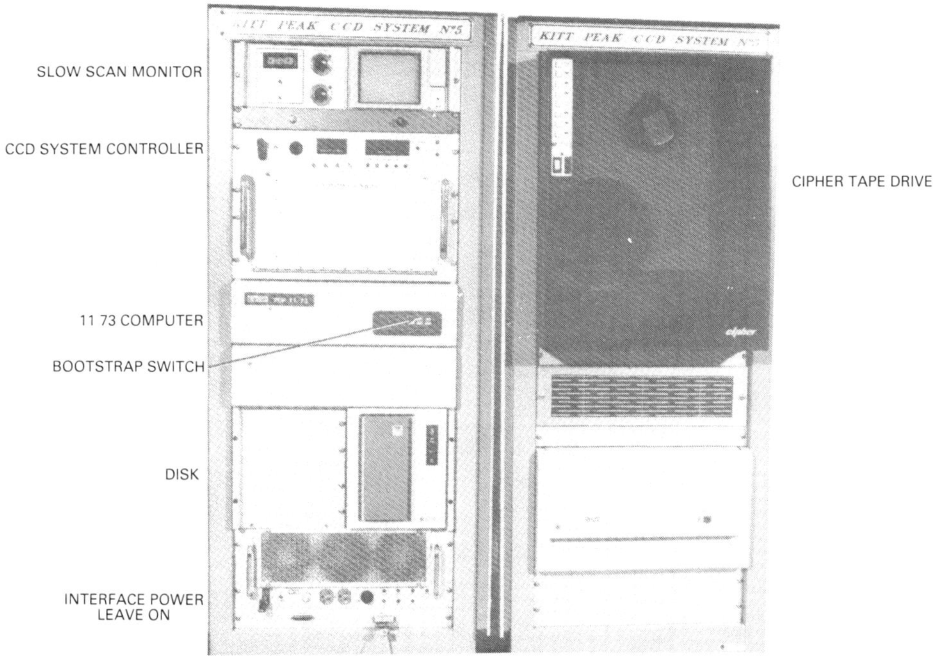

Fig. 6.6 — (b) A photograph of how the Kitt Peak system appears physically.

identical controllers are used to operate the four different CCDs in use there. A CCD identification code 'hardwired' into each croystat allows the controller to select the correct set of clock waveforms from menus stored in EPROMs erasable by exposure to intense UV light. All of the controller software is stored in EPROM† in fact, since this allows testing of the CCD with an oscilloscope, say, without the aid of the data computer. The CCD controller receives commands on a high-speed (4 MHz) serial link from the data acquisition computer. Commands include CCD identification request, erase, transfer of one or more rows to the serial CCD register followed by readout of one row via the A/D to the computer, and start-up of a timed integration, with the shutter either open or closed. Each command is echoed, but when a row of data is requested, a complete row of successive 16-bit numbers is transmitted at a fixed rate and stored in computer memory by direct memory access (DMA) without additional commands from the computer. Maintaining a uniform readout rate eliminates voltage drift or 'jitter' in the amplifier which can add to the readout noise. Special readout modes can be requested, with on-chip summing or 'binning' of data from two or four rows and/or columns (section 6.3.2). In the Lick system, the controller can also operate a CCD in the standard TV mode, in which half of the CCD image is rapidly moved to a covered frame store 'on the CCD' itself so that a second exposure can take place while the data from the first exposure are read out to the computer.

†Software stored in PROM is usually termed 'firmware'.

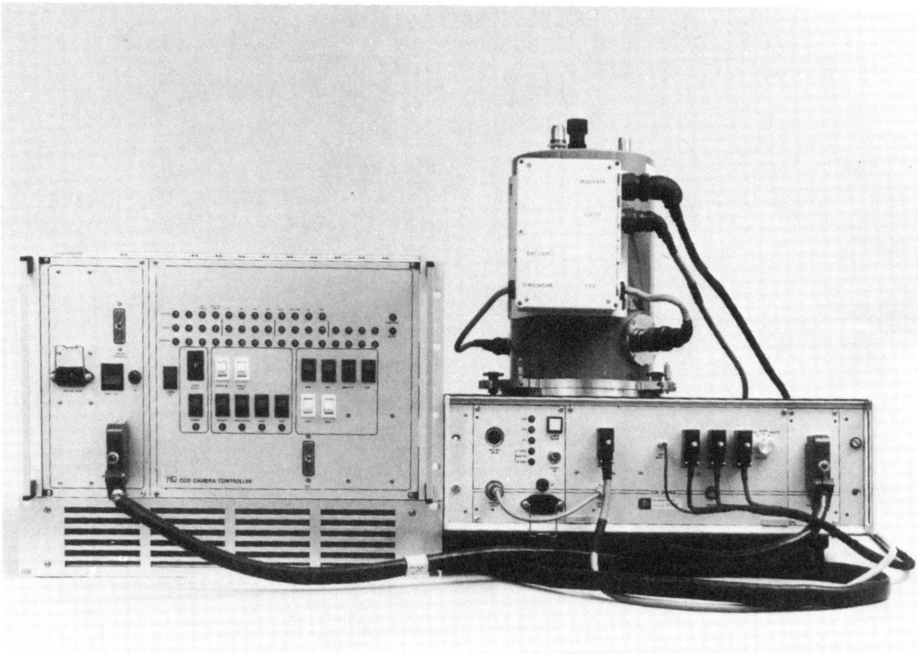

Fig. 6.7 — The RGO CCD cryostat and electronic system controller.

Several of the major manufacturers of optical and infrared solid-state imagers have invested in a very expensive but totally programmable system by a company called Pulse Instruments which enables every clock and bias level to be controlled.

This has been a very rough cross-section of CCD controllers. For more detailed technical descriptions the reader is recommended to write for engineering manuals available from the national astronomical facilities.

6.3 SIGNAL PROCESSING AND DIGITISATION

CCD outputs were discussed in section 4.2.4 and the concept of readout noise was introduced in section 5.5. Here we consider CCD signal processing in more detail, and in particular explain the crucially important technique known as **correlated double sampling** or **CDS.**

6.3.1 Correlated double sampling

As each charge packet arrives at the output node it produces a voltage change which must first be amplified and then digitised by an A/D. This cannot be done instantaneously, it requires a finite amount of time — hence the term **slow-scan** — if a high degree of accuracy is required such as can be achieved with a 16-bit A/D; a 16-bit A/D divides a specified voltage range, e.g. 10 V, into 65536 (2^{16}) parts and therefore each voltage interval is 152.5 μV in size (i.e. 0.000 152 5 V)! The A/D circuit matches up the actual voltage to the nearest number on the scale of 1–65536.

To measure the voltage of each charge packet we need a 'reference' voltage. We could use ground but, since it is important to reset the output capacitance back to some nominal value on each readout cycle — otherwise we would be differencing one charge packet from the previous one while drifting away (in voltage) from the ideal operating point of the MOSFET — there is another way. The output capacitor can be recharged to a fixed voltage by briefly pulsing the gate of another transistor, called the **reset transistor** (see Fig. 6.8(a)), to briefly turn that transistor on (like 'closing' a switch) so that current can flow from a power supply to charge up the output node to the supply level. When the reset pulse disappears the reset transistor is turned off (like 'opening' a switch) and the output becomes isolated to await the next charge-dump from the horizontal register. As a capacitor is charged to a certain voltage level (call it V_{RESET}) it rises steeply at first and then levels off to approach its final value shown in Fig. 6.8(b). But, again due to random thermal agitation of electrons in the material, there is 'noise' or uncertainty on the mean value and so the final voltage can lie anywhere within a small range, given mathematically by

$$\text{Reset noise} = \sqrt{(kT/C)} \text{ in volts or}$$
$$\frac{\sqrt{(kTC)}}{e} \text{ in electrons}$$

In this expression, k is Boltzmann's constant (1.38×10^{-23} Joules/K), e is the charge on the electron, T is the absolute temperature of the output node in Kelvin and C is the node capacitance. If C is expressed in picofarads (i.e. 10^{-12} Coulombs/V) then this noise uncertainty (called the **reset noise** or, from the formula, **kTC** (kay-tee-cee) **noise**) is simply $400\sqrt{(C)}$ electrons at room temperature and about $250\sqrt{(C)}$ at 120 K (−153°C); for the EEV/GEC device this would yield 80 electrons of noise, whereas for the RCA CCD it would correspond to 195 electrons. In both cases the reset noise greatly exceeds the readout noise of the MOSFET alone and so some means must be found to remove it.

Fortunately, removal of reset noise is quite straightforward due to the fact that whatever the final reset voltage actually is, and it must be in the range $V_{RESET}-\sqrt{kT/C}$ to $V_{RESET}+\sqrt{kT/C}$, it will get 'frozen' at that value because the leakage of current through the switched-off reset transistor is exceedingly slow (its **RC time constant** is seconds compared to the microseconds between arrivals of discrete CCD charge packets) and hence, if this uncertain reset level is sampled by the A/D just prior to a charge packet being dumped at the output, and then again after the charge has been added, it will have *exactly* the same value in each sample. Forming the difference of these two signals therefore automatically eliminates this voltage level — without ever knowing exactly what it was! This technique is known as **correlated double sampling** (or CDS) because two samples of the CCD output voltage are taken per pixel and the offset due to the reset or kTC noise in each sample is the same, i.e. it is correlated.

In practice it is advantageous to integrate or sum the signals rather than merely 'spot sample' them. Integration smooths out high-frequency noise and limits the effects of $1/f$ noise for frequencies lower than the overall signal-processing time. A schematic version of how a full, integrating CDS circuit works is depicted in Fig. 6.9 and a practical layout is given in Fig. 6.10.

Signals from the CCD output transistor are amplified by a low-noise solid-state

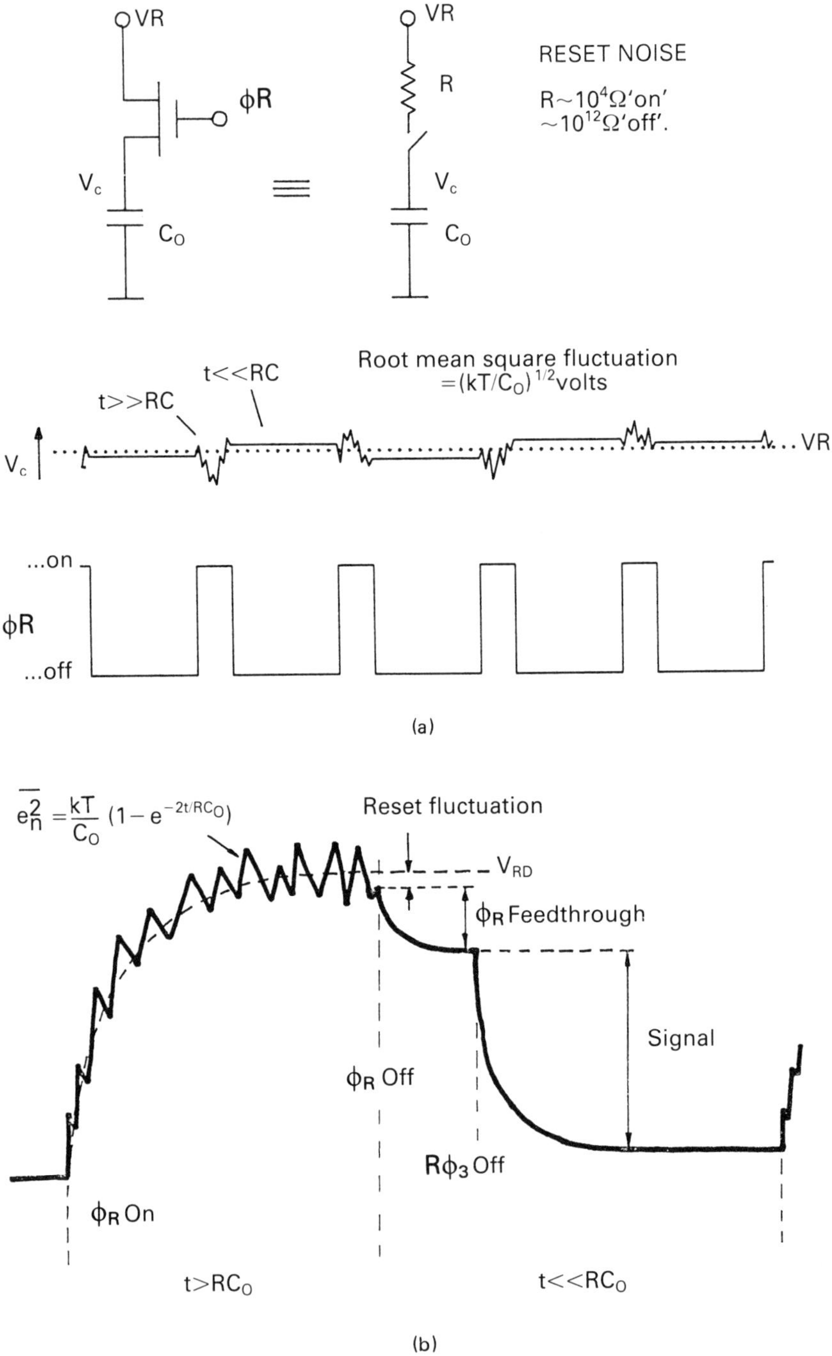

Fig. 6.8 — (a) An equivalent 'switch' circuit to explain the operation of the reset transistor. Note that the resistance, and therefore the RC time constant, is very different between the off and on states. (b) The charging profile of the output of a CCD when reset. Notice that the curve is 'noisy' but when the reset pulse disappears the last value of the signal becomes frozen.

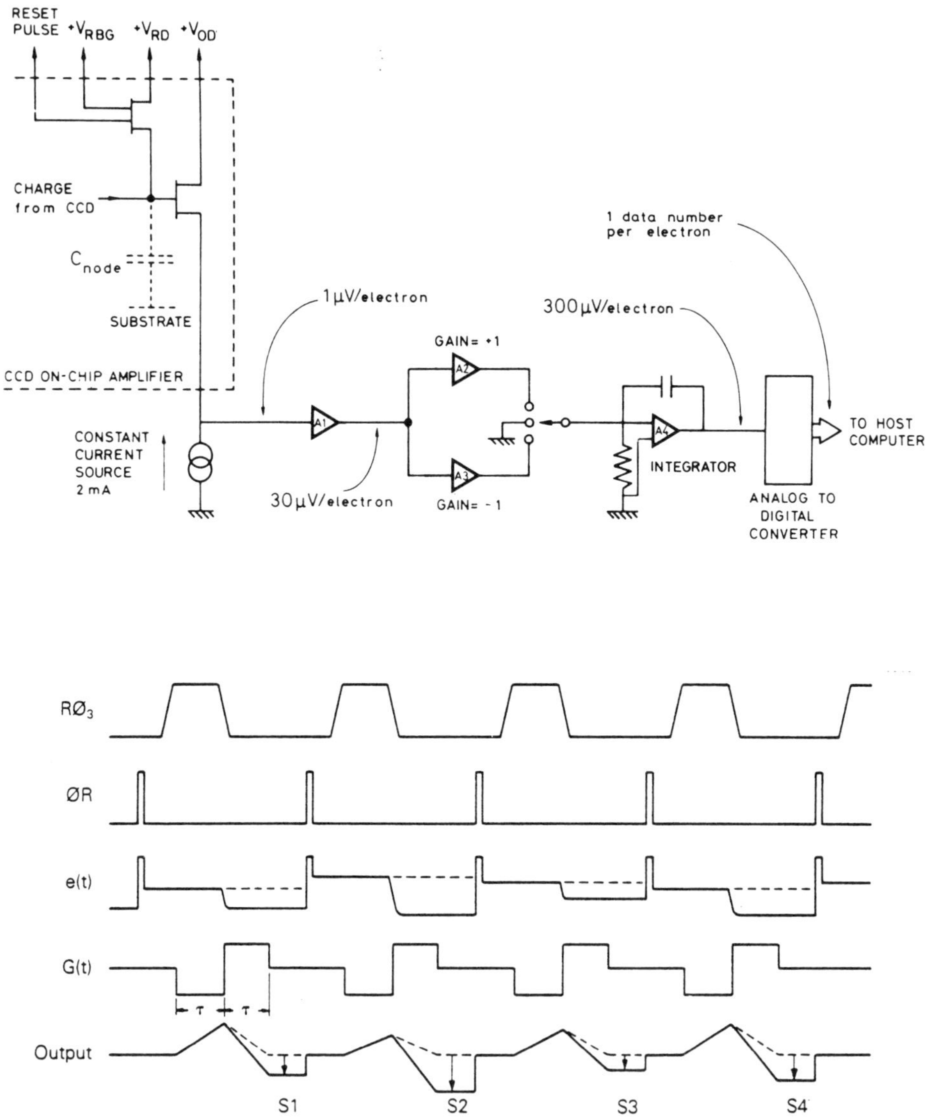

Fig. 6.9 — A block diagram of the principle of the correlated double sampling (CDS) method of removing reset noise before the signal is digitised and sent to a computer.

operational amplifier (A1 in the diagram) and fed to the integrating amplifier A4 (with the capacitor in its 'feedback loop' from the output to the input) by way of a three-position switch. One of the three positions is ground, and the other two contain a *non-inverting* and an *inverting* amplifier respectively each with a gain of ×1; in other words in one case the signal voltage is multiplied by 1 and in the other case by −1. Amplifier A1 is connected to A4 via A3 (times −1 gain) for a precise period of

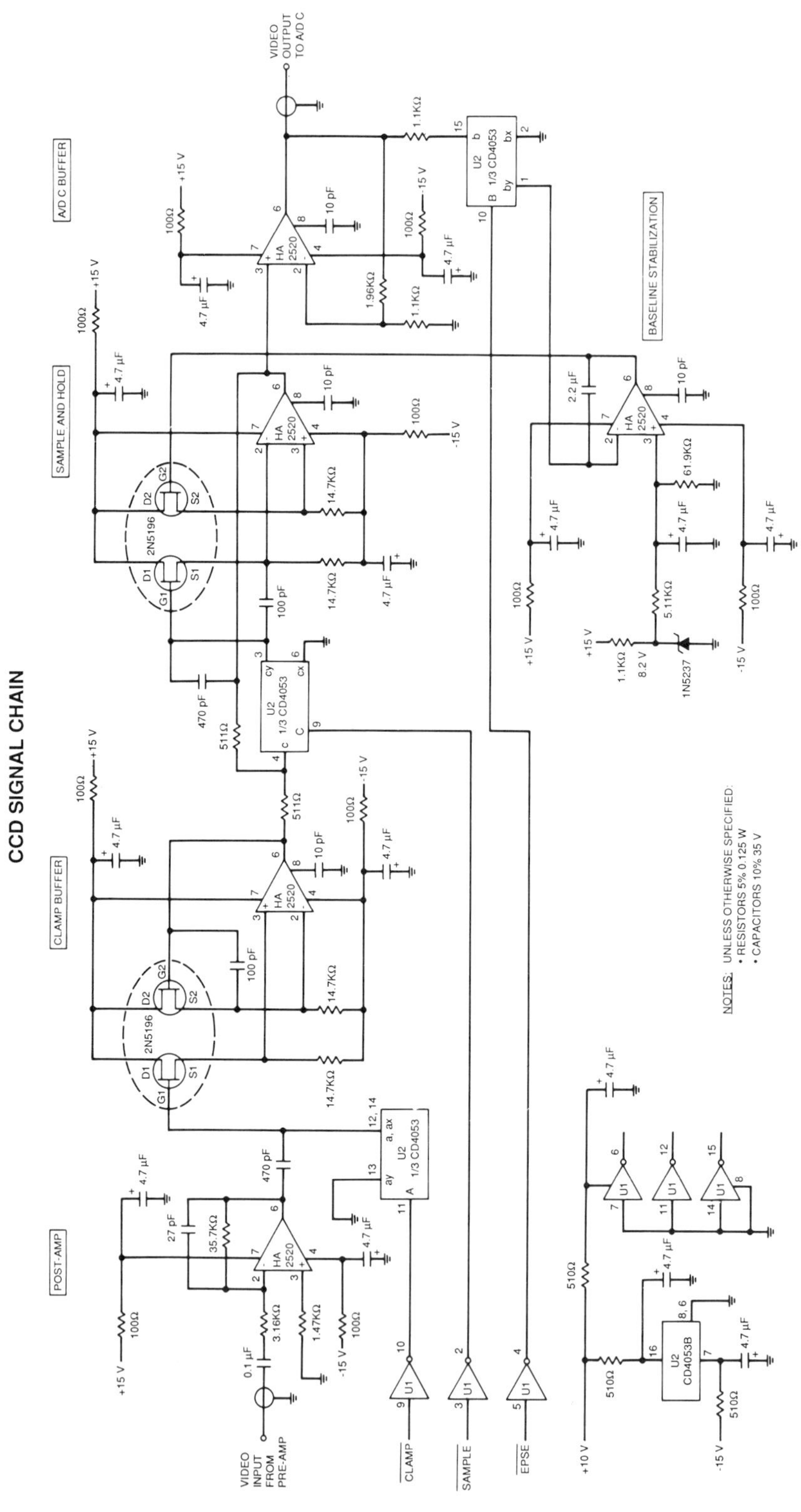

Fig. 6.10 — A practical CDS circuit used by Jim Janesick at JPL and the associated 'timing' diagram.

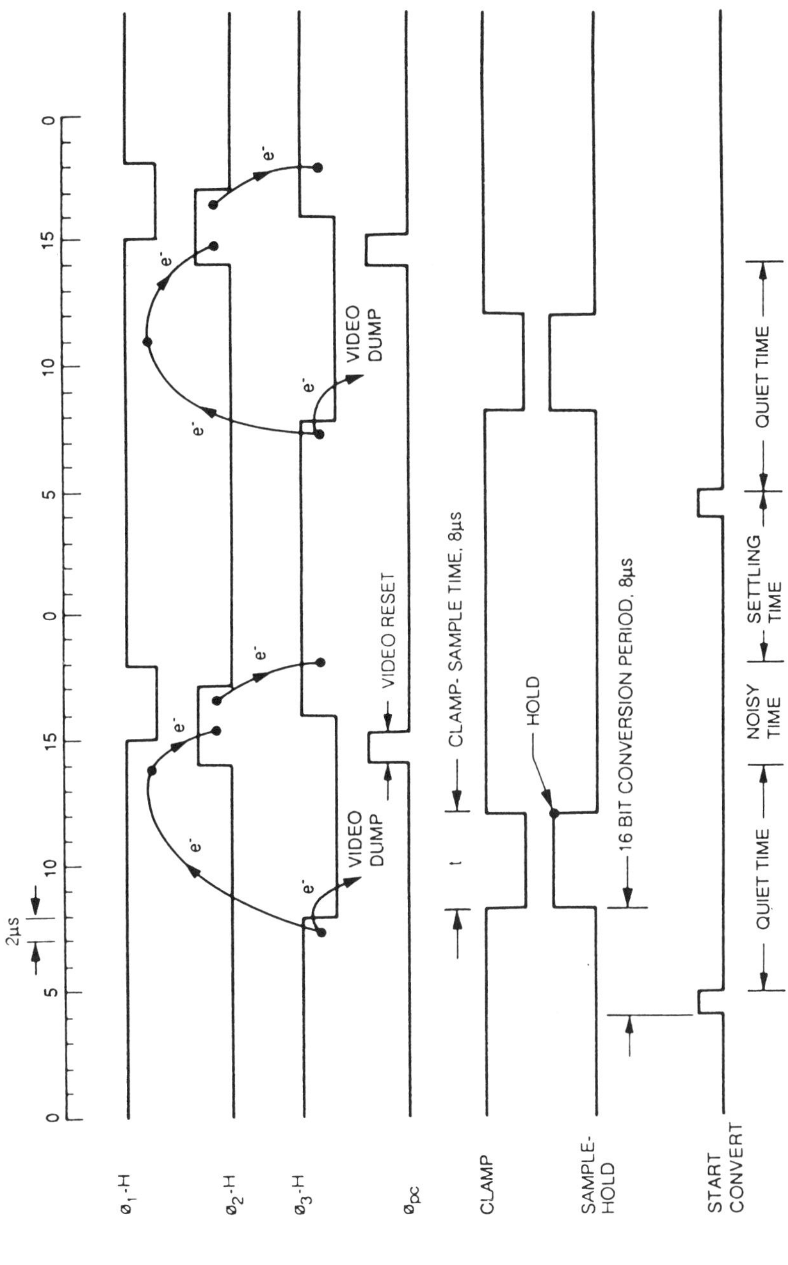

Fig. 6.10 — A practical CDS circuit used by Jim Janesick at JPL and the associated 'timing' diagram.

time, for example 20 μs, during which the reset voltage level is being integrated (negatively) onto the capacitor on A4. Next, the switch is connected to ground for a few microseconds which isolates the integrator and allows the signal charge packet to be transferred from the serial horizontal register onto the CCD output node. Then the switch is thrown to the non-inverting amplifier A2 ($\times$1 gain) and the 'reset plus voltage' level is integrated (positively) onto the same integrating capacitor on A4 for exactly the same period of time. Since A2 and A3 give identical but opposite signals in identical times, the resulting output of A4 is the real *difference* signal, i.e. the signal due to the charge packet itself. Reset noise is eliminated! The output of the integrator is fed to a sample and hold (S/H) circuit which 'freezes' the value and then to a fast A/D (often in the same commercial package) to convert the voltage to a number. The parallel output (typically 16 wires) of the A/D goes directly to the data-input part of a computer.

Choosing the correct (pre-)amplifier gains can be important. To work at very low signal levels, the total system **gain** should be such that the root-mean-square (rms) readout noise corresponds to a few 'bits' (also called data numbers or DN, and ADUs for A-to-D units) of the A/D converter. The maximum or **saturation** signal from a CCD is less than 500 000 electrons, but some devices have a readout noise of only five electrons; the ratio of these two numbers is called the **dynamic range**. In this case the dynamic range is 100 000:1 whereas a 16-bit A/D can only give a range of 65536:1. If we choose a gain to give one electron per DN then the readout noise will be 5 DN and the A/D will saturate at 65536 DN which is only 13% of full well. If we use five electrons per DN then the readout noise will be 1 DN (and not well determined due to digitisation noise) but we can utilise 65% of the well. A more realistic noise figure for many systems is 10 electrons, which gives a maximum dynamic range of 50000:1 which is within the 16-bit A/D range of 65536:1. Since most CCD applications in astronomy are for low light levels then it is better to use a high gain and work only in the lower part of the wells, e.g. five electrons/DN in the last example yields a readout noise of 2 DN and the A/D saturates at 65% of full well. A 10-V 16-bit A/D has a resolution of 152 μV per DN or 30.5 μV per electron if five electrons/DN are required. But a typical CCD output yields only 1 μV per electron and so an amplifier gain of 30 is needed to match up to the A/D.

As CCD readout noise levels continue to improve it will become necessary to develop more than 16-bit A/Ds or to install switchable (or variable) amplifier gain settings. For example, Photometrics offer variable gain settings by changing the dual-slope integration time. Any method of changing the gain introduces another calibration step for astronomers.

The CDS integration, S/H delay and A/D time will determine the pixel readout rate; typically this is 10–50 kHz. In principle, the longer the processing time spent on each pixel the better the final noise performance. The CCD output noise from the CDS is inversely proportional to S_v, the sensitivity of the CCD output node (μV/e$^-$) and proportional to the CDS-filtered noise voltage spectrum $V_n(f)$ of the on-chip amplifier; $V_n(f)$ has two components, a white noise and a $1/f$ term. Readout noise will decrease approximately as the square root of the CDS integration period until correlation is lost by $1/f$ noise. A plot of readout noise versus CDS integration time for three CCD types studied at JPL is shown in Fig. 6.11; the Thomson CCDs lie

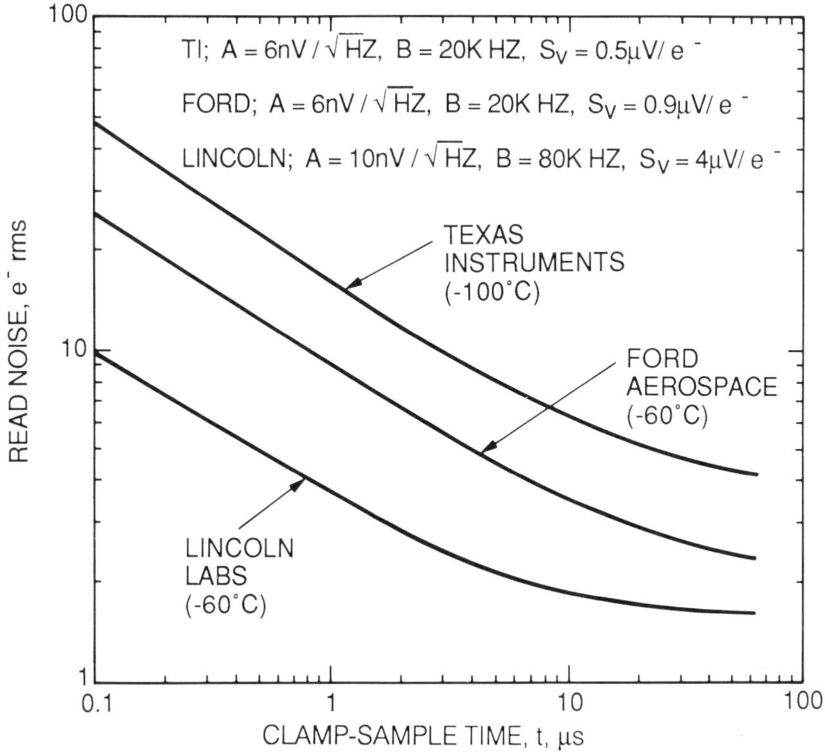

Fig. 6.11 — The theoretical readout noise performance of three CCD types as a function of CDS integration or clamp-to-sample time. Although these plots were derived from an equation they were found to be in good agreement with experimental data obtained at JPL.

between the Ford and Lincoln devices. With a CDS integration time of 8 μs the Lincoln Laboratory CCD gave two electrons noise.

For a typical pixel time of 20 μs a 512×512 CCD requires a total readout time of 5.2 seconds. For the GEC CCD, which has extremely low $1/f$ noise, a longer pixel time of 80 μs, is possible and with 385×578=222 530 pixels, it takes almost 18 seconds to read out the CCD!

Low-noise performance depends on the use of very low-noise transistors such as the 2N5564 JFET in the JPL-style preamplifier and on good operational amplifiers in the signal processing chain, for example, Harris 2520 or National Semiconductor type LF357. The pixel clock timing should be synchronised with the cyrstal-controlled microprocessor clock and with a non-varying (in time) instruction sequence in the micro processor to avoid noise due to time 'jitter' or changes in the amount of signal induced by accidental coupling from the microprocessor circuits. It is essential to maintain a precisely uniform readout rate since, due to the settling times of various components, temporal jitter can produce small variations in the steady or 'baseline' voltage of the amplifiers which the CDS cannot cancel and so the noise will increase.

6.3.2 On-chip binning

Although CCD architectures vary from one manufacturer to another, all frame transfer devices incorporate a structure which permits **on-chip binning.** On-chip binning of charges is the process of adding or combining charge packets from rectangular groups of pixels at a 'summing point' — the output node capacitor of the CCD. These groups of pixels effectively become single larger pixels, sometimes called 'superpixels'. On-chip binning is achieved as follows.

Any number of rows can be co-added together in the serial register simply by omitting the horizontal clocking sequence which usually occurs after each vertical transfer. Likewise, when the horizontal register is clocked, the charges from any number of pixels can be combined on the output capacitor by simply omitting the reset pulse which usually occurs between each horizontal transfer. By not running the horizontal (serial) clocks after a vertical transfer, the next vertical transfer will co-add the next row of charges into the horizontal register. Similarly, by not resetting the output transistor after a pixel has been read out, the charge associated with the following pixel will be co-added with the first. In this way rows and columns can be 'binned' together (Fig. 6.12). The final picture has smaller dimensions and less

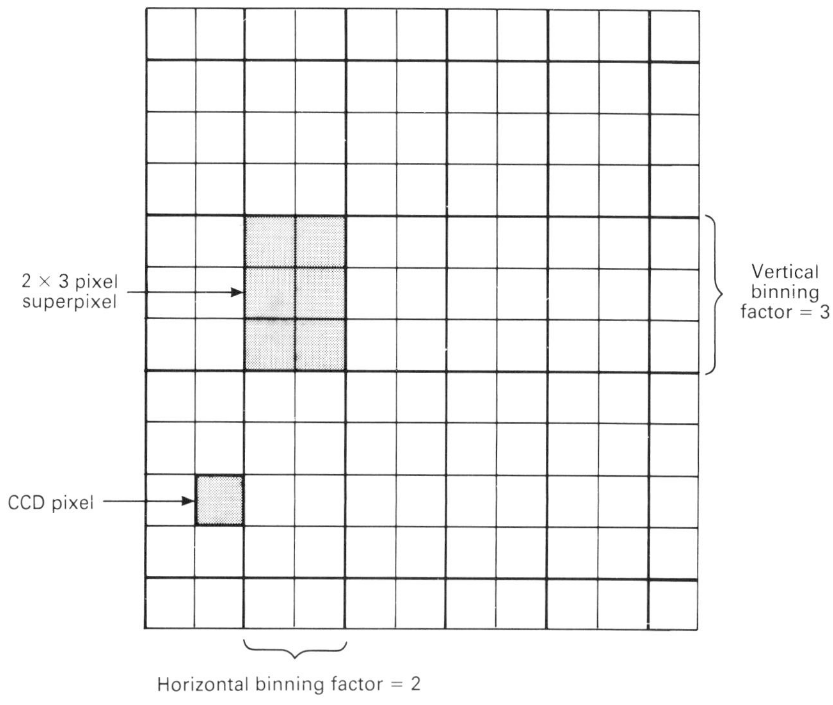

Fig 6.12 — The principle of on-chip charge binning.

resolution, but it may have a higher signal-to-noise ratio because more signal was combined 'on-chip' before the noise of the readout process was added. Of course, care must be taken to avoid saturation. An 800×800 CCD using 4×4 pixel on-chip binning (i.e. omit four horizontal transfers and four reset pulses) will produce an image with only 200×200 pixels. This feature is most useful in high resolution echelle or coudé (stationary) spectrometers where binning is usually slight (×2 say) or not at all along the spectrum so as not to lose spectral resolution, but substantial (×15 or more) at right-angles to the spectrum since it allows more signal per readout and so a much better signal-to-noise ratio is obtained in a faint spectrum. This technique is also used for optimising the resolution in variable seeing conditions or to obtain a preliminary look at a field containing very faint sources.

A clock sequence which has these properties can be stored in PROM and 'called up' when required.

6.3.3 Overscan

Inevitably, when a CCD is read out rapidly with essentially a zero exposure time, a picture appears on the screen which, although random, has an average value which is not zero. This value — the **bias** value — must be subtracted from every image before any other arithmetic manipulation of the data can be performed.

A bias frame can be obtained by closing the shutter of the CCD camera and setting the exposure time to zero. Alternatively, the bias level can be obtained by pretending that the CCD array is larger than it really is — say 10 pixels in each direction — so that the electronics continue to 'clock' even though there is no longer any physical correspondence with points on the CCD; 10 more vertical transfers are requested and 10 more horizontal transfers are included each time the serial register is activated. Again, a PROM-based controller can deal with this request by simply calling-up a different 'repeat count' for each type of clocking sequence. When the resulting image is displayed it will now have a 'border' on two sides which contains no true signal, only the voltage corresponding to the bias level (Fig. 6.13). Many pixel values in this **overscan** area can be averaged to get a good estimate of the bias level. Thus, no extra frames are required. In Chapter 8 we will return again to a discussion of bias frames when considering the detailed calibration of CCDs. Finally, just as you can send more pixels than are required, you can also opt to read out a subsection of the CCD, usually by controlling the A/D unit with a (pulse) command to inhibit digitisation except within the sub-area specified.

REFERENCES

Gunn, J. E., Emory, E. B., Harris, F. H., and Oke, J. B. (1987) The Palomar Observatory CCD camera, *Publ. Astron. Soc. Pacific,* Vol. 99, 518–534.

Janesick, J. R., Elliott, T., Bredthauer, R., Chandler, C., Burke, B. (1988) Fano-noise-limited CCDs, SPIE, Optical and optoelectronic applied science and engineering symposium; *X-ray Instrumentation in Astronomy,* San Diego, August.

Optical Engineering, guest ed. J. R. Janesick, three volumes, Aug.–Oct. 1987.

Raymer, J. T. (1988) Aspects of the development of a 1-5μm infrared camera for astronomy, *Ph.D. Thesis,* University of Edinburgh.

White, M. H., Lampe, D. R., Blaha, F. C., and Mack, I. A. (1974) Characterisation of surface channel CCD image arrays at low light levels, *IEEE J. Solid State Circuits,* Vol. SC-9, 1–12.

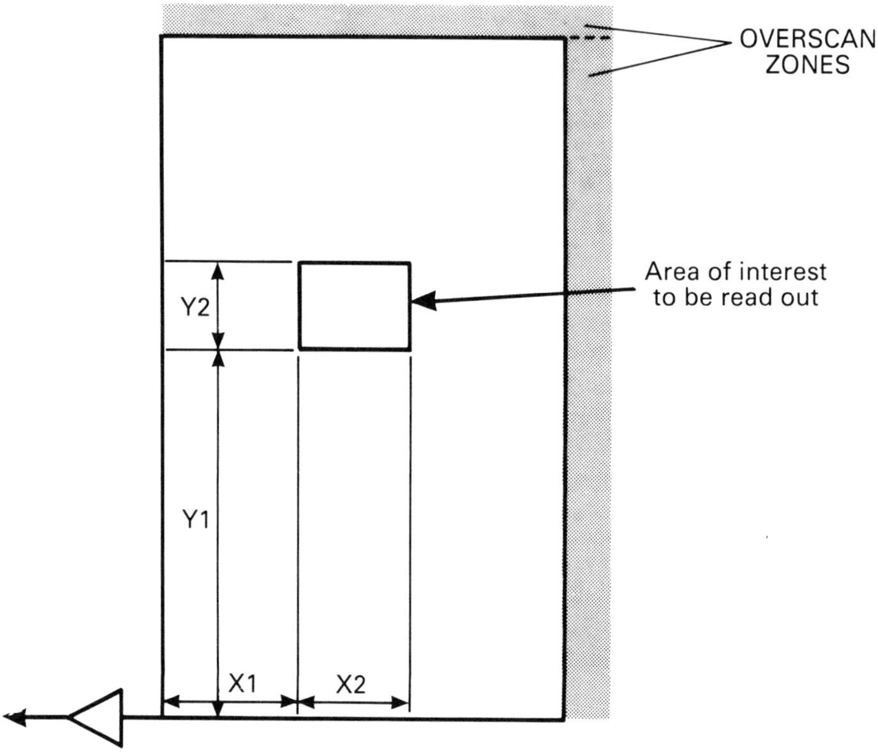

Fig. 6.13 — The overscan region of an image is shown as well as an arbitrary sub-image.

7

Image displays and computers

As we have seen, an integral part of all modern astronomical instrumentation is a computer and an image display. The computer provides the immense degree of flexibility and reliability demanded by astronomical researchers, and is an essential tool in handling the digital information from CCD detectors and similar devices. Computers are also needed for later analysis of the data once the astronomer has left the mountain observatory. There is little doubt that CCDs have done much to force the subjects of ground-based optical and infrared astronomy towards a greater reliance on computers and more awareness of progress in computer technology.

It would be impractical to describe in detail each and every image processing system used in astronomy, and it is certainly pointless — in a rapidly changing market — to make recommendations regarding a 'shopping-list' of good purchases. The aim of this chapter is to give the reader a strong impression of how widely computers and image displays are used and needed in observational astronomy, and that this resource is of vital importance if we are to benefit from other technological advances. I will try to explain some of the jargon, briefly describe the most vexing problems facing astronomical computing, and introduce some general guidelines and principles.

7.1 DATA RATES, FORMATS AND PROTOCOLS

The 'data rate' is the number of digitised pixel values transferred per second to the host computer, or possibly an intermediate buffer or control computer, from the astronomical instrument. For a CCD, or any solid-state array detector, this rate is largely decided by the electronic configuration, although the actual application and the device physics may also be factors. Examples of each of the latter two considerations would be high-speed CCD photometry and detector saturation. For instance, detector saturation occurs easily in the infrared due to the higher background light levels.

Since the integrating correlated double sampling (CDS) method described in section 6.3.1 is almost always used with CCDs then the interval of time required for digitisation of a pixel value is typically about 20 to 100 μs — corresponding to readout rates of 50 kHz and 10 kHz respectively. For a 20-μs pixel time and digitisation to 16 bits (65 536 levels), the maximum output data rate is 800 000 bits/s. Note that this level of digitisation is mandatory if one is to capitalise on the large dynamic range of CCDs. Eight bits make a **byte**, and two bytes (16 bits) commonly make a **word**. So this data rate is 100 000 bytes/s or 100kbytes/s. Data rates are often specified as a **baud rate**, which is the number of times per second that there is a transition on the signal line from a 0 to a 1 or from a 1 to a 0. For normal communication between a computer and a terminal/keyboard the baud rate is usually not larger than 19 200 bits/s, but data links can operate at 8 million transitions per second or 8 Megabaud, so that a 16-bit data transfer can be accomplished in 2 μs! Digitisation itself may require 5–10 μs depending on the A/D; 2-μs A/D units for 16 bits are now on the market, but there may be an additional 2 or 3 μs needed to allow the electronics to sample and hold the signal for digitisation.

The transfer of data from the A/D unit to the computer can be accomplished in two ways. Either by a high-speed **serial** link which is a single line along which signals are sent in a sequential pattern, or by a multiway cable called a **parallel** link which sends all 16 bits at once. Most modern computers handle the flow of data by using **direct memory access** (DMA). In several systems the digital data are stored first in a large buffer memory called a **frame store** or **external memory**, and in a few systems the data go directly onto magnetic tape. In general, however, digital data from a CCD are initially saved on a large magnetic disk and later 'archived' to magnetic tape.

For compatibility between two devices, several parameters must match besides the rate, in particular, the voltage levels, the timing, the format, the 'code' by which numbers and characters are digitised, and the rules or 'protocols' by which data transfers are acknowledged — called 'hand-shaking' — and by which computer operations are tracked — called 'housekeeping'.

Is there a universal standard for digital electronic data acquisition systems? The answer is no, but there are several systems which are quite common.

Among the older systems still in astronomical service, is a modular system known as CAMAC. Functional circuit-board units such as counters, voltage-to-frequency converters, A/D boards etc., are constructed as 'plug-in' modules which go into a central power supply unit called a **crate** which contains a set of connections called a **data bus** which link each unit to a central crate 'controller' module. With each of the units under digital control through an external host computer connected to the crate controller, the CAMAC system is capable of performing a wide range of control and data manipulation functions once it has been 'programmed'. To the uninitiated, this is a daunting task because a very basic instruction set known as 'machine-language' for handling individual bits must be used. In practice, the program is written mainly in the well-established languages of FORTRAN 77 or C with appropriate 'calls' to machine-language subroutines.

A more recent type of data-acquisition system uses the IEEE-488 standard which is based on the Hewlett–Packard general-purpose interface bus (GPIB). Input–output data paths are 8 bits wide (parallel) and so transfer of a 16-bit word consists of two serial bytes. Each unit on the bus is independent and connections to the bus are by means of a standard 24-pin connector which allows other connectors to be added on to it to create a 'stack' of several devices on the same bus. Instruments connected to the bus are designated **talkers, listeners,** and **controllers**. Again, a computer program is required to control the modules on the bus, but the support for this system is very widespread and many commercially available pieces of apparatus come with an IEEE-488 interface.

A more recent possibility, at least for data transfer, is the use of a high-speed, single-cable serial link called **Ethernet** which can also be used as a general means of interlinking several computers in what is termed a **local area network** or LAN. The idea of linking computers in LANs is very topical and it is one way of arranging computer resources. In the Ethernet approach (due originally to Xerox, DEC and Intel) messages and data are broadcast along a coaxial cable from the sender station to the receiver station. To send a message, the sending station senses the cable to see if it is free. If it is free the station transmits but it also continues to sense in case some other station on the net transmitted at the same time. If this were to happen a 'collision' would have occurred and the transmission would be unsuccessful, so both stations stop sending, wait a random time interval and then try again. This is called **carrier sense multiple access** with **collision detection** or CSMA-CD. The hardware components of the Ethernet consist of the coaxial cable, a tap which connects the transceiver to the cable, and a twisted-pair interconnecting the transceiver and the controller. An individual Ethernet consists of a 500-m cable segment with up to 100 stations; the stations can be a wide range of devices from complete image-processing workstations to terminals. Ethernet, which derives its name from the legendary all-pervasive medium — the ether — once thought to be the carrier of electromagnetic radiation, is gravitating towards an Electronic Industries Association standard.

For electronic data acquisition, an alternative 'bus' system capable of very high speeds and extremely compact hardware receiving widespread commercial support is the **VME bus**. The VME bus is designed for input–output paths which are 32 bits wide and compatible with the current generation of microchip central processing units (CPUs) such as the Motorola 68000 series.

Optical fibres are also rapidly emerging as a high-speed, high-capacity, low-error communications link for data transmission, and they are already beginning to appear in some astronomical situations. Optical fibres are insensitive to electrical interference.

7.2 COMPUTERS AND IMAGE DISPLAYS — BENEFICIAL OR BEWILDERING

Of paramount concern to many groups of astronomers in university departments and observatories throughout the world is the question of 'portability' of good software for astronomical image processing and data analysis from one institute and computer system to another. Unlike in the commercial world, this kind of software is usually freely available within a nation or an international partnership — regardless of the cost of programming effort — because it is often developed under the auspices of the national science funding agency for the benefit of all. Major software packages are also exchanged between nations on a reciprocal basis. Although, given that astronomy today is very international in character, much more could be accomplished in that regard. As in other branches of science and aerospace, several countries find it much more cost-effective to collaborate to construct and operate major facilities — often in remote areas of the world where observing conditions are excellent — and so today, more than ever, compatible data formats and compatible software are important.

The decisions facing any astronomy group or department with a need to purchase computing systems for CCD-type image processing are:

(a) which manufacturer?
(b) which operating system?
(c) which applications system?

Performance of a scientific computer devoted to astronomical image processing and the analysis of digital data — as distinct from data acquisition and instrument control which can be handled extremely well by a combination of fast electronic hardware, one of the modular bus systems and an efficient microprocessor — can be discussed in terms of technical quantities such as **MIPS** and **megaFLOPS** which describe the efficiency of the machine, vector and parallel-processing capability, and the operating system. These performance criteria must be weighed against the capital cost, the cost of maintenance and support (if any), and the likely longevity of the computer vendor in a highly competitive, bewilderingly innovative, rapidly changing market.

Perhaps the most important and certainly the most controversial item in any CCD system is the **image display**. It is crucially important to ensure that the astronomer can view the CCD image on a screen as soon as it is obtained at the telescope in order to evaluate progress, as well as during data processing and analysis later on. Recall that the cosmic source may be quite invisible on a TV acquisition and guide camera, it may be at an optically invisible wavelength, and the physical measurement of the light may pertain to the polarisation state or the velocity distribution or something other than a simple image. Many times he/she will use some form of visualisation to make a tactical decision as to the next move (see colour plates for typical computer-generated image displays). It will often happen that blemishes, cosmic rays, flat-field effects or the sheer faintness of the object will necessitate some image processing at the telescope before the displayed image can serve its purpose. Visualisation of physical parameters at the telescope is crucial (see Fig. 7.1 for a typical arrangement of displays at the telescope). The concept of 'Visualisation in scientific computing' is a hot topic; a report of the same title was recently submitted to the US National Science Foundation and discussed in *Computers in Physics* (Vol. 2, No. 3, 1988). It is alleged that the lack of sufficiently good computer graphics is effectively wasting resources by impeding data interpretation.

Image display systems are very numerous. The range of options has increased dramatically in the last 10 years and costs vary greatly; it is a minefield of trouble for the non-expert. You have to think carefully about the needs of your own instrument, the needs of the observatory and what the rest of the astronomical community is doing. If you are an instrument builder, it is worthwhile to visit many places and see different hardware in action. Some of the actual systems in use are named in the discussion below.

Fig. 7.1 — Image and TV displays found at a typical large observatory.

7.2.1 Image processing systems

The art of image processing is a topic with boundaries well beyond astronomy. For most astronomers, image processing follows from the act of 'data reduction' and is largely associated with visualisation of a set of measurements. By image processing we mean the mathematical manipulation of the two-dimensional (x,y) array of *numbers* stored in the computer to represent the intensity on each pixel of a CCD camera. Some of the required manipulations are straightforward, for example, add, subtract, multiply or divide each pixel value by a constant, or by the value at the same pixel location in another digitised image, such as a dark sky, or flat-field frame. Software is also needed for numerous other manipulations, including logarithms, statistics, smoothing, noise suppression, image enhancement, contouring and profile fitting. There are commercial sources of software to do these manipulations, e.g. IDL, NAG, but there is a tendency for academia to develop its own. In almost all cases it is essential to 'display' the manipulated image on a screen.

Just 10 years ago, colour image display and image processing systems were large and expensive and mainly connected with the space programmes. Today all that has changed. There are now numerous image display systems on the market and the trend for faster and cheaper associated computing power for these displays continues at an incredible pace. Of course, not all the available image display systems are suitable for scientific use — important markets are CAD/CAM (computer-aided design and computer-aided manufacture), business machines, military and civilian radar and medicine — but the remaining options are still so numerous as to be confusing.

Many 4-bit displays are now on the market for use in CAD/CAM for engineering applications.

For astronomical applications most people are content with an **8-bit plane** image display. In such a display the true data values are re-scaled so as to lie in the range 0–255 (2^8=256, hence 8 bits) and a colour or a shade of grey is allocated to each interval for display on a special high-resolution TV monitor. These monitors have at least twice the fine detail or resolution of a normal TV tube and are usually of the RGB type which means that the intensity of each of the three (red, green and blue) electron guns can be controlled from the computer to produce any colour or hue. When the distribution of brightness in an image is represented by arbitrary colours it is known as a **false-colour** representation. The computer can also construct 'true-colour' displays by suitable scaling or 'level-balancing' of CCD images taken through colour filters and then software to control the image display so that red objects appear red and blue objects appear blue. Since the human eye cannot easily distinguish more than about 32 (2^5 or 5 bits) brightness levels then the effect of an 8-bit image display is pleasing. It is not generally necessary to use a 16-bit plane image display for the image itself.

Sometimes it is helpful to have extra planes for other things however, such as annotation, overlays (contours, vectors, moveable apertures, patterns or charts) or for rapidly 'blinking' between two images — a powerful technique for visualising small areas of difference between two data sets. Machines which provide these features are obviously more expensive. Since all that is required at the telescope (in conjunction with the instrument) is a display which can be labelled with axes, graphs, letters and numbers it is possible to do this by the method of 'overwriting' the displayed image; the required overlay is done by setting a pixel value to the last bit of the 8-bit display. Note that the real data are in no way affected, only the 0–255 level representation.

The **MIPS** (millions of instructions per second) rating is used to measure speed of integer (whole number) calculations, and the **megaFLOPS** (millions of floating-point operations per second) rating measures how efficient the computer is in dealing with real (non-integer) numbers of arbitrary size. Provided the convention is adopted of measuring the MIPS against a *standard* program relative to performance on, say, a Digital Equipment Corporation (DEC) VAX11/780 computer with a MIPS rating of 1, then the term can be useful; but MIPS is dependent on the action being implemented. Computers designed for commercial applications do not normally rely on good floating-point performance but, since this quantity is independent of the computer architecture, it is a useful measure for the scientific user. The VAX 11/780 has a double-precision MFLOPS rating of 0.14 when equipped with a special hardware unit called a **floating-point accelerator**. Many computer magazines carry articles from time to time which explain the standard performance 'benchmarks' and which compare leading computers and new products; two references are given at the end of this chapter.

Vector computers can perform floating point operations on vectors (quantities with a magnitude and a direction) as well as scalars (ordinary numbers), and **parallel** computers can perform simultaneous operations on several different scalars and vectors. Not many of these machines are in use in ground-based optical astronomy yet. Of equal importance in many applications will be the efficiency with which a computer can perform input and output actions — referred to as I/O — such as 'writing' to disk or 'reading' from magnetic tape.

Certainly the most widely used computer type in astronomy in recent years has been the DEC VAX series of computers running an operating system known as VMS. These machines, and others like them, can accommodate many terminals and many different 'users' running lots of different programs. Often, an independent image display and store is 'added' to the overall system. Unfortunately, unless the main computer is very powerful, all this 'loading' results in a relatively slow system. Some changes in this situation are becoming apparent. As Steven Grandi of the National Optical Astronomy Observatories says, '. . . the proper computer solution for three to five years ago . . . is by no means the proper solution for today' (Grandi, 1988). Most pundits would agree that four broad classes of computer systems can be distinguished as follows. **Desk-top** image processing **workstations** without a magnetic disk but linked to disks elsewhere, **desk-side** image processing workstations with powerful large disks and tape drives, small departmental systems with add-on image displays or diskless workstations supported over Ethernet, and large departmental systems involving mini-supercomputers with ample vector processing used, for example, for theoretical work.

Vendors for the first class include DEC, Sun Microsystems and Apollo and examples would be the DEC MicroVax 3000 series (3 MIPS, 0.5 MFLOPS) with an 8-bit 1024×864 display, the Sun 3/60 (3 MIPS/0.1 MFLOPS) machine with a 1152×900 8-bit display and the Apollo DN4000 (4 MIPS/0.2 MFLOPS) system with a 1024×800 8-bit display. Many astronomy departments currently operate slightly older (slower) machines in this class such as the DEC MicroVax II and the Vaxstation 2000. All of these computers, although diskless, can be configured so that they can access data stored on the large disks attached to one of the higher classes below. These systems can also be used at the telescope. Prior to the advent of these 'computer+image display' workstations, homemade or commercial 'scan-converters' were attached to the instrument computer at the telescope, or single-board graphics cards were inserted into free slots in the computer to enable the CCD image to be seen immediately upon readout. Actually, some of the latter techniques are still used.

In the market for desk-side systems might be the DEC VAXstation 3500 (3 MIPS/ 0.4 MFLOPS) and the Sun 3/260 plus floating-point accelerator (4 MIPS/0.4 MFLOPS) or the Sun 4/260 (10 MIPS/1.1 MFLOPS) which is a new machine based on a concept known as reduced instruction set computers or RISC! Systems in this range are not necessarily 'single-user' computers but they can be obtained for much less than $100 000.

Examples of computers in the third class — once the arena of the VAX 11/780 — could be the DEC VAX 8250 or the Sun 4/280. In the past, image display add-ons have been common with such machines, e.g. the Sigmex ARGS (in the UK), the International Imaging Systems I^2S and the Grinnell (in the USA) and the Gould-DeAnza (USA and Europe) image display systems; some are even available at the telescope.

In the fourth class we find some very powerful machines such as the Convex C1/

XP (3 MFLOPS) and the vector/parallel Alliant FX/8 (8 MFLOPs). The Alliant and Convex systems have been adopted by some major radio astronomy centres as replacements for VAXes, and Sun workstations like the Sun 3/60 are being widely adopted in North American establishments, although MicroVAXes are still very popular.

Finally, it is worth considering the relatively recent development of 'clustering' of less expensive computers to lighten the load on any particular one. Many astronomical establishments already operate Vaxclusters.

7.2.2 Which operating system?

All computers employ special software supplied under licence by the manufacturer to allow the machine to operate and execute programs — this is called the 'operating system'. Ideally, one would like to use an operating system which is independent of the actual computer hardware. In practice this is not possible. The only operating system which is claimed to come close to this ideal in portability is called UNIX, a development and registered trademark of Bell Labs AT&T. Since VAX computers have been the norm in the scientific arena for many years, most astronomers have become familiar with their DEC operating system called VMS. More recently, VAX users have begun to utilise UNIX, in one of its forms, and this is also the preferred system for the Sun workstations. As with many things in life, these competing systems have their ardent supporters and their vociferous antagonists! Undoubtedly this is one of the most widely debated, hotly contested issues in the computing world which actually has a direct bearing on astronomy. It seems probable that these systems will exist side by side for some time to come.

7.2.3 Applications software — a unified approach

During the past decade astronomers and national institutions around the world have become aware that computing facilities and good software are of the utmost importance to the successful use of modern detectors such as CCDs. The lesson has been learned slowly, and there is still a reluctance on behalf of funding organisations to support astronomy in this area. It can be a serious problem for scientific development if observations are made but their publication, in final form, is delayed because subsequent, so-called 'off-line', computing facilities are inadequate. Some of this reluctance can certainly be traced to a genuine concern about capital expenditure on items which have a very short life-time before they are superseded!

Some excellent attempts to reach common standards in astronomical computing have been made.

Credit must go to Don Wells, who from 1974 to 1978 at the Kitt Peak National Observatory (NOAO) was responsible for the development of a system known as the Interactive Picture Processing System or IPPS which was to begin the move toward workstations and to common-user software.

In the late seventies an ambitious project called STARLINK was initiated in the UK when in mid-1978 a committee, chaired by Prof. Mike Disney (who invented the name STARLINK), began to look into urgent complaints by the optical/UV astronomers of a dreadful lack of image processing machinery, and then struck it lucky when, within 6 months, an underspend in the Science Research Council budget allowed the project to get off the ground and go into operation in early 1980. From the outset STARLINK was to be a nation-wide linked network of VAX computers from DEC covering all of the major astronomy centres in the UK as well as at overseas observing sites in Australia, Hawaii and La Palma. Similar peripherals (such as the Sigmex ARGS display or, more recently, the Digisolv Ikon display) are established at each site and identical software provided to each site. Moreover, applications programmers at all sites contribute to the common store of software. A headquarters for the project was established at the Rutherford-Appleton Labs (RAL) near Oxford, from where each new software package — all written to the same standard specification — is scrutinised before general release. So everyone contributes and everyone benefits. STARLINK supports 900 UK users, covers every wavelength regime used in astronomy, supports numerous packages for astronomical data anlysis of all kinds and has been widely exported (Fig. 7.2). A decade later STARLINK is about to undergo a metamorphosis with major hardware upgrades and the widespread adoption of a new software environment called ADAM. Most of the original 11/780 VAXs will be replaced with MicroVAX 3500-based local area VAX clusters.

STARLINK applications software is accessed through an 'environment' which overlays the DEC VAX VMS operating system. This environment is known as ADAM — for Astronomical Data Acquisition Monitor — a system originally developed for data-taking and the control of instruments at the telescope, but now extended in a grand unification to data reduction; in other words, the observer at the telescope will 'see' the same computer environment as he does at his home institution. The ADAM environment offers menu-style interaction as well as the more familiar single-line command input, and a programmable command language called ADAMCL. The menu style is considered very useful by beginners to the system. Conceptually, this approach and this degree of generality offer many advantages if the community using these facilities is large. For example, the reduction process looks the same to all, the system used at the telescope can also be used 'back home', more complex algorithms can be acquired at no extra cost and standards of disk and tape storage formats become established such as the FITS or Flexible Image Transport System. One disadvantage of the STARLINK approach is that generalised software packages always work, or in the jargon 'run', more slowly than more specific programs, which means that more powerful (costly) machines are required. But in this era of decreasing cost for increasing computing power, this does not seem like a major problem.

Of more concern, perhaps, is the time needed to develop and document a universally approved astronomical data analysis system. But the alternative of do-it-yourself schemes leads to an immense amount of duplication and no lasting product, which may be just as costly!

A similar general-purpose scheme for astronomical data reduction has been set up in Europe under the auspices of the European Southern Observatory (ESO) which operates several telescopes on La Silla in Chile on behalf of a large consortium of European nations; the ESO headquarters are in Garching near Munich in the Federal Republic of Germany. The ESO scheme is called MIDAS — Munich Image Data Analysis System — and as the name implies it is principally intended for imaging data (which includes CCD spectroscopy). Apart from the VAX computer, ESO hardware incorporates Gould-DeAnza image displays (Fig. 7.3). Again, as

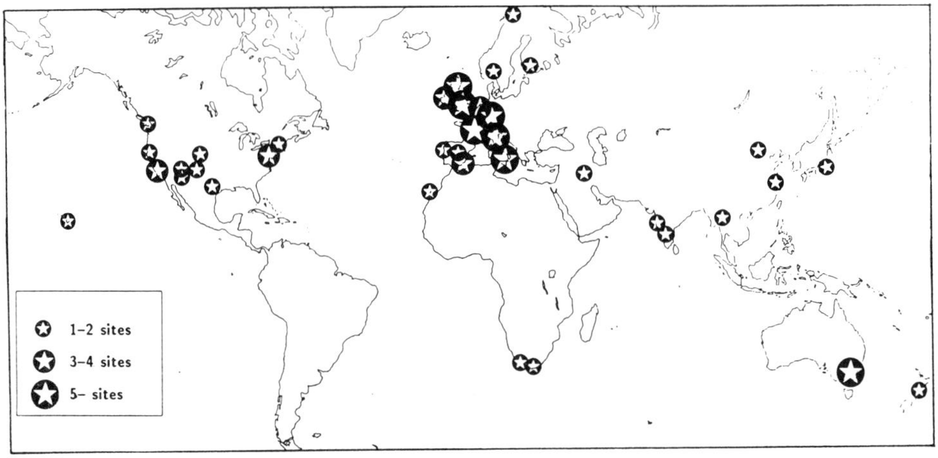

(a)

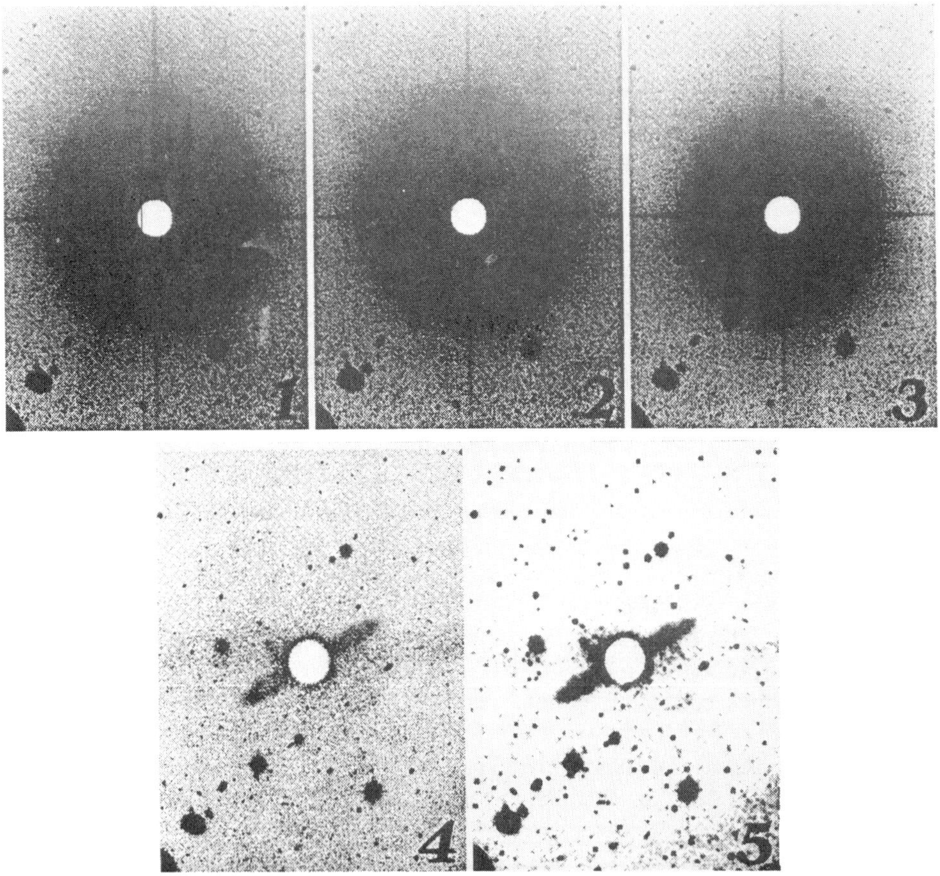

(b)

Fig. 7.2 — The distribution of STARLINK software worldwide (a) and an example of 'image processing' (b).

Fig. 7.3 — An operator at a MIDAS terminal and image display workstation at the ESO headquarters in Munich. Courtesy ESO.

with ADAM, within the MIDAS environment there is a command language (MCL) and a large set of general software packages for image processing, plotting, spectral analysis and I/O functions.

In the USA many astronomy institutes also use VAX computers and a major new initiative supported by the National Optical Astronomy Observatories (NOAO), called the IRAF (for Image Reduction and Analysis Facility), has been under development in recent years. Headed by its originator Doug Tody, the IRAF project began formally in late 1981 at the Kitt Peak National Observatory. The IRAF system was implemented initially on a VAX 11/750 computer running Berkeley UNIX — a development of the AT&T UNIX undertaken by a group at Berkeley, some of whom formed SUN Microsystems — but from its inception was intended to be transferable to all major scientific computers. In December 1983 the Space Telescope Science Institute in Baltimore selected the IRAF Command Language to host their Science Data Analysis System (SDAS), and an initial transfer or 'port' of the IRAF to VAX/VMS was made in 1984/1985. The first public release was in 1986; it is now extremely widely used and the NOAO currently supports the IRAF on an incredible variety of different computers.

IRAF goes beyond the already well-established AIPS (Astronomical Image Processing System) at the National Radio Astronomy Observatories (NRAO), in the sense that every effort has been made to make this software system as portable and as device-independent as possible. Clearly, the American system bears many similarities to the British STARLINK system and the European MIDAS system but because of the importance placed on portability of the software much more emphasis has been given to the design of the command language. The IRAF also carries with it its own programming language for maximum portability and was developed more comfortably in the UNIX operating environment. This departure from VMS is considered an important development because it maintains software compatibility for reduction and analysis of data from instruments at national facilities without restricting the owner of the data (or their institute) to purchasing a DEC VAX or any other computer. The IRAF team do, however, also support a VMS version of their system. On Kitt Peak mountain near Tucson, Arizona, the IRAF CCD reduction package is available at the telescope on a Sun Microsystems Inc., workstation and similar facilities are now available at the CFHT in Hawaii.

There are two other image/spectral processing systems in fairly wide use. These are VISTA (Lick Observatory, USA) and FIGARO (developed by Keith Shortridge at CalTech after he left Alec Boksenberg's group at University College London, and further enhanced from his current position at the Anglo-Australian Observatory). Keith has pushed the idea of treating data acquisition as a subset of data reduction since the earliest days of ADAM. FIGARO is the official STARLINK software package for spectral reductions.

These huge investments, in terms of cost and time, in general-purpose astronomical software should be encouraged up to a point, and that is the point at which they cease to improve the rate at which astronomers can eliminate massive data bottlenecks. It is also essential that such developments are as international in character as possible, and as well-managed as possible. To be honest, it is by no means clear that the astronomical world-community has 'got it right' yet. But these moves are in the right direction *provided* the designers and writers of all general-purpose software systems interact intensively with practising astronomers, and the data analysis systems themselves satisfy the following goals often expressed by the users of astronomical software:

(a) they must run on the user's computer and display system without any difficulties,
(b) they must be reasonably easy to use and learn,
(c) they must be well-equipped with a wide range of computational procedures which are properly documented and supported.

It is for funding agencies, astronomy departments and national facilities to tackle the problem of good computing facilities seriously and rigorously to ensure that the flow of results from the new generation of powerful astronomical instruments is as rapid as it should be.

7.2.4 Software development methodologies

Enormous advances have occurred during the last 20 years in the size, speed and reliability of computer hardware. Main memories have increased from kilobits to megabits, external storage has increased from kilobytes to gigabytes, and processing speed is now measured in MFLOPS or MIPS.

As Rudi Albrecht of the Space Telescope European Coordinating Facility (ST-ECF) has put it. 'The question is, have the hardware improvements been paralleled by equivalent improvements in the software area?' Many, including Albrecht, believe that the answer is no. Software, at least in the scientific arena, is not being developed orders of magnitude faster than before, and software 'quality control' has not progressed much either. Too often, instruments and subsystems, and even entire telescopes, are developed and delivered without adequate software, where by adequate we mean a product which has been specified in writing, developed to professional standards, documented fully, tested for functionality and consistency with the computer hardware and ancillary systems of the environment into which it is to be incorporated and, finally, has undergone acceptance tests.

It is interesting to compare the large-scale development of software in commerce and industry with that in academic research. In industry, attempts to enforce rigid methods upon what is essentially a creative process have led to software development becoming a management-intensive exercise. In fact, software is being produced using the management methodologies developed in a hardware-producing environment. This is perfectly justifiable in projects with well-defined goals or markets. Academia has succeeded in retaining more of the intellectual challenge. However, there is a definite tendency always to start from scratch, thereby unnecessarily duplicating the efforts of other people. The 'Not invented here' syndrome thrives. Of course there is a reason for this: most software systems developed in academic research environments — including astronomy — are of such poor engineering that they barely survive upgrades of the host operating system, cannot be used on other computers, and usually die with the departure of the original author! This may be a little severe, but without a determined effort to coordinate actions and improve standards an inordinate amount of resources will be wasted. In the past, all investments in software usually had to be written off at the end of a research project.

Advances in hardware technology are mainly caused by the fact that each new generation of devices is used to design and build the next generation — semiconductor microelectronics has led to CAD/CAM by which even more complex chips can be designed. This development is only now being paralleled in software technology with CASE — computer-aided software engineering. CASE is widely supported and used by the 'big' companies. In astronomy however, experience indicates that 'software engineering' is not yet recognised as a genuine discipline. After all, the programmer only needs to know a programming language, the computer operating system, and some very vague statements of requirements — right! When the product is delivered (predictably, late), it is accepted that 'bugs' will have crept into the code. What is being overlooked here is the fact that it is very hard to deal with the hundreds of small details in a complex system unless time is allotted for a thorough design and evaluation analysis — just as must be done for hardware. The basic approach has to be:

(1) plan and review the software development process,
(2) obtain a complete specification of requirements,
(3) model the system in several ways,
(4) provide good programming tools and a support environment.

What about 'expert systems', an application of artificial intelligence (AI), which has occupied the creative energies of computer scientists for quite some time now? They have not been used to any significant extent in astronomy. An expert system 'knows' that apples and oranges cannot be compared and will not let the user do it. Furthermore it will infer that apples cannot be compared with other fruits either, thus exhibiting 'intelligent' behaviour. This is also known as 'knowledge processing'. The nearest thing to an expert system is what is called 'black-box processing'. This is a 'raw data in, answers out' software package which follows a very well agreed standard reduction procedure for certain well-understood astronomical topics. The difficulty is that there are always subtleties! Therefore developing an expert system for astronomy would be virtually impossible, but for limited-scope projects it might be tractable.

7.2.5 Data archives

To permanently store or 'archive' CCD data can be a monumental task. All of the national facilities in the UK, USA, Europe and elsewhere are committed to data archives. Major archives include the Strasbourg Data Centre, the Space Telescope Science Institute and the Canadian Astro Data Center. But even temporary storage of digital images on computer magnetic disks requires enormous space. For example, the 800×800 TI CCD with 16-bit digitisation yields 1.28 Mbytes per readout. With a 1024×1024 CCD a single frame occupies 2.1 Mbyte of disk space which for a 300-Mbyte disk implies room for only 143 images — ignoring the need for extra space for informational data called 'headers'. It is by no means unreasonable to expect to obtain about 50–100 images per night at the telescope, making 250–500 images in a single observing run, and during image processing and analysis the number of intermediate images will often be four to five times as many — at least temporarily. And CCDs are getting larger. So computer disk space is crucial, and large observatories possessing an entire suite of CCD instruments will need tens of thousands of megabytes of storage capacity.

On a nightly basis, the digital image data stored on magnetic disks is transferred to magnetic tape. A standard magnetic reel-to-reel tape can accommodate 1600 bits per inch (bpi) along the tape with 8+1 bits across the tape, i.e. nine-tracks; the ninth bit is known as a parity bit. This yields 1600 bytes per inch, but 1 block is 8192 bytes (5.12 inches) and there is an interblock gap of 0.75 inch. Therefore a standard 2400 ft reel can hold about 40 000 000 (40 Mbyte) 8-bit words or 122 images from the RCA 512×320 CCD; when allowance is made for extra informational data stored ahead of the pixel values in the header block, the capacity is reduced by about 30%. More recent tape drives are capable of the higher density of 6250 bpi which yields a capacity of 157 Mbytes per tape or 479 images; about one tape per run. Note that for the Photometric–Ford 512×512, the TI 800×800, and the Tektronix 2048×2048 a single image requires 0.52 Mbytes. 1.28 Mbytes and 8.4 Mbytes respectively! For the TI chip you can store 122 images per tape and use one tape per night, and for the Tektronix chip ... well, tapes become a nuisance.

Magnetic tape data archives are on the verge of becoming impractical, especially since the information on the tapes can become lost or unreadable after a period of five to seven years unless the data are re-copied. It seems very likely that the next generation of astronomical data archives will be based on high density tapes such as

VHS or on optical laser disks of the write once-read many (WORM) kind. Typically, a double-sided 12-inch disk will hold 2 Gigabytes=2000 Mbytes.

7.3 PCS AND MICROPROCESSORS IN ASTRONOMY

The scientific use of personal computers (PCs) is widespread for word-processing, spread-sheets and project-planning, but PCs are also being used for instrument control and even for modest data reduction uses.

For example, both Wright Instruments and AstroMed Imaging Systems use powerful PCs such as the IBM PC/AT as the host computers in the CCD camera systems. The IBM AT uses the Intel 80286, a full 16-bit microprocessor with 8 expansion slots for use with peripherals. In the Wright Instruments CCD Camera (Fig. 7.4) an 80287 maths co-processor is also fitted. A minimum of 512 kbytes of

Fig. 7.4 — The use of an IBM PC to run a CCD camera system is demonstrated by this picture by Jonathan Wright.

RAM are installed, together with a 1.2-Mbyte floppy disk and a 20-Mbyte hard disk. In Jonathan Wright's 'AT1' system he offers standard 0.5-inch magnetic tape using the RC500M Overland Data interface board giving 1600 bpi. For the display of images, a proprietary 16-bit image store is provided but an alternative is to install the 8-bit IDE Associates 'Ideagraph' colour graphics board. The software runs under the standard IBM PC-DOS operating system. Software supplied with the Model 1 Camera and IBM PC/AT uses IBM's 'Professional' FORTRAN as produced by the Ryan-McFarland Corporation in conformance with the FORTRAN 77 standard. Astromed Imaging Systems CCD 2200 cameras are normally supplied with a host IBM PC/AT running under the UNIX operating system. This allows for portability of the same software environment to, for example, a DEC MicroVax II. Operation under MicroVMS is also supported. Numerous ancillary hardware options are offered, but the basic system comprises the IBM PC/AT (or compatible), 8/10 MHz clock speed with keyboard, Hercules graphics card and TV monitor, 1.2-Mbyte floppy drive, 30-Mbyte hard disk, parallel serial board, additional 2-Mbyte memory board, Astromed/NCR Image display card, colour monitor and graphics digitiser tablet.

Photometrics have a 'lab' system with a small computer that runs FORTH or UNIX, has a scan-converter and image store, and there is a high-capacity tape drive as well.

These systems are ideal for applications in which long-term data storage is not required, laboratory detector evaluation systems or real-time applications. For scientific use these camera systems can be linked into major (faster) computers with lots of disk space.

Other PCs such as Macintosh, Commodore Amiga, Apple II, Atari 1040ST and MEGA do find applications in astronomy as intelligent terminals, word-processing systems and occasionally for instrument control screens.

7.4 AUTOMATED DATA REDUCTION

Given the high data rates from large CCD cameras and spectrographs, there is a need to do more than develop general-purpose software packages. Automated data reduction is needed, that is, reductions which do not require the continuous interaction or intervention of the astronomer — this is the 'black-box' system, a modest use of an expert system. Earlier in the book we encountered this problem when describing plate-measuring facilities like COSMOS, and survey telescopes such as the CTI. The data handling chain for COSMOS is shown in Fig. 7.5 and the principle of the CTI archive is shown in Fig. 7.6. Computer hardware is up to it, so what about automated CCD data reduction? One such package is FOCAS which stands for Faint Object Classification and Analysis System. This software package was developed over a period of seven years at AT&T Bell Labs by J. F. Jarvis, J. A. Tyson and F. Valdes (now with the IRAF team). Basically, it is a collection of image analysis programs and special programs for the automatic recognition and classification of patterns in images. Catalogues of detected image properties like brightness and shape are produced automatically and may be compared with other catalogues of different spectral regions, or sifted for certain correlations or unusual trends, all without much intervention by the astronomer. Run time for FOCAS on a VAX 11/780 computer, without an array processor, for a fairly deep 2-h exposure with the 320×512 RCA CCD is about 39 minutes. On the other hand, using a dedicated workstation and fast array processor with 10 MFLOP capability would allow, for example, the statistical correlation plots for 100 000 objects to be obtained within an

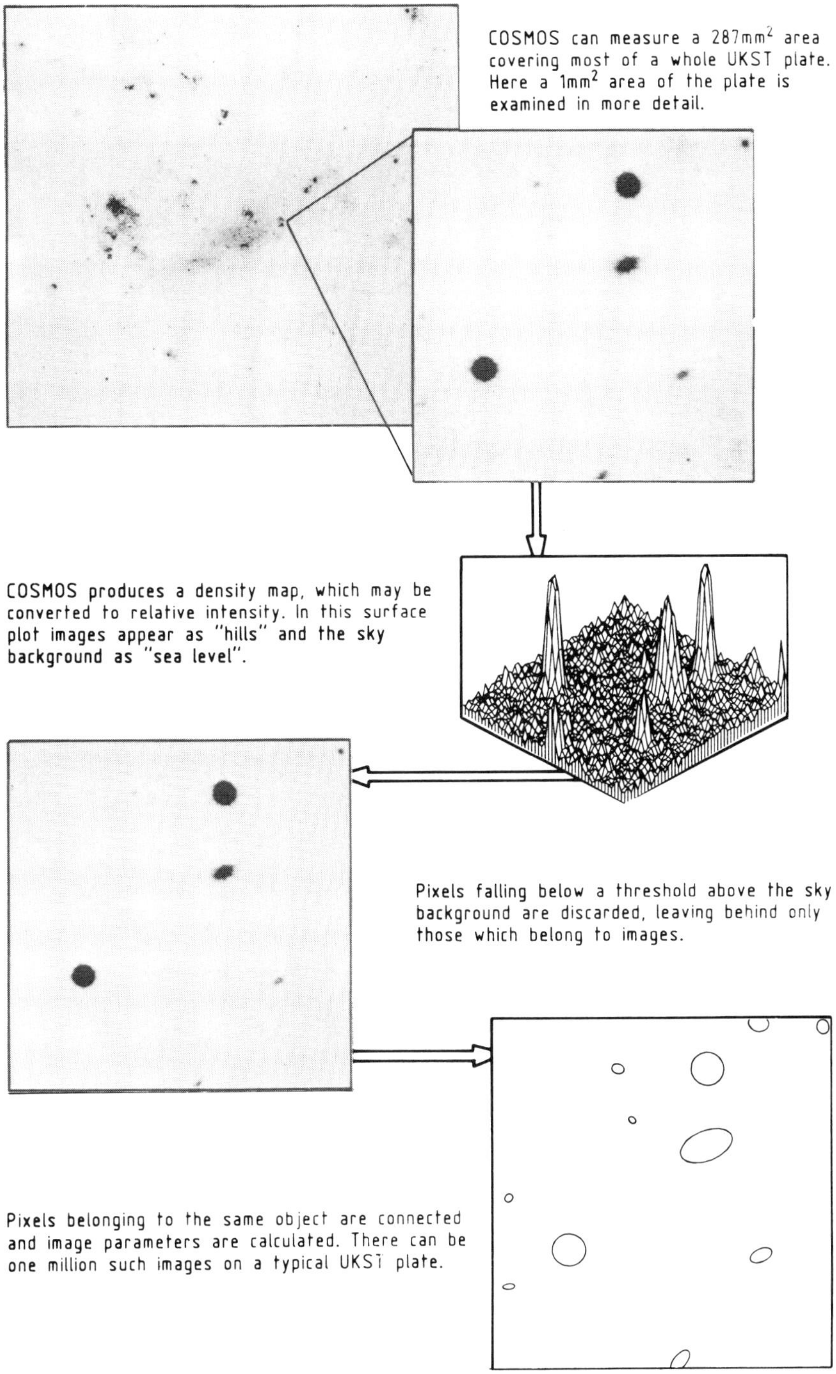

Fig. 7.5 — The COSMOS data processing chain for automated computer analysis of Schmidt plates.

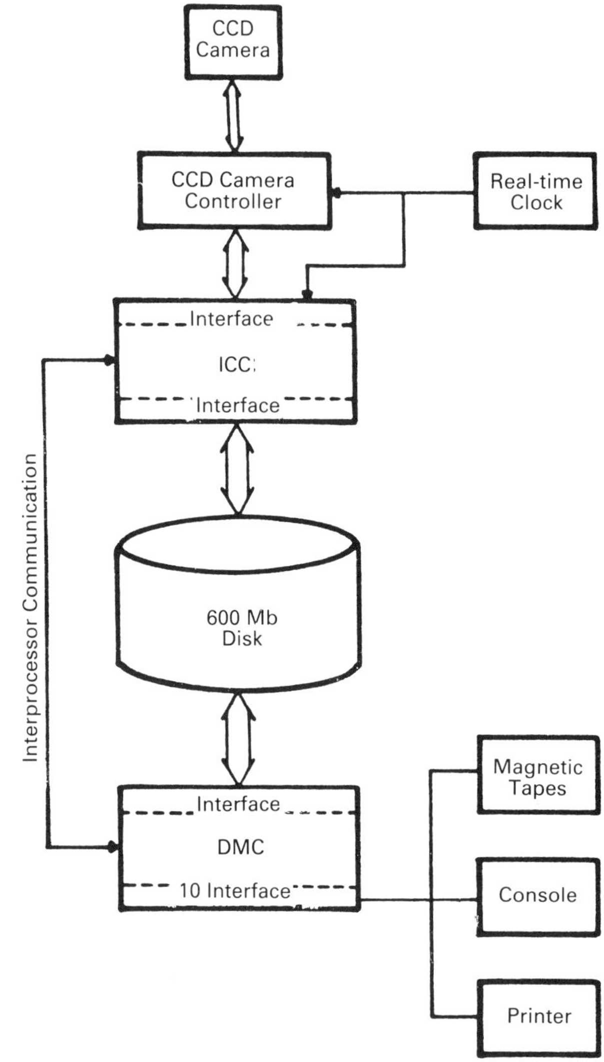

Fig. 7.6 — The University of Arizona CCD Transit Instrument (CTI) requires a sophisticated data analysis pipeline to keep pace with the efficient and deep-sky survey.

hour. Commercial 'hardwired' boards are now available that outperform software packages if a single 1-bit pattern is being searched for.

Tony Tyson is now working on a camera with a mosaic of large CCDs giving about 4000×4000 pixels, and a 200–900 MFLOP computer to handle and preprocess the data has already been completed! It would be nice to see more situations like this one. More developments in the areas of automated and black-box data reduction of CCD images and spectra are likely in the years ahead.

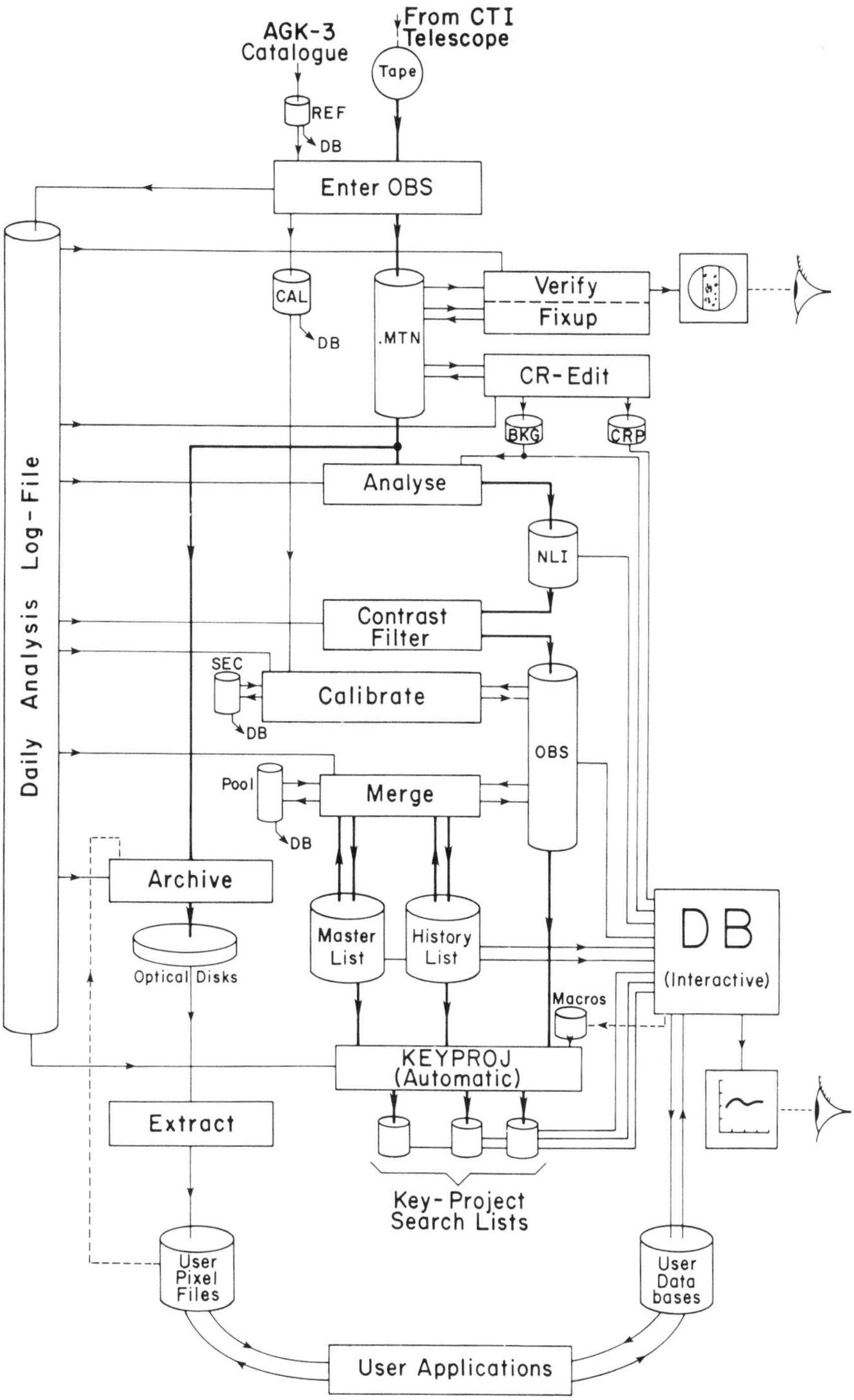

Fig. 7.6 — The University of Arizona CCD Transit Instrument (CTI) requires a sophisticated data analysis pipeline to keep pace with the efficient and deep-sky survey.

REFERENCES

Starlink Bulletin Issue No. 2 (1988), Rutherford Appleton Laboratory, UK.
Mini-Micro Systems Vol. 21, No. 1 (1988) Cahners Publishing Company, USA.
Macworld Vol. 5, No. 2 (1988), PCW Communications Inc., USA.
Computers in Physics Vol. 2, No. 3 (1988), American Institute of Physics, USA.
Computers and the Cosmos (1988) by the editors of Time-Life Books, USA.
Albrecht, R., (1986) in Instrumentation in astronomy VI, ed. D. L. Crawford, *Proc. SPIE,* Vol. 627.
Grandi, S. (1988) in *Instrumentation for ground-based optical astronomy; present and future.* Ed. L. B. Robinson, Springer-Verlag, New York.
FITS, the Flexible Image Transport System, is described in Wells, Greisen and Harten 1983, *Astron. Astrophys. Suppl. Ser.* **44,** 363.

8

Characterisation and calibration of CCD instruments

This chapter deals more quantitatively with CCD measurements and calibrations. Although the bias of the presentation is toward optical astronomy, many of the characterisation and calibration techniques are just as valid for non-astronomical imaging applications. The same principles also apply to solid-state detectors for UV and infrared wavelengths. The aim of this chapter is to introduce the serious CCD user to the steps needed to transform a 'camera' into a quantitative scientific tool of immense power.

8.1 FROM PHOTONS TO MICROVOLTS

The observed quantity in a CCD experiment is the stream of photons of light, but the detected quantity is a small voltage (V_0) which is amplified and digitised. If N_P photons per second are absorbed then (QE) N_P electrons per second are detected corresponding to a voltage A_{SFD} $\{(QE)\ N_P\ e\}/C$ volts at the output pin of the CCD. In these expressions, QE is the quantum efficiency, e is the charge on the electron, C is the capacitance of the output node of the CCD and A_{SFD} is the amplification or gain of the CCD output amplifier. The first step therefore is to determine the QE.

8.1.1 Quantum efficiency

In principle, QE can be determined in a laboratory provided a stable and well-designed calibration system is constructed to properly illuminate the CCD through a known passband. One example of a suitable experimental set-up is shown in Fig. 8.1. Either an incandescent lamp or a monochromator — a grating spectrometer — can be used as a source of illumination. After passing through a device called an **integrating sphere** which randomises the light rays and produces a uniformly illuminated source, the light passes through a shutter and a filter holder. The light is then split by mirrors; part is directed towards the CCD cryostat and part toward a calibrated detector called a photodiode. Exposures are taken at the desired wavelengths and recorded along with the signal from the calibrated photodiode. One

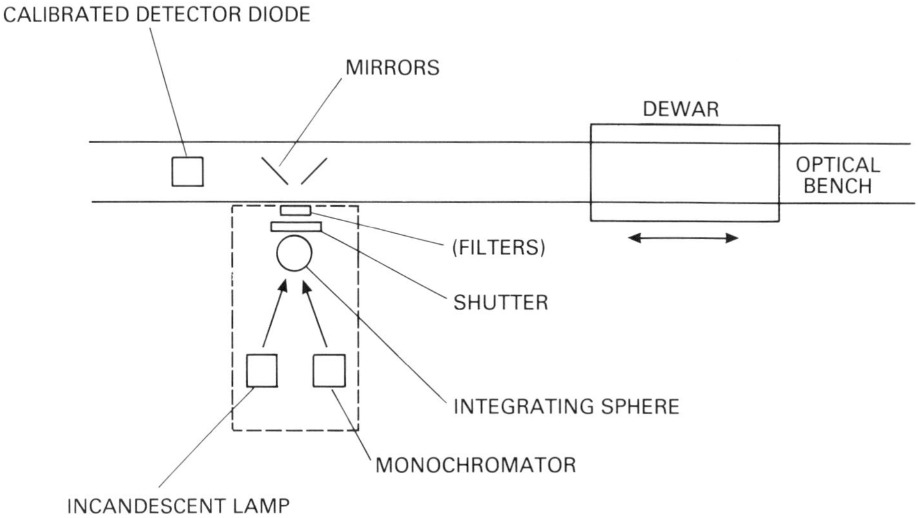

Fig. 8.1 — Laboratory arrangements for calibration and characterisation of CCDs as used at National Optical Astronomy Observatories in Tucson.

advantage of this arrangement is that since the CCD camera unit is on an optical bench it can be moved closer or farther from the light source in a very controlled manner. This allows the experimenter to use the **inverse square law** for light as a way of changing the illumination on the detector. For some wavelengths and passbands it is also possible to use non-wavelength-dependent attenuating filters called 'neutral density' filters since their attenuation can be determined fairly accurately with the calibrated photodiode. With this set-up it is easy to obtain the relative QE as a function of wavelength, but to convert this to an absolute QE at any given wavelength requires a precise calibration of the illumination, the exact transmission or 'profile' of the filter passband at each wavelength and an accurate determination of the solid angle on the source subtended by a pixel. The latter is usually achieved by using a well-defined geometry and filter profiles are measured in commercial spectrophotometers. An illumination source called a **black-body,** operating at a known temperature, will radiate an energy spectrum according to **Planck's function** which states that the energy at any given wavelength is determined only by the temperature.

In practice in astronomy, laboratory systems such as the one above are generally reserved for obtaining the relative QE as a function of wavelength and this is tied to an absolute scale by observing the signal from a star of known brightness and energy distribution. One reason for taking this approach is that the solid angle (on the sky) of a telescope is very well-defined. Stars such as Alpha Lyrae, or fainter stars of the same temperature which are not too distant and therefore are not **reddened** by interstellar dust, are commonly used. Fainter stars called **white dwarf** stars which have almost featureless spectra are also utilised as flux standards. The observed QEs of several CCDs are compared in Fig. 8.2.

When discussing systems which exhibit readout noise, as opposed to systems with

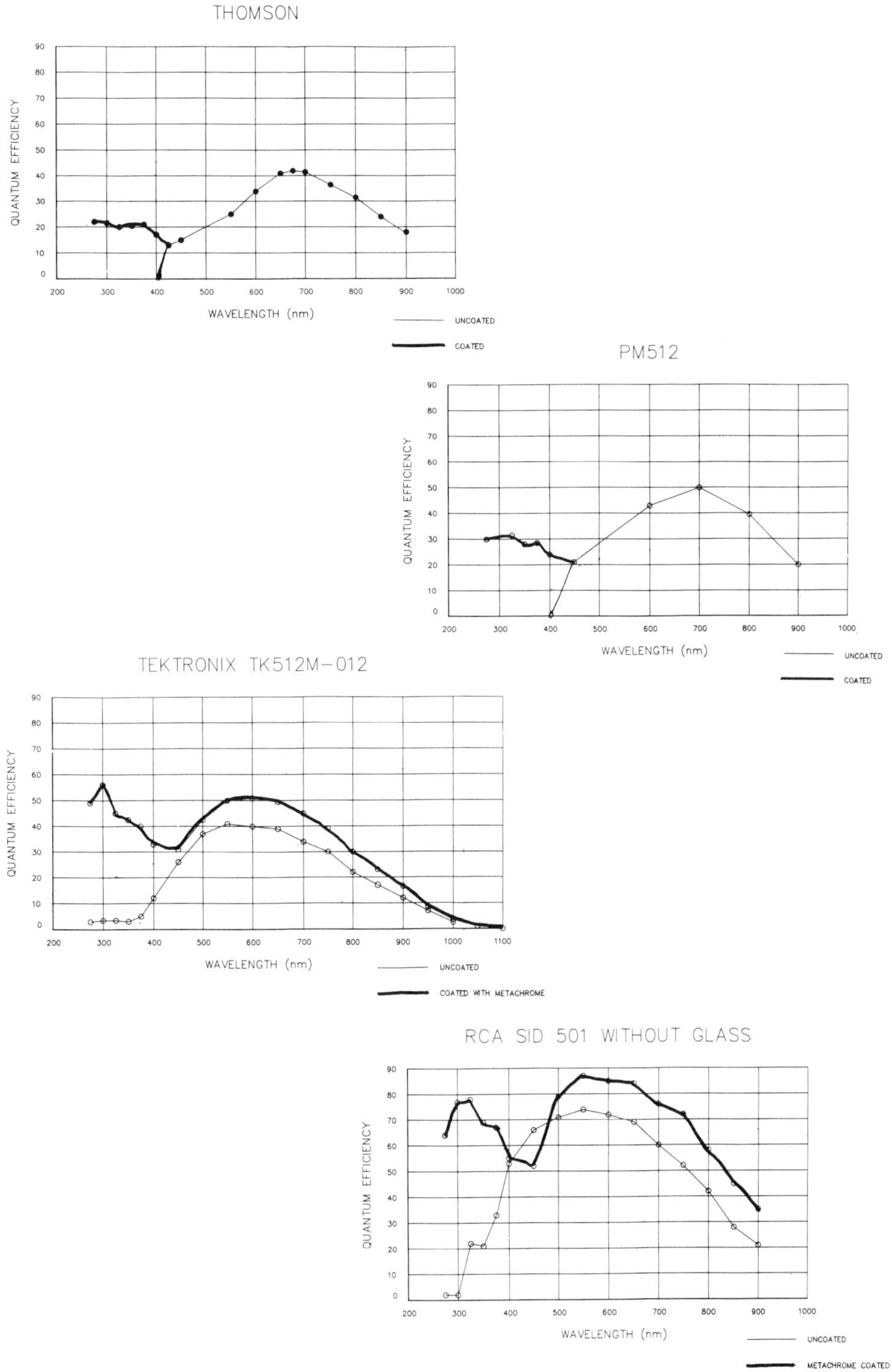

Fig. 8.2 — Curves of the detected quantum efficiency (DQE) of a variety of CCDs. Courtesy CFHT.

pure photon-counting detectors (PCDs), such as the Boksenberg IPCS, it is useful to introduce the concept of detected quantum efficiency or DQE. The DQE is defined as the QE of an idealised imaging system with no readout noise but which produces the same **signal-to-noise ratio** as the actual CCD system in question. Formulae for estimating the signal-to-noise ratio and the corresponding signal from a star of known magnitude are given in Appendix 4.

The signal-to-noise ratio for a CCD pixel is given by

$$S/N = q\, I/[qI+R^2]^{\frac{1}{2}}$$

where q is the QE at the wavelength of concern, I is the total number of photons incident on the pixel in the exposure time, and R is the root-mean-square† (or rms) value of the readout noise. An idealised detector with no readout noise would have QE equal to q' and a noise (N), given by Poisson statistics, of the square root of $q'I$. Equating the S/N ratios yields an expression for the DQE q' of

$$q' = q\{1/[1+(R^2/qI)]\}$$

which shows that q' is less than q, and that the DQE of a CCD is dependent on the signal level (Table 8.1). To keep the DQE within 10% of the QE then I must be

Table 8.1 — The detected quantum efficiency (DQE) as a function of readout noise (R (electrons)) and incident photon flux (I (photons/s)) for two values of the QE (30%, 60%)

R	I: 1	10	100	1000	10 000	100 000
0	30(60)% independent of I					
1	6.9	22.5	29.0	29.9	30.0	30.0
	(22.5)	(51.4)	(59.0)	(59.9)	(60.0)	(60.0)
10	0.1	0.9	6.9	22.5	29.0	29.9
	(0.4)	(3.4)	(22.5)	(51.4)	(59.0)	(59.9)
100	0.001	0.009	0.1	0.9	6.9	22.5
	(0.004)	(0.215)	(0.4)	(3.4)	(22.5)	(51.4)

greater than $10\times R^2/q$. For a good modern CCD with five electrons noise and a QE of 0.5 (50%), I must be greater than 500 electrons which, as shown in Appendix 4, corresponds to a star of visual (V) magnitude 22.0 in a 10-second integration time for direct imaging on a 4-m telescope. On the other hand, a spectrometer might require an integration time a thousand times longer than this. Clearly, this equation puts

† The term root-mean-square derives from the fact that to estimate the noise one forms the difference between any value and the mean value for a set of measurements, squares each difference, finds the mean value and then takes the square root to estimate the noise.

stringent demands on CCDs for low noise *and* high QE, and without both of these simultaneously there will be a need for intensified photon counting systems despite their problems of stability.

8.1.2 Photon transfer function

The digital signals actually recorded by the CCD system — usually called **data numbers (DNs)** or A/D units (AD Us) — need to be turned back into microvolts and then into electrons and finally to photons in order to calibrate the system. The relation between DN and microvolts at the CCD output depends on the 'gain' of the amplifiers in the system, and the conversion between microvolts at the CCD output and an equivalent charge in electrons requires a knowledge of the capacitance (C) of the output node of the on-chip amplifier. The actual data counts or DNs produced by the CCD camera system are linearly related to the numbers of electrons in the charge packets being read out by the following algebraic expression

$$S_0 = G\,N_e + b$$

where S_0 is the recorded output signal in DN or counts, N_e is the number of electrons in the charge packet, G is the gain or **transfer** factor expressed in counts (or DN) per electron (the inverse quantity is $g = 1/G$ in electrons/DN), b is the electronic offset or **bias** level (in DN) for an empty charge packet. The value of b can be controlled by the designer and is usually substantially above zero, by ten to one hundred times the readout noise (expressed in counts), to ensure that there are no problems with the A/D unit when it receives zero or slightly negative signals by chance. (Bipolar A/D units can be used which will properly record negative and positive signals.) Typical values of b would be in the range 50–500 DN depending on the particular CCD and its associated readout noise.

There are two ways to derive the transfer factor g in electrons/DN:

(a) by calculation, knowing the overall amplifier gain and the capacitance of the CCD;
(b) by a series of observations of a uniformly illuminated scene at different brightness levels.

We will describe each method in turn.

Let V_{fs} be the full-scale voltage swing allowed on the A/D unit, and n be the number of **bits** to which the A/D can digitise. The full-scale range is therefore subdivided into 2^n parts, the smallest part — the **least significant bit** or LSB — is simply 1 DN. Thus the voltage corresponding to 1 DN at the A/D unit is $V_{fs}/2^n$; as an example, suppose the full-scale voltage is 10 V and the A/D is 16 bits then 2^n is 65536 and so the ratio is 0.000 152 5 V, that is, 152.5 μV at the A/D is 1 DN.

But we need the number of microvolts at the CCD rather than at the A/D, so we must divide this number by the gain A_g of amplifiers in the system. To convert this number of microvolts to an equivalent charge of electrons we must then multiply by the CCD capacitance (C) and divide by the value of the charge on the electron (e). Therefore,

$$g = [V_{fs}C]/[2^n A_g e]$$

where $e=1.6\times10^{-19}$ coulombs.

The value of C varies with the type of CCD. For example, the RCA CCDs typically have $C\approx0.6$ pF whereas for the GEC devices C is approximately 0.1 pF. Thus for the GEC CCD we get that $g=95.3/A_g$. The term A_g is the product of two components, namely the amplifier gain of the 'on-chip' output transistor which is typically about 0.7, and the gain of the 'external' electronics such as the preamplifier and any additional amplifier.

If the value of g ($1/G$) is too small then large signals produced by the CCD will be above the maximum input level of the A/D unit. On the other hand, if g is too large then errors due to the process of conversion of analogue to digital signals — called quantisation errors — will significantly add to the system noise. A good rule of thumb is to arrange for g to be about 0.5 to 0.25 times the CCD readout noise expressed in electrons. So, for a 10-e device we might choose g to be 5 e/DN implying an external gain of about 27. Notice that a full-scale reading of the A/D corresponds to 65536 DN in this case which is equivalent to 327680 electrons; this is smaller than the physical full well of some CCDs but larger than the well-size of others.

In practice C is not well-known and it is important to determine g experimentally.

The latter experiment can be performed on the same **flat-field** as used to remove pixel-to-pixel response differences. There are essentially two approaches. Obtain several exposures and examine the mean signal and noise from each and every pixel independently before adjusting the illumination level — which must be absolutely constant during the measurements — or take fewer exposures and examine the mean signal from a small array of pixels (in a good clean area of the CCD) which have been flat-fielded to remove the dispersion (or noise) in the mean due to their variations in sensitivity. The latter method will not work unless the field has been flattened.

From either data stream X_i, one obtains a mean, bias-subtracted signal (S_m) in counts and a **variance** (V_m) — the square of the standard deviation (or noise) of a single observation from that mean — by using the well-known formulae:

$$S_m=\Sigma\, X_i/n \text{ and } V_m=\Sigma\, (X_i-S_m)^2/(n-1)$$

where the summation indicated by Σ is over all n pixels in the data set.

Now, if all is well, there should be only two noise sources, namely

(1) photon noise on the signal photoelectrons (p),
(2) readout noise from the CCD output amplifier (R).

Since these two noise sources are independent and random they add together in a way known as **quadrature,** so that the total noise is given by

$$(\text{noise})^2=p^2+R^2 \qquad (1)$$

The noise-squared $(\text{noise})^2$ is the variance. It is important to realise that this expression applies to photoelectrons and not to A/D counts.

But the measured quantities (S_m and V_m) are in DN. To convert from electrons to DN in equation (1) we need to divide each noise term by g (electrons/DN) or alternatively multiply by G ($1/g$) DN per electron, to give

$$(G\,\text{noise})^2 = (G\,p)^2 + (G\,R)^2$$

This is much simpler than it looks because the left-hand side is just V_m, the observed variance, in DN. Also, the mean number of photoelectrons is S_m/G and the photoelectron noise (p) on this number is simply the square root of S_m/G, so $p^2 = S_m/G$ which gives

$$V_m = G \times S_m + (G\,R)^2 \quad (2)$$

If we rewrite this equation in the simple form of

$$y = mx + c$$

where $y = V_m$ and $x = S_m$ are the 'observed' quantities (noise and signal) as the illumination changes, then we can see that this is the equation of a *straight line*. The gradient (slope) is m and the value of c is given by the y-value at the point where the line cuts the y-axis (i.e. when $x=0$). In other words, equation (2) is a straight line of gradient (or slope) $m = G = 1/g$ and intercept $c = (G^2R^2)$ (see Fig. 8.3(a)). (Graphs of noise versus signal on a logarithmic scale (Fig. 8.3(b)) are also used.) This method of deriving g and R is known as the 'variance' method or the 'photon transfer method'.

At the lowest signal levels the noise (variance) is dominated by the fixed readout noise whereas for larger signals it is dominated by photon noise — the statistical fluctuations in the photon detection rate — and finally, at the largest signals, the CCD begins to saturate and the noise actually falls.

In practice this technique requires some care. Four or more flat-field exposures are taken at each of many illumination levels, usually increasing by factors of two, and a dark frame is taken at each of the exposure levels. Sub-areas of the CCD that are free from artifacts and show minimal non-uniformity are selected. For each illumination level the mean picture is computed and this is subtracted from the individual images to obtain pixel deviations. (Alternatively, the mean, dark-subtracted image can be used to flat-field the individual, dark-subtracted images.) The variance for each pixel in the clean sub-array is calculated and averaged over the sub-area, and the mean unbiased signal value is derived from the difference between the averaged picture and the dark frame. A least-squares straight-line fit can then be made to the data set to derive the slope and intercept and hence the parameters in equation (2).

8.2 NOISE, BIAS AND DARK CURRENT

In almost all CCD systems it is necessary to purposely bias the video signal slightly towards the positive so that the readout noise never drives the A/D input into the negative. Some systems employ bipolar A/Ds which allow negative-going signals to be properly handled but again, a 'no-signal' condition will always correspond to some positive offset or bias signal. This electronic instrument signature is therefore known as the **bias level** and can be determined easily by taking an unexposed frame (zero exposure time, shutter closed), also alluded to as an **erase frame,** as shown in Fig. 8.4. Several bias frames can be averaged to reduce the random readout noise while

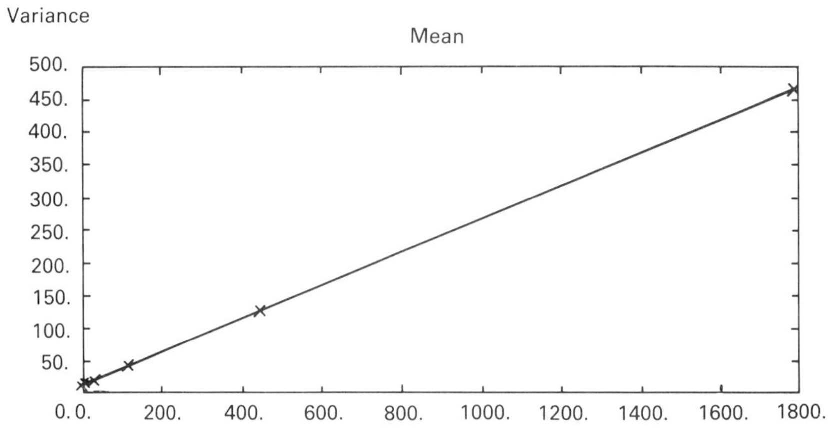

Least Square fit to variance/mean calibration statistics
gain calibration = 3.9611 electrons/adu
noise level = 14.0010 electrons rms
correlation coeficient = 1.000

(a)

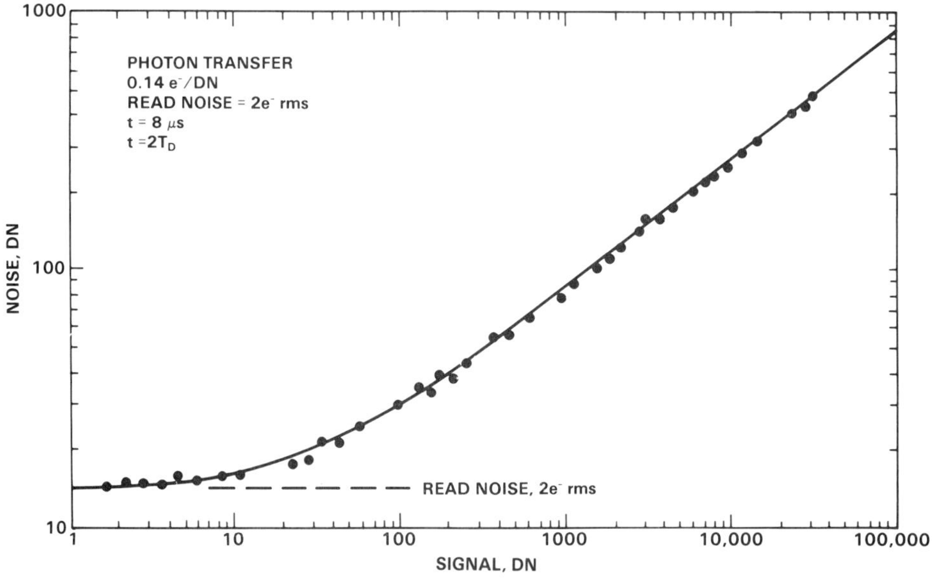

(b)

Fig. 8.3 — The photon transfer curves of noise versus signal for different CCDs. (a) A variance plot; (b) a logarithmic plot.

preserving any spatially coherent noise and the pixel-to-pixel bias levels. A mean bias frame must be subtracted from each exposed frame as the first step in data reduction. For some applications it may be sufficient to subtract a single, mean bias

Fig. 8.4 — A good bias frame showing no serious amplifier fixed-pattern noise or ground-loop interference.

number from each pixel rather than a bias frame, but the normal practice in astronomy is to take many bias frames and use the **median** of that set.

Bias information can also be conveniently obtained in many systems by **overscanning.** That is, the sequencer is told to send more clock pulses than are actually required to vertically and horizontally read out the real CCD (e.g. 522×522 instead of 512×512). The entries in the image so obtained which lie outside the physical area of the CCD should only contain bias level signals — provided the CTE of the device is good. This latter point is important, since overscanning a uniformly illuminated CCD is one way of probing the CTE (section 5.4). Charges which are trapped during the primary transfer will slowly leak out and will therefore appear at the output as a result of the overscanning. Usually, the median bias frame is computed and subtracted from the object frame, then the overscan in the basis frame is compared to the overscan in the astronomical object frame to derive any small offset which is averaged and subtracted as a single number (scalar) from the object frame.

For CCDs requiring a preflash to alleviate global charge transfer problems, the preflash must also be applied to the bias frame.

If the system is working perfectly then a bias frame should contain very little fixed-pattern structure; it should be dominated by random readout noise variations. The standard deviation ($\sqrt{V_m}$) of a good-sized patch of the CCD therefore gives an immediate estimate of the readout noise in data numbers without recourse to the full 'photon transfer curve' experiment described previously. If there is some unavoidable fixed pattern in the bias frame then it is straightforward to take the difference between two bias frames and divide the resultant noise by $\sqrt{2}$.

In several CCDs it is necessary to turn off, or reduce to a very low level, the current through the output transistor *and* the clock voltage swing in the horizontal register during long exposures. If this is not done then an immense accumulation of charge can occur in the corner of the chip and along the rows nearest to the horizontal register. The horizontal register and output transistor need to be switched on again in sufficient time to allow them to be swept clean and re-stabilise before the integration time ends.

Dark-current levels are determined by long exposures with the CCD shutter kept closed. It is advisable to perfom these measurements at night or in a darkened room or in the actual telescope environment. Exposure times of 1 hour will be needed to accurately determine the dark-current levels. Several identical exposures of this duration will also enable cosmic ray and radioactivity events to be isolated and counted. Thinned CCDs or those with an epitaxial (or thin) layer are less prone to cosmic ray events. Sometimes 'hot pixels' cause the dark current to be non-linear, then for the highest accuracy many dark frames are required at the same exposure time as used for the object frame.

8.3 FLAT-FIELDING STRATEGIES

As already described in section 5.6, CCDs do not have a uniform response to light across their surfaces. Pixel-to-pixel variations in sensitivity (QE) arise due to physical differences between pixels as the result of fabrication processes and due to optical attenuation effects such as microscopic dust particles on the surface of the CCD. Although these variations can be as small as a few per cent of the mean sensitivity they must nevertheless be reduced much further because such variations result in a 'noisy' image at a level corresponding to a few per cent of the sky brightness.

A common practice is to observe the inside of the telescope dome (if it is matt white) or a huge white card on the dome. In both cases the objects are so close that the telescope is actually completely out of focus, which helps to ensure that the field is uniformly illuminated (flat), but the beam passing through the instrument is filled in virtually the same way as when observing objects in the sky. Dome illumination is usually done with tungsten lamps which do not remotely mimic the spectrum of the night sky, which is a complex composite of a thermal continuum, scattered light plus a number of practically monochromatic (and variable) night-sky emission lines. Optical filters often exhibit 'red leaks', that is, an unwanted transmission to light at much longer wavelengths — typically double the original centre wavelength. Although the light leak is at a very low level (less than 1%), the sensitivity of the CCD is so good that even this is a problem. The solution is to 'block' the red light with an additional filter. Since tungsten lamps are much redder than the sky, dome flat-fields with such filters would be a poor match. For faint objects it is the light of the sky which dominates and so it is better to try to use the sky itself as a flat field, whereas for brighter objects it is their own intrinsic colour which matters and that matches neither sky nor dome. Most workers have found it desirable to establish a set of narrow, rather than broad, passbands for imaging so as to limit the effect of colour-dependent non-uniformity.

Since, to first order, CCD detectors are quite linear, then only a simple arithmetic division, pixel by pixel, by an image of a uniformly illuminated scene — the flat-field — of the appropriate colour, followed by re-scaling, is required.

Let I_{FF} be the uniform illumination of the flat-field source on a pixel whose quantum efficiency is q_{ij}. The observed signal from that pixel in DN is

$$(X_{ij})_{FF} = G\, q_{ij}\, I_{FF}$$

where i,j are the 'coordinates' of the pixel in terms of rows and columns.

The mean signal in the flat-field is obtained from averaging X_{ij} over all the rows and columns; call this S_{FF}.

$$S_{FF} = G\, q_{m}\, I_{FF}$$

where q_m is just the mean QE over the entire CCD.

For the image scene — which of course is not uniformly illuminated — we have a similar expression for the recorded signal on the pixel on the ith row and jth column.

$$X_{ij} = G\, q_{ij}\, I_{ij}$$

To eliminate q_{ij} we form the ratio of image scene to flat-field

$$X_{ij}/(X_{ij})_{FF} = I_{ij}/I_{FF}$$

and finally we rescale this ratio to the mean of the flat-field to give

$$S_{FF}\, X_{ij}/(X_{ij})_{FF} = I_{ij}\, Gq_{m}$$

Thus the flat-fielded, re-scaled image constructed by the operations on the left-hand side of this equation differs from the true image scene I_{ij} by a single, constant scale factor Gq_m which can be determined by observations of a source of known brightness — a flux standard.

In practice, the same exposure time 'on-chip' is used for the flat-field, but many repeated flat-fields are averaged to increase the accuracy of the flat-field itself and to remove from the flat-field various artifacts such as cosmic ray events. Also, the colour of the flat-field should be as good a match as possible to that of the image scene since q is a strong function of wavelength and the cancellation of q_{ij} described above may not be perfect if the two scenes are significantly different in colour.

To detect the very weakest signals, particularly of extended images, it is necessary to correct for residual non-linearity by more complex means such as obtaining a series of flat-fields at different exposure levels and determining the response of each pixel individually by means of a polynomial-fitting routine.

As already mentioned, for many applications CCD users have found it convenient and acceptable to use the inside of the telescope dome as a flat-field. For very high accuracy work or for observations on objects fainter than the night sky itself, various systematic errors tend to dominate over the expected random errors from photon arrival statistics. The first advance in counteracting low-level systematics in the detector was the **drift-scan** technique developed by Craig Mackay and Johathan Wright; this approach is based on the idea of **time delay and integration (TDI)** used in infrared detector sytems. The CCD charge pattern is transfered along columns very slowly — i.e. the clocking rate is slow — while the image from the telescope is physically 'scanned' along with it in precise synchronisation, using a moveable motor-driven stage. In this way the image keeps up with the charge pattern and every pixel along a column contributes to the final image of the object (Fig. 8.5). This

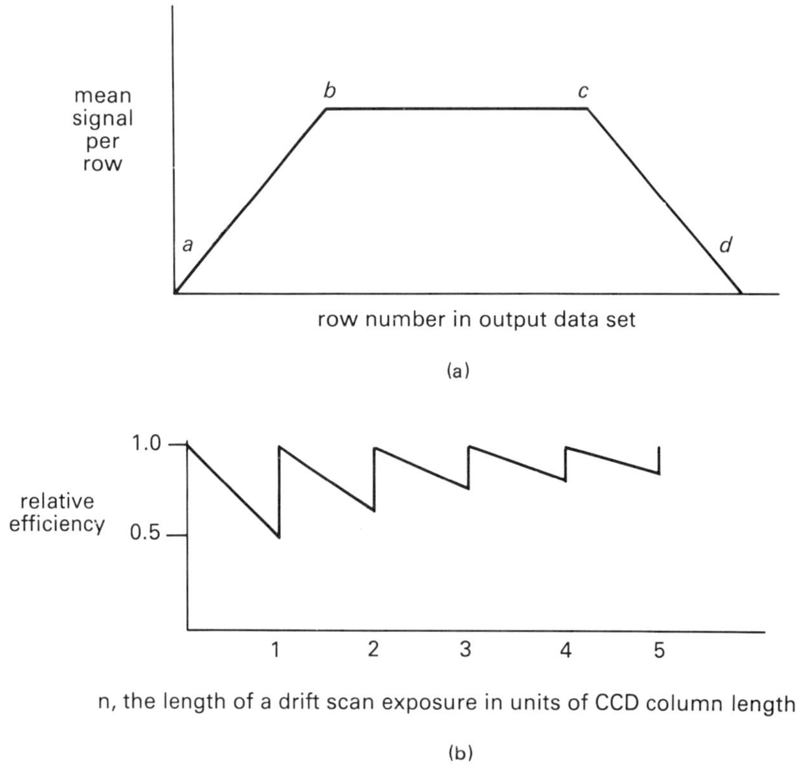

Fig. 8.5 — The ramp-up and ramp-down of the mean signal from a given column with the drift-scan technique for flat-fielding. (a) The mean signal in each row of a drift scan image. The shutter is opened at *a* but it is not until *b* that sky elements have passed over the entire length of the CCD. All pixels until *c* do this where the shutter is shut and the rest of the image quickly read out until *d*. (b) The efficiency in actual observing time of a drift scan exposure compared with a conventional exposure of the same area of sky.

technique averages over some of the fixed pattern noise on the CCD (at least in one dimension) and is effective in reducing systematics to below 1% of the night-sky background level due mainly to the fact that the 'colour match' in deriving the flat-field from the sky itself is much better than from an artificial lamp illuminating the inside of the telescope dome.

There are also variations on this technique, such as the 'short drift scan' in which the drift is over a much smaller number of pixels than in a whole column.

Tony Tyson and others have proposed, and demonstrated, a number of steps which can be taken to reduce systematic errors to a level of at least 0.1% of the sky background without the need for a mechanical stage, and which can enable a large 4-m-class telescope to reach its theoretical limit of 27th magnitude in a six-hour exposure.

The dominant sources of error are the mismatch between calibration flat-fields and the actual night-sky background and, for some thinned CCDs, interference **fringing** due to unblocked night-sky emission lines. Generation of a master flat-field and sky frame from *the object frames themselves* has been shown to remove systematic effects to better than 0.03% of night sky. Recently, the author and several colleagues have obtained at least as good a performance using this approach with the new generation of infrared array detectors (see Chapter 9). Essentially, this powerful technique involves numerous observations of a piece of relatively blank sky but with the telescope pointing to a slightly different position on the sky (displaced by, say, 5 arcseconds) for each CCD exposure. Positions can be chosen randomly or in a simple pattern. When the sequence of displaced exposures is examined later, and the frequency or number of times a certain value or signal is found in a given pixel is plotted against those signal values, it is found that just one signal value (or a small range) is most favoured. This value is just the signal corresponding to pure background night sky. An occasional faint object will fall on the given pixel very infrequently because of the semi-random displacement between exposures. The mathematical statistic used to determine the most likely sky value from such a data set is called the **median** since it ignores a value discrepant from all the rest, including values due to bad pixels; if sufficient frames are taken it is possible to calculate the **mode** which literally is the most frequently occurring value. Clearly, the disadvantage of this technique is that it will not work on object frames which are crowded, e.g. the image of a large galaxy or nebula. A combination of dome flats and sky flats is often recommended in that instance. However, with a telescope of the right scale and a 2048×2048 CCD, Harold Ables and colleagues at the US Naval Observatory in Flagstaff managed to apply this technique to the field shown in Fig. 8.6; the Tektronix chip was an early 'set-up' device not intended for scientific use!

As a matter of practicality, it is essential to 'normalise' the various flat-fields before applying a median filtering algorithm, since a drift in the mean level of the illumination — while not disasterous for the overall technique — will affect the calculation of the median value.

CCDs exhibit both additive and multiplicative effects which require calibration. In the 'additive' class we have

(1) electronics pattern noise,
(2) charge skimming or trapping,
(3) interference fringes,
(4) LED activity.

And in the 'multiplicative' class there are the effects of

(1) quantum efficiency variations across the CCD,
(2) transmission of optics and coatings,
(3) thickness variations of thinned CCDs.

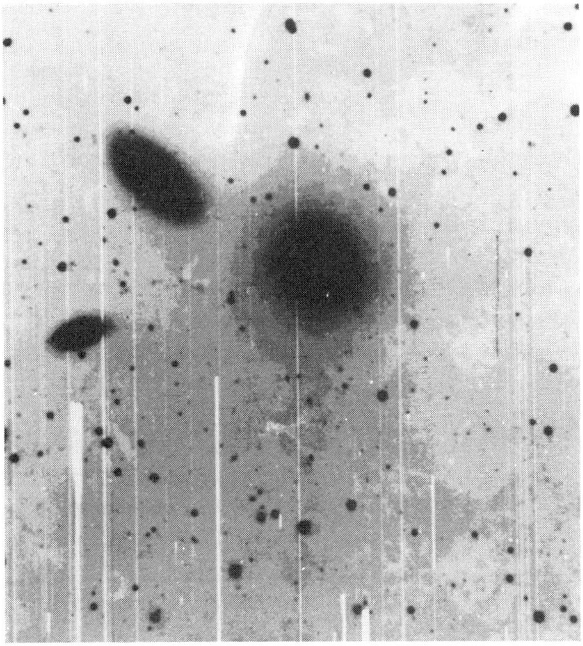

(a)

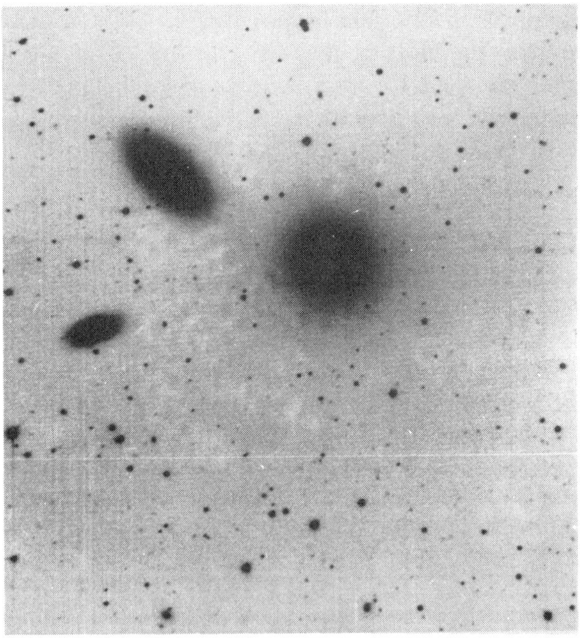

(b)

Fig. 8.6 — (a) A raw CCD image with lots of defective pixels; (b) same image flattened by using median sky flats by shifting the images. Courtesy Harold Ables, US Naval Observatory.

Clearly, it is important to devise steps to remove these effects and to perform these corrections in an optimum sequence. In general, the best recipe for reduction and calibration of raw CCD images is to follow the following pattern:

(1) **Subtract bias and bias structure:** A 'bias structure' image, obtained by averaging many bias frames, is subtracted and a further small scalar subtraction to eliminate any offset in bias between programme and bias frames is also made using overscan data. For CCDs using preflash the bias must also use preflash.
(2) **Subtract dark:** Most CCDs exhibit some dark current or a low-level LED activity, leading to pollution during long exposures. To correct this, the 'median image', rather than the 'mean image', of many long (bias-subtracted) dark exposures, obtained as recently as possible, must be subtracted from the object frames.
(3) **Divide by flat-field:** Many exposures on a diffusely illuminated screen are normalised (adjusted to the same mean) and averaged, for each filter or spectrograph setting. Exposure levels in each flat-field and the illumination colours must be similar to the astronomical exposures. For spectroscopic applications the colour of the flat-field is less important. The bias-corrected, dark-subtracted, normalised flat-field master averages are then divided into the dark-corrected object frames to calibrate for QE variations (and other effects such as out-of-focus dust spots) from pixel to pixel. Alternatively, the master flat-field is derived from a series of sparsely populated sky frames each of which has been displaced (semi-randomly) from the other. The **median** or the **mode** (the most frequently occurring value) of this set of images will be a frame composed purely of sky.
(4) **Subtract fringe frame (sky subtraction):** Night-sky emission lines limited to a very narrow band in wavelengths can cause **interference fringes** on some back-illuminated CCDs bonded to glass, in sky-limited exposures. A 'fringe frame' derived from an adaptive modal filter routine (see below) must be scaled and subtracted from the already flat-fielded object frame derived from Step 3. This step effectively performs a sky-subtraction.
(5) **Interpolate over bad pixels:** Either a bad-pixel map for the particular CCD in use is supplied as input to a median interpolation routine or, if the two-dimensional displacement or offsetting technique (see above) is used during observing, then bad pixels at low levels will automatically get filtered out of the final image. Remaining bad pixels can be replaced by good data from other frames for that part of the scene.
(6) **Remove cosmic ray events:** A computer algorithm can be used to identify non-starlike point sources or the random offset technique can be used to eliminate (by median filtering) cosmic radiation events greater than several times the sky background noise.
(7) **Registration of frames and median filtering:** If the random offsetting technique is used, re-registration of data frames to a fraction of a pixel using interpolative software routines, followed by median-filtering and edge trimming, is used to produce the final cleaned image.

8.4 FRINGES AND SKY EMISSION

The occurrence of interference fringes with backside-illuminated thinned CCDs was described earlier (section 5.6). Obviously enough bias, dark and flat-field frames must be taken to ensure sufficient accuracy in the median-processed images. Many astronomers unaccustomed to CCD methods balk at the thought of spending so much time on calibration data-taking. Although CCDs are improving steadily, it is nevertheless absolutely essential to invest effort in calibrations if one is to extract the faintest signals and to use a telescope with a CCD instrument at the limits of their capabilities. There is usually plenty of opportunity to take such calibration data during daylight hours.

Fringe removal can be performed by a technique called **adaptive modal filtering.** This routine computes the absolute difference between the mean and the median of values associated with a pixel over all the images in a set and rejects deviant values until this difference falls below a certain value or a maximum number of values have been rejected. A given pixel is then median-filtered over all the images. Of course, this technique fails with large, extended objects such as nearby galaxies and nebulae, but it is very suitable for relatively uncrowded fields in which two-thirds of the actual CCD area is occupied by sky. This fringe frame must then be scaled by trail-and-error and subtracted from the object frame until a patch of sky on the object frame is entirely flat.

Fig. 8.7 shows a 500-s CCD exposure on a 4-m telescope before and after fringe removal. The effect is remarkable!

8.5 LINEARITY

CCDs, if operated properly, are quite linear detectors over an immense dynamic range. That is, the output voltage signal from a CCD is exactly proportional to the amount of light falling on the CCD to very high accuracy, often better than 0.1% of the signal. The good linearity of the CCD makes it possible to calibrate observations of very faint objects by using shorter, but accurately timed, exposures on much brighter photoelectric standard stars. Fig. 8.8 demonstrates some linearity data.

Since it is difficult to vary light intensity in the laboratory with high accuracy — recall that one possible method is to keep the source output constant and alter the distance from the detector in as controlled a manner as possible — linearity curves are usually derived by observing a constant source with various exposure times. This method assumes that exposure times can be very accurately controlled — which is generally a good assumption — and that $1/f$ noise sources are negligible. Simply put, if the exposure time is doubled the (dark+bias)-corrected signal should also double.

Non-linear behaviour from CCDs can occur if incorrect voltages are applied. Care must be taken to ensure that the output transistor is operating in its normal linear regime and, as with the RCA device (shown in Fig. 8.8), it is essential to use the correct clock voltages to ensure the CCD pixel is fully inverted. A comparison of clock voltages is given in Appendix 5.

8.6 PHOTOMETRY

Photometry is the process of obtaining quantitative (numerical) values for the brightness of objects. Sometimes relative brightness is adequate and this may be in

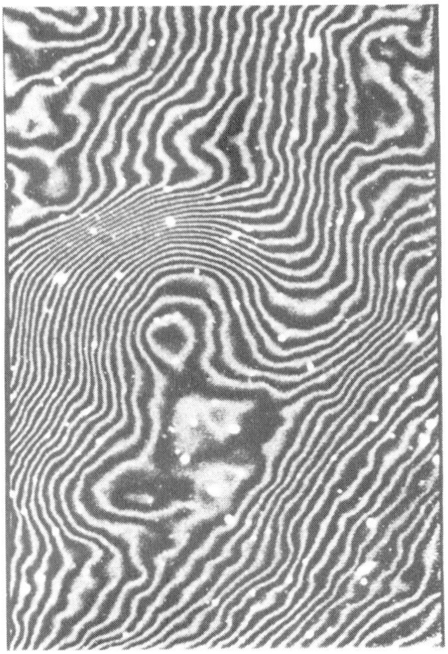

(a)

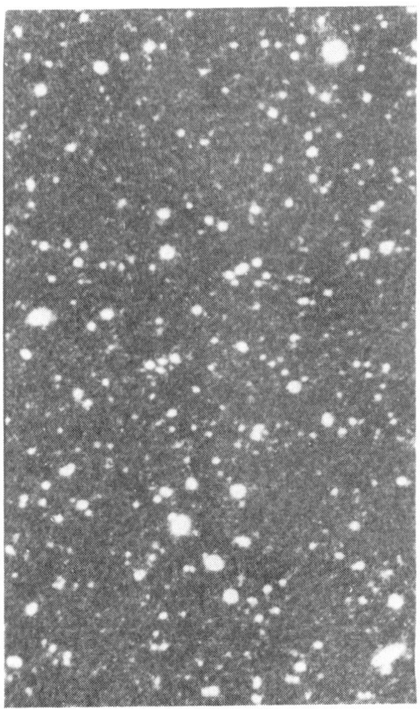

(b)

Fig. 8.7 — (a) A severe fringe effect on a deep 4-m telescope CCD image before processing. (b) The same image after computer processing to remove the fringe pattern.

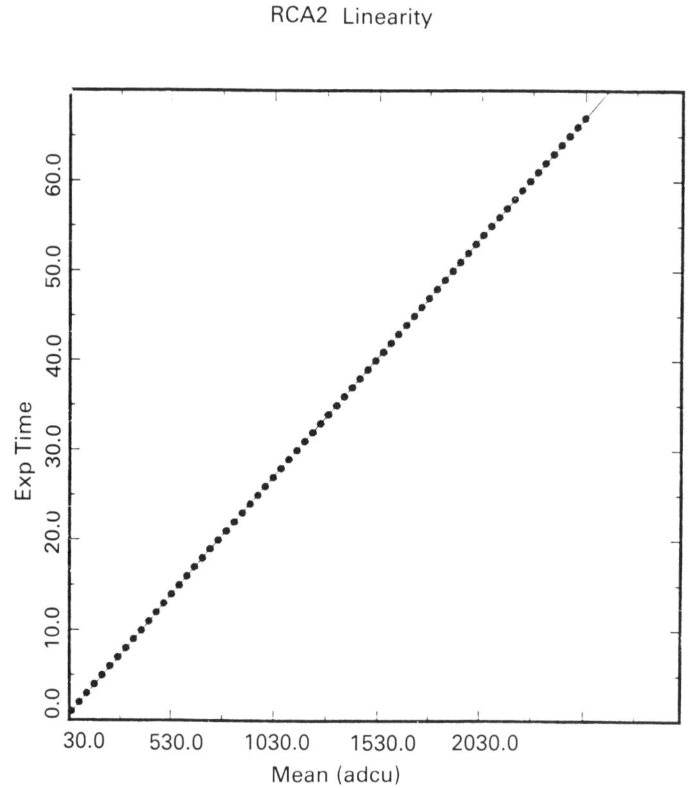

Fig. 8.8 — The exceptional linearity of a CCD is demonstrated in this graph for an RCA chip; see also Fig. 5.10.

relation to other objects in the same field of view or nearby in the sky. For other studies it is imperative to derive the true or absolute amount of radiant energy reaching the earth.

In general, the accuracy with which brightness can be measured on CCD frames is extremely high, and repeatability from one CCD image to another is excellent. Limitations on accuracy are really only introduced when comparing CCD photometry with the results of classical photoelectric photometers which employed photomultiplier tubes. Calibration of CCD systems against the old phototube systems to high accuracy (0.5% or 0.005 magnitudes or better) is very difficult due mainly to the gross mismatch between CCDs and classical photoelectric photometers in terms of the huge span in wavelengths to which CCDs are sensitive and the effect which this has on the filter profile. To a lesser extent, small systematic errors can be introduced by the measuring method during CCD data reduction and analysis. In general, CCDs perform better, and can be calibrated more easily, if the spectral bandwidth is limited to a few hundred ångströms (i.e. a few hundredths of micrometres). Unfortunately, this is not standard, and does not easily permit a comparison with classical photomultiplier tube photometry. It has been said that classical photoelectric photometry was 'a great idea at the time' but now that we have CCDs we should do everything over again!

Fig. 8.9(a) shows the standard UBVRI filter bands and Fig. 8.9(b) shows some effects of their 'red leaks'.

Two basic procedures for obtaining photometric data from CCD images are

(a) aperture photometry,
(b) profile fitting.

Each is discussed in turn, and both rely on computer processing of digital data.

(a) **Aperture photometry:** Conceptually this is analogous to photoelectric photometry. After the CCD image has been obtained, a computer program is used to reconstruct the signal which *would* have been obtained from an object in the field of view if the light had passed through a physical aperture (usually circular) of a certain diameter in arcseconds. The imaginary aperture is called a **software aperture**. Although this approach sounds quite straightforward there are subtle practical difficulties which can influence the results. For example, the procedures used to estimate the background level, the centre of the optical image, and the shape of the image itself are important.
(b) **Profile fitting:** Also called **point-spread-function** (PSF) fitting, where by PSF is meant the actual recorded shape on the detector of an unresolved point source, this method relies on modelling the image rather than summing over the image. Mathematical curves are 'fitted' to the real data using computer programs until a good match is obtained. This method is much more time-consuming for the astronomer, and for the computer! Various programs are available at major institutes for photometric reductions including such packages as DAOPHOT, ROMAFOT and STARMAN. Information on these software packages is available from NOAO (USA), ESO (Germany) and STARLINK (UK) respectively, or from the original writers.

The use of CCDs for photometric measurements is founded on two basic assumptions, namely

(1) The response of each pixel is a well-defined function of exposure level, optical bandpass, and CCD architecture and control. Considerable engineering effort may be required to fully optimise and stabilise CCD control for any given application.
(2) The incident signal from the astronomical source can be calibrated or transferred to the desired 'standard' system.

For all techniques, whether profile-fitting or apertures, an understanding of the image profile and the centroid of the image is important since centring errors and inappropriate apertures or fitting parameters can lead to systematic effects. There are four other important issues: (1) passband mismatch of the filters — even narrow band filters; (2) red/infrared 'leaks' in the filters (Fig. 8.9(b)) which complicate flat-fielding; (3) the finite opening and closing times of electromechanical shutters; and

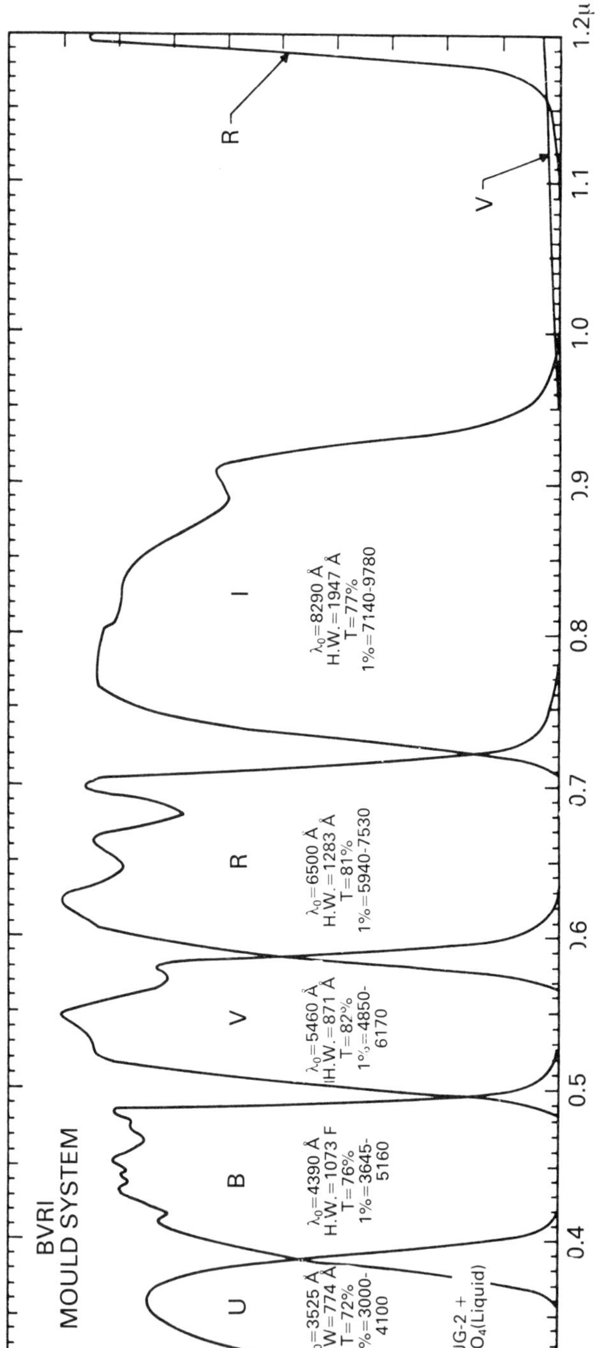

Fig. 8.9 — (a) Standard filter bandpasses used with CCDs.

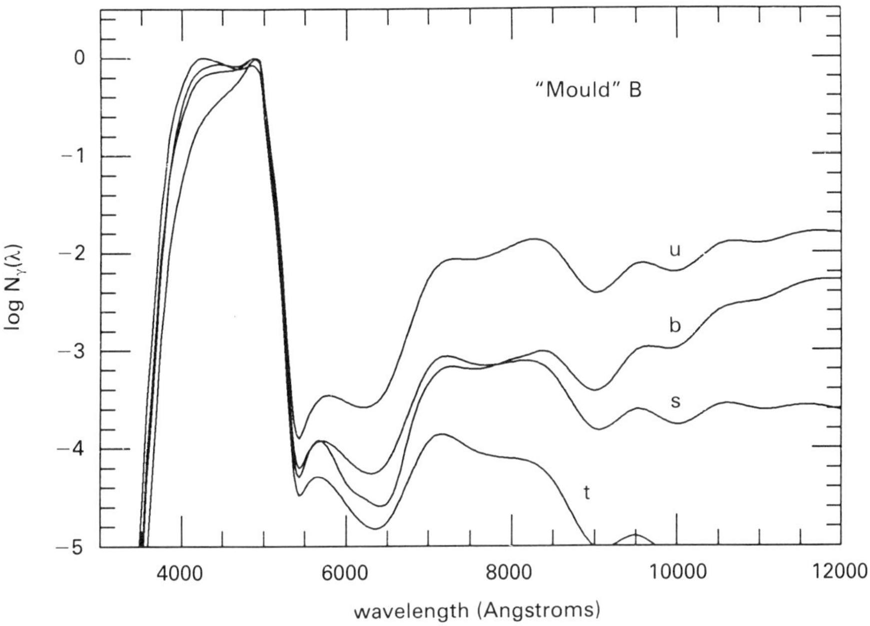

Fig. 8.9 — (b) The effect of (accidental) imperfect blocking of the B(lue) filter is a 'leak' of red photons to which the CCD is very sensitive. The consequence depends on the spectrum or colour of the source; u and b are balanced (filtered) and unbalanced artificial lamps illuminating the telescope dome, s is a typical solar spectrum and t is the twilight sky. Courtesy Peter Stetson.

(4) changes in the atmospheric attenuation or 'airmass' for long on-chip CCD integrations. Items (1) and (2) need to be eliminated by design.

It is usually assumed that the basic CCD calibration procedures of bias subtraction, dark subtraction, flat-fielding and de-fringing have been applied correctly, that the effects are small or 'infinitely well-known'! This is not always true. Extreme care and patience at the telescope is required to ensure that all necessary calibration data are obtained with very high accuracy so as not to limit the photometric determinations, and that nothing has happened 'on the night' which might render reductions difficult, e.g. an accidental ground-loop, detector temperature instability, electronic failures, or inadvertent saturation which may result in latent images in older types of CCDs.

Shutter-timing errors arise from the finite opening and closing time of electromechanical shutters. If T is the requested integration time, then $T+\delta$ will be the actual integration time and, since shutters are normally of the iris type, δ will be greatest near the centre of the CCD. It is straightforward to map δ. For example, take the sum of 10 1-s flat-field exposures and compare this with one 10-s exposure by forming their ratio $r(x,y)$ for each pixel which, provided the flat-field is constant, gives

$$r(x,y)=\frac{10(1+\delta(x,y))}{10+\delta(x,y)}$$

from which we get

$$\delta(x,y)=\frac{10(r(x,y)-1)}{10-r(x,y)}$$

The effect is typically about 7–11 ms with a TI 800×800 CCD (see Fig. 8.10).

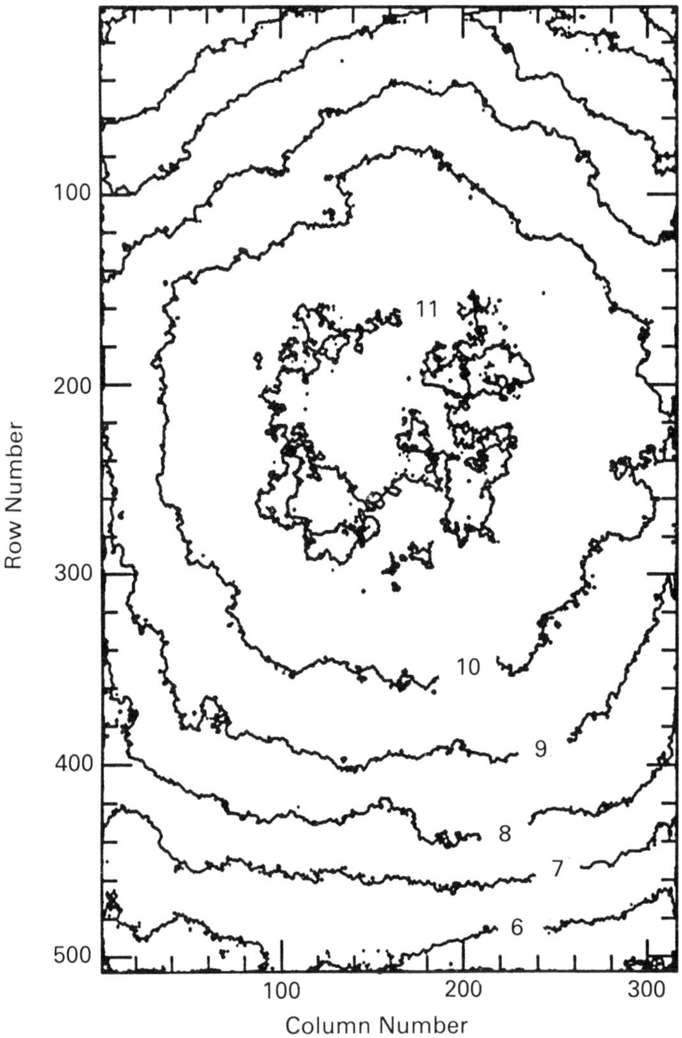

Fig. 8.10 — Shutter exposure time error map. An 'iris' type shutter is open longest in the centre of the array. Courtesy Peter Stetson.

Absolutely critical for photometry is the derivation and/or calibration of any non-linearity, either at high brightness levels or at very low brightness levels. Both laboratory and telescope experiments are usually required to calibrate non-linearity.

Fortunately, CCDs are exceptionally linear when operated correctly, that is, in inverted mode and well within the manufacturer's specification of full-well charge capacity. Some of the current generation of infrared array sensors are not CCDs and are intrinsically less linear. To convert from the CCD filter system to one of the classical systems implies that observations must be made of an object common to both systems — a so-called 'standard star'. It is also preferable to observe the standard star under exactly the same conditions as the programme object.

Modern large telescopes are being used more and more in the 'common-user' mode which means that the average visiting astronomer and graduate students have neither the time nor the training to become completely familiar with device characteristics, and observatory staff are often faced with the dilemma of a working system which must be kept working at all costs and the need to explore a huge ensemble of physical parameters which may or may not affect the end results. The only practical approach is vigilance and scepticism, check and double-check!

Once photometric values are obtained they must be compared or calibrated against well-measured 'standard' sources so that different data sets can be intercompared, or the values related to theoretical predictions. With hindsight, the classical photometric systems such as the UBV system, may not be the most rational for use with CCDs, but such comparisons are inevitable. Photoelectric standard stars must be observed over a wide range of **airmasses** — given essentially by the secant of the zenith angle of the object, which is proportional to the thickness of atmosphere through which the light passes — and cover an appropriate range of spectral types (or temperatures) to obtain colour coefficients. The instrumental magnitudes† are calculated from

$$-2.5 \log_{10} (\text{counts in e/s})$$

and these can be compared on excellent nights with the standard stars to produce a fiducial set. The parameters to be determined are:

(1) **The 'zeropoint'**, i.e. the magnitude corresponding to 1 count/s for a star of zero colour-term above the Earth's atmosphere.
(2) **The 'colour equation'** relating the CCD photometric system to the older photoelectric systems.
(3) **The 'extinction' factor** or light-loss through the Earth's atmosphere per unit airmass; at the zenith the airmass is unity whereas at an altitude of 30° above the horizon the airmass is 2.

The resulting equation for the magnitude (m) of a star is given by

$$m = -2.5 \log_{10} [\text{counts(e/s)}] - \alpha \times (\text{AIRMASS}) + \beta \times (\text{COLOUR}) + \text{ZP}$$

For example, for one of the CCDs (known as RCA1) at the CFHT the following parameters are derived; g=28.6 e/DN R=75 electrons(e)

†The astronomical 'magnitude' system is a logarithmic scale of brightness. It is explained in more detail in Appendix 3.

Filters	U	B	V	I
α	0.26	0.15	0.11	0.04
β	0.12(U−B)	0.21(B−V)	−0.05(B−V)	−0.01(V−I)
ZP	23.70	26.35	26.00	24.95

In this example, large values of β — the colour term — for U and B arise because the passband of the blue filter and the response of the CCD differ from the classical system established by Johnson.

Peter Stetson, who developed the DAOPHOT package, has shown that during a long CCD exposure when the airmass has changed from X_0 at the beginning to X_1 at the end the best estimate is a 'weighted' mean given by

$$\overline{X} = \frac{(X_0 + 4X_{1/2} + X_1)}{6} + \mathrm{O(e)}$$

where $X_{1/2}$ is the airmass at the mid-point of the integration and O(e) is a small error of about 1 part in 10000.

8.7 SPECTROSCOPY

Spectroscopic calibrations proceed much as with imaging. A flat-field is required to remove optical interference effects caused by the near-monochromatic light and variations in the thickness of thinned backside-illuminated CCDs. Bias frames must precede the derivation of the flat-field as before, but now, because of the much weaker signals and longer integrations, dark current may be more serious. Many dark frames may need to be averaged and subtracted from the object frame. The flat-fielded spectra must be 'sky-subtracted'. To do this it is normal to collapse or sum together all rows of the spectrum containing source flux and all rows containing sky spectra. After allowing for the fact that there will probably be more rows of sky spectrum than source spectrum, the pair of flat-fielded, summed spectra are subtracted. Next, the relationship between pixel number and wavelength is determined using arc-lamp (emission-line) spectra containing numerous lines with accurately known wavelengths. Sometimes bright planetary nebulae like NGC 7009 are used instead (Fig. 8.11)! Atmospheric extinction corrections are applied to the intensity and correction to absolute flux levels is accomplished by forming the ratio of the observed spectrum to that of a flux standard.

Spectrographs are large instruments and inevitably, even in excellent designs, some small amount of mechanical flexure may occur as the telescope tracks an object across the sky. For this reason it is advisable to perform flat-fields and arc-lamp wavelength-calibration images at fairly regular intervals.

One feature of CCDs of special use to spectroscopy is **on-chip binning.** In this mode the extension of the spectrum along the length of the entrance slit, and therefore at right angles to the dispersed spectrum itself, can be summed into one

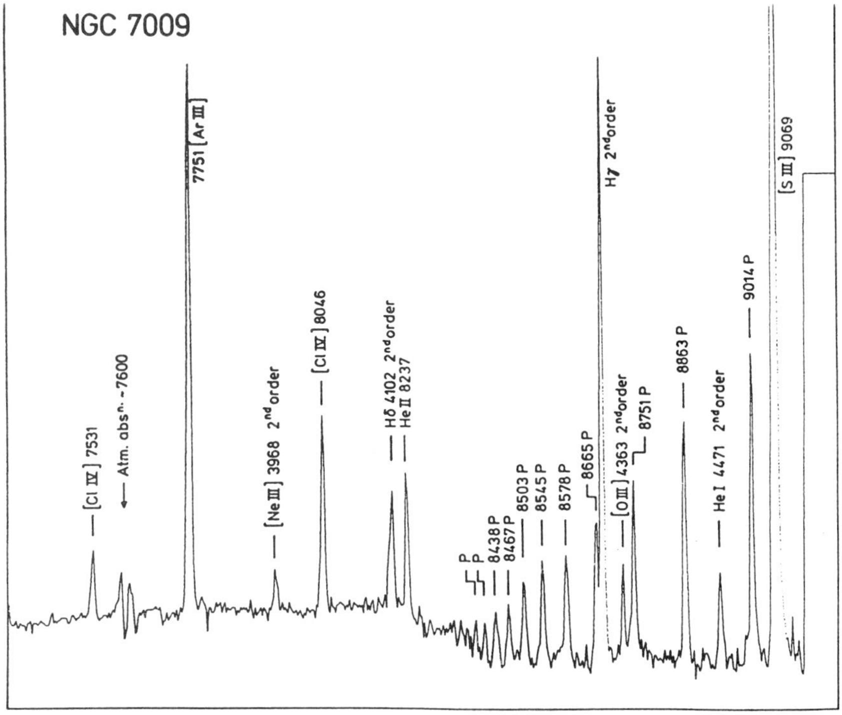

Fig. 8.11 — The emission line spectrum of NGC 7009 obtained with the Cambridge EEV/GEC CCD attached to the AAT spectrograph.

pixel while still on the CCD by a suitable clocking sequence (see section 6.3.2). The readout noise associated with this new binned pixel will be approximately the same as before even though the charges from several pixels have been summed together. Thus, apart from compressing the size of the image frame to be handled by the computer, a gain in signal-to-noise ratio is obtained. Care must be taken not to saturate the CCD by summing too many charges.

Similarly, CCDs used in polarimeters and Fabry–Perot imaging spectrometers are all reduced in a similar way to achieve a basic set of clean frames before any other operations (e.g. determining Stokes parameters, velocities) are carried out. In the case of polarimeters and Fabry–Perot interferometers, a set of flat-fields corresponding to different positions of the polarimeter waveplate or the Fabry–Perot etalon spacing would be needed respectively.

It is usually of some importance in all spectroscopic applications to ensure that the calibration arc lamps illuminate the spectrograph collimator with the same focal ratio as does the telescope.

8.8 COMPARISON OF CCDS

To conclude this chapter a comparison of the properties and operating conditions of a wide selection of CCDs in astronomical use is given.

In Table 8.2 the primary properties of nine types of CCD are given. The

Table 8.2 — Comparison of the main CCD types used in astronomy

	TI	TI	RCA	RCA	GEC	TH-CSF	TEK	PHOT-FORD
Format, pixels	800×800	800×800	320×512	640×1024	385×578	384×576	512×512	516×516
Pixel size (μm)	15	15	30	15	22	23	27	20
Technology	3PCCD	VPCCD	3PCCD	3PCCD	3PCCD	3PCCD	3PCCD	4PCCD
Type	thin	thick	thin	thin	thick	thick	thick	thick
Array size (mm)	12×12	12×12	15.4×9.6	15.4×9.6	8.5×12.7	8.8×13.2	13.8×13.8	10.3×10.3
Quantum efficiency: peak (%), λ peak (μm)	69, 0.55	50, 0.6	70, 0.62	85, 0.50	42, 0.74	38, 0.70	54, 0.60	—
Range (μm)	0.33–1.1	0.33–1.03	0.33–1.03	0.33–1.03	0.35–1.02	0.4–1.02	0.35–1.03	—
Coated/UV flashed	yes	no	no	UV enhanced	yes	no	no	—
Read noise (e^-)	7	25–150	35–55	50	5–7	5–10	8	6
Dark current (150 K) e^-/pix/min	<0.07	<0.07	<0.2	1.0	<0.07	<1.5	0.1	<0.07
Full well capacity (e^-)	50 000	100 000	650 000	110 000	200 000	170 000	500 000	320 000
Dynamic range	7000	700–4000	12 000	2200	33 000	22 000	62 500	53 000
Sensitivity (μV/e^-)	0.35	0.57	0.1	0.1	1.1	1.0	0.5	0.8
Charge transfer								
horizontal	good	good	variable	good	good	good	good	good
vertical	good above 30–40 e	good above 100 e	good	good	good above 25 e	good above 100 e	good	—
V-H threshold	—	—	0	0	300 e	400 e	—	—
Traps, residual image effects	few, no	—	some, yes	some, yes	some, yes	few, yes	some	—
Cosmic rays (event/cm²/min)	low	—	5.5	—	2.3	2.8	>5	—
Uniformity	fair	good	variable	variable	good	very good	—	—
Fringes	yes	none	severe >0.7 μm	yes	none	none	none	—

information in this table is meant to be representative rather than specific since there is considerable variation from CCD to CCD within the same type. However, all of the tabulated data were obtained from published results pertaining to particular astronomical systems. Since different groups evaluated different CCDs there is some incompleteness in this table.

REFERENCES

Argyle, R. W., Mayer, C. J., Pike, C. D., and Jorden, P. R. (1988) *A user guide to the JKT CCD camera,* Royal Greenwich Observatory.

Christian, C. (1987) *CFHT CCD user's manual,* Canada–France–Hawaii Telescope Corporation.

Hall, P., Mackay, C. D. (1984) Faint galaxy number-magnitude counts at high galactic latitude, *Mon. Not. R. astr. Soc.,* Vol. 210, 979–992.

Janesick, J. R., Hynecek, J. and Blouke, M. M. (1981) A virtual phase imager for Galileo. In Solid state imagers and astronomy, *Proc. Soc. Photo-Opt. Instr. Eng.* Vol. 290, 165–173.

Schoening, W. (1988) *Operation of the CCD direct imaging camera for the 2.1 meter telescope,* Kitt Peak National Observatory, National Optical Astronomy Observatories.

Stetson, P. B. (1987) DAOPHOT: A computer program for crowded-field stellar photometry, *Publ. Astron. Soc. Pacific,* Vol. 99, 191–222.

Willmarth, D. (1988) *Instrumentation operation manual Coude spectrograph–CCD camera,* Kitt Peak National Observatory, National Optical Astronomy Observatories.

9

Electronic imaging at other wavelengths

9.1 INFRARED

It has been intimated several times in this book that CCDs have revolutionised optical astronomy, but as we shall see, that same pronouncement would seem like a gross understatement for **infrared astronomy**!

In the past six or seven years there has been a tremendous explosion of activity in this discipline. Solid-state electronic imaging systems in computer-controlled instrumentation have recently had an astonishing impact on ground-based near- and mid-infrared astronomy. The analogy with optical astronomy in the early seventies when silicon CCDs appeared is self-evident. Of course, the dust has not yet settled — an appropriate metaphor for infrared astronomy — but one cannot help experience a sense of *déjà vu* as the number of pixels in the new **infrared array** devices are increased, and experience in their usage is gained. Much of the same semiconductor physics, electronic control schemes and computer image processing facilities as already discussed in this book can be applied. Therefore, fluid though this 'new era' in infrared astronomy may be, it rightfully deserves a place in this book.

9.1.1 Optical or infrared — the elusive boundary

For many years the infrared part of the spectrum was considered to be the region just beyond the red limit of sensitivity of the human eye, at a wavelength of about 0.7 μm. With the advent of semiconductor silicon technology for imaging applications at very low light levels, optical astronomy extended its territorial claims to about 1.1 μm — the cut-off wavelength for detection of light imposed by the fundamental band-gap of silicon (section 4.2). Then came along lower band-gap solid-state arrays enabling astronomers to image at wavelengths as long as 10 μm. So where is the optical–IR boundary? In reality it occurs at a wavelength of about 2.5 μm. At about this wavelength there is a marked and fundamental change in the nature of the 'background' light entering the camera system. For wavelengths shorter than about 2.5 μm the background light comes mainly from faint emission produced by Sun-induced photochemical reactions in the Earth's upper atmosphere, whereas at wavelengths longer than 2.5 μm the dominant source of background radiation is the infrared or **thermal** emission from the atmosphere, telescope and optical components themselves! This emission is very strong indeed, as can be seen from Fig. 9.1.

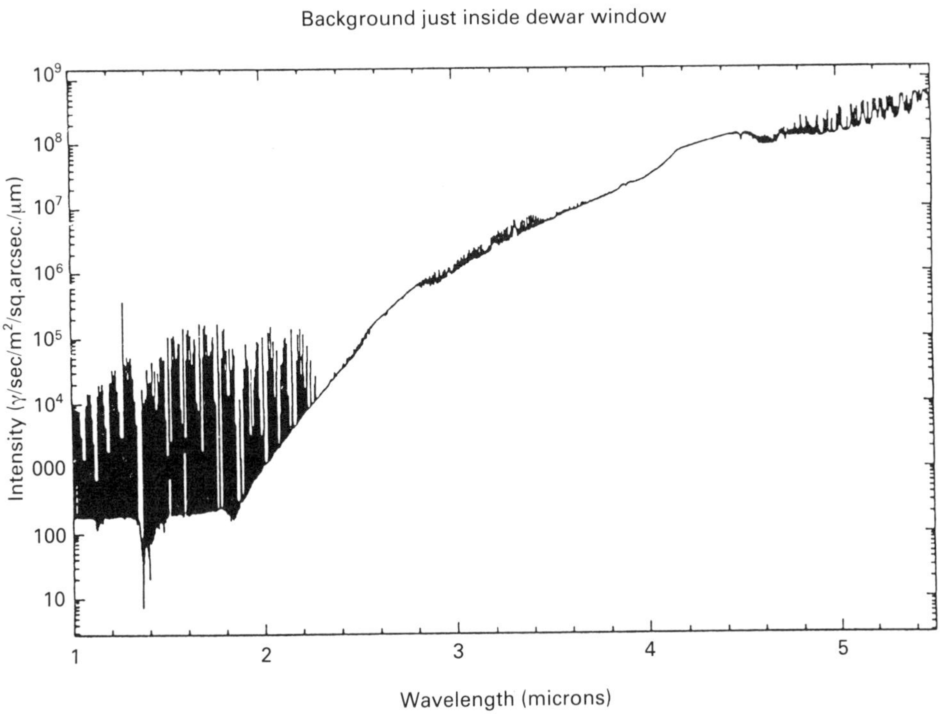

Fig. 9.1 — The rapid increase with wavelength in the number of background photons entering an infrared camera from a small patch of sky. This plot is based on a model developed by Mark McCaughrean of the UKIRT infrared camera team. Backgrounds in the infrared are hundreds to thousands of times larger than at optical wavelengths.

For practical reasons however, associated with the ubiquity of the silicon CCD, the dividing line is usually set at 1.1 μm.

For our purpose we can define the domain of infrared astronomy as extending to longer wavelengths from a wavelength of about 1.1 μm, at which silicon CCDs become insensitive. **Near**-infrared is generally taken to be the interval from 1 to 5 μm, **mid**-infrared extends from 5 to 20 μm and **far**-infrared stretches from 20 to 200 μm. Wavelengths longer than about 350 μm are nowadays referred to as **sub-millimetre**, and although sub-millimetre astronomy is closely allied with infrared wavelengths in terms of the objects and regions of space which are studied, many of its techniques are more akin to those of radio astronomy. The progression in wavelength from near- to far-infrared is essentially a progression to cooler and cooler objects and regions of space; eventually, in the microwave region, we come across the cool remnant of the Big Bang itself at a mere 3 degrees above absolute zero. The hotter an object the more electromagnetic energy it radiates at all wavelengths, but the more the peak amount of energy occurs at shorter (toward the ultraviolet) wavelengths.

Amazingly enough, infrared astronomy is not a new subject born out of the high technology era of the post World War II years ... like radio astronomy. Infrared astronomy had an early, albeit somewhat accidental, origin when, as early as 1800 in a series of papers, Sir Willian Herschel noted that a thermometer placed just beyond the red end of the optical spectrum of the Sun indicated an increase in temperature. Herschel called these unseen radiations **calorific rays** and proved that they were refracted and reflected just like ordinary light. In a way, his discovery of an unseen form of radiant energy heralded the beginning of the understanding of the existence of a remarkable, continuous form of radiant energy which, thanks to the brilliant work of the Scottish physicist James Clerk Maxwell (1865), we now know to be electromagnetic radiation. Extending from gamma-rays through optical to radio, the only distinguishing feature is the 'wavelength' of the radiation. The region of the spectrum discovered by Herschel was only the tip of the iceberg.

Why is infrared astronomy important? Observations at infrared wavelengths offer three important advantages over optical measurements, namely,

(1) Improved seeing. At longer wavelengths the distortion of starlight caused by the Earth's atmosphere decreases. It is therefore easier to approach the ultimate or **diffraction-limited** performance of the telescope. This is an important consideration for the next generation of large telescopes and for very high-resolution imaging.
(2) Less attenuation by interstellar dust. For example, the **extinction** of light from the centre of our Milky Way galaxy amounts to about 30 **magnitudes** at ordinary visible wavelengths, that is, the light is dimmed 1 million million times. Even at the relatively short infrared wavelength of 2.2 μm however, the extinction is reduced to only 3 magnitudes; a dimming of only 16 times. Quite an improvement! Near-infrared wavelengths open up a host of observations which cannot be pursued at all by optical astronomy because the objects are obscured by **interstellar dust**. For example, regions of star formation, or the nuclei of active galaxies.
(3) Highly **red-shifted** objects become infrared sources. Distant objects in the universe are receding from us at velocities which are a large fraction of the speed of light. The Doppler effect from this motion displaces the radiant energy emitted at optical and ultraviolet wavelengths towards much longer wavelengths. The redshift (z) is expressed as the 'change in wavelength divided by the original wavelength' so for objects with z greater than 2.0 optical emission becomes observable only in the near infrared. Quasars are known with redshifts greater than 4, and recently, partly due to the progress in infrared instrumentation, galaxies have also been discovered at redshifts greater than 3 (Lilly,

1988)†. The light from such distant sources started its journey to Earth when the universe was less than one-sixth its present age — comparatively young!

Therefore, in summary, infrared wavelengths provide us with a unique way of 'seeing' into optically opaque stellar nurseries for a peep at newborn stars, make it possible to study the enigmatic content of the very centre of our own Milky Way Galaxy and can also provide the ultimate tool for studies of the early universe.

9.1.2 Infrared detectors

The many possible ways of detecting infrared radiation are reviewed in detail in a number of classic texts.

Infrared detectors are classified as **photon** detectors or **thermal** detectors. In photon detectors, individual incident photons interact with electrons within the detector material. For example, if a photon frees an electron from the material the process is called **photoemission**; this is how a photomultiplier tube works and it is also the underlying principle of the Schottky-Barrier diode except that the free electron escapes from the semiconductor into a metal rather than into a vacuum. When absorption of photons increases the number of charge carriers in a material or changes their mobility, the process is called **photoconductivity**. These devices are usually operated with an external voltage across them to separate the photogenerated electron–hole pairs before they can recombine. If absorption of a photon leads to the production of a voltage difference across a junction between differently 'doped' semiconductors the process is known as the **photovoltaic effect**. The interface is normally a simple **pn junction** which provides an internal electric field to separate electron–hole pairs. More often the pn junction or **diode** is operated with an externally applied voltage to cause a reverse bias. In thermal detectors, the absorption of photons leads to a temperature change of the detector material, which may be observed as a change in the electrical resistance of the material as in the **bolometer**; the development of a potential (voltage) difference across a junction between two different conductors, as in the **thermopile**; or a change of internal dipole moment in a temperature-sensitive ferro-electric crystal, as in a **pyroelectric** detector.

By cooling these detectors and their local environment with **liquid nitrogen** and **liquid helium** two important results are achieved. Firstly, the heat or infrared emission from all the filters, lenses and support structure surrounding the infrared detector is eliminated. Secondly, operating the detector at a low temperature greatly reduces the detector's own thermal background or 'dark current' and results in a huge increase in sensitivity.

Despite its early start, the major breakthroughs in infrared astronomy did not come until the 1950s when solid-state photoelectric semiconductor detectors were perfected. The lead sulphide cell (see Chapter 4) was used by Harold Johnson, the father of photoelectric photometry, and by Neugebauer and Leighton for a huge

† Simon Lilly's discovery that the object known as 0902+34 was a galaxy at the astonishing redshift of 3.395 is a splendid example of what modern astronomical instrumentation can achieve. The RCA1 CCD on the CFHT was used for the optical imaging, a TI 800×800 CCD for the University of Hawaii's faint object spectrograph, and the SBRC 62×58 array in the UKIRT IRCAM for infrared imaging.

2-μm survey of the sky — albeit at the very low resolution of 4 arcminutes. For wavelengths beyond 3 μm something else was needed. One important event was the invention of the gallium-doped germanium bolometer in 1961 by a young physicist called Frank Low working for the Texas Instruments Corporation. This detector opened up much longer wavelengths to astronomers, and with the aid of rocket, balloon and airplane surveys and, later, the development of large ground-based infrared observatories such as the UK IRT, the NASA Infrared Telescope and the Wyoming Infrared Observatory, the subject 'took off'. David Allen recounts many of the events of that exciting period in his book *INFRARED; the new astronomy*. Lead sulphide detectors were later superseded by indium antimonide photodiodes for the near-infrared, while doped silicon photoconductors and improved bolometers became prevalent at longer wavelengths.

Thirty years later Frank Low played another key role in the highly successful Anglo-American-Dutch Infrared Astronomical Satellite (IRAS) mission, which gave astronomers their first deep all-sky survey of the infrared sky.

With the launch of IRAS in 1983, the science of infrared astronomy took another quantum leap. Unfortunately, IRAS had a finite lifetime in space because it was a cryogenic (or cooled) telescope (23 in or 0.57 m in diameter) and once the on-board supply of liquid helium coolant was exhausted then the telescope and its detectors warmed-up and thereby lost their sensitivity. So successful was IRAS that follow-up space missions involving 'observatory class' satellites are planned by both ESA and NASA. The European project is called ISO (Infrared Space Observatory) and the American project is called SIRTF (Space Infrared Telescope Facility) — see *Satellite astronomy* by John K. Davies in this series for more details. The current launch-dates for these missions are in the 1990s. Due partly to the small size of these telescopes, partly to their limited lifetimes and partly to the fact that they must use 'old' detector technology — i.e. the detectors available now, rather than those which may be available five years from now — it is very important to pursue infrared astronomy from (very much larger) ground-based telescopes.

Easier said than done! For a number of reasons it is by no means easy to perform infrared astronomy from the surface of the Earth. To begin with, the atmosphere is not uniformly transparent at infrared wavelengths as it is in the visible part of the spectrum. Water vapour and carbon dioxide do an efficient job of blocking out a lot of infrared radiation, which for life on Earth is just as well, of course! Happily, there are breaks or 'windows' in this safety blanket which improve if the observatory is at a high altitude and, moreover, there are very important astronomical features to be seen at the wavelengths corresponding to the atmospheric windows.

But, as we have already hinted, there are other difficulties! Everything on the Earth is warm, even the air and the telescope, and it is all a lot closer than the stars! Objects with a temperature of 300 K (27°C) emit a spectrum of radiant heat energy which has a peak value at a wavelength of about 10 μm and is still very strong even at a wavelength as short as 3 μm. In other words, there is an immensely bright background light at infrared wavelengths and the faint celestial sources must be extracted from this glare. Early infrared astronomers found a solution to the problem of a bright sky . . . it is called 'chopping' and it involves measuring first the object plus bright sky and then changing quickly to a nearby piece of bright sky with no object in view, and forming the difference. Of course, the question arises 'where to look for a

Table 9.1 — Infrared windows in the Earth's atmosphere

Wavelength (μm)	Designation	Width (μm)
1.25	J	0.30
1.65	H	0.30
2.22	K	0.42
3.50	L	1.05
4.8	M	0.4
11.0	N	1.0
20.0	Q	5.0

blank piece of sky?'. Well, one could look at a picture. And that brings us to the third difficulty which gound-based infrared astronomy faced until recently, namely, the nature of the available detector systems. The kinds of infrared detectors available to astronomers, although very sensitive, were not really conducive to the making of extensive, fine-detail infrared pictures.

Until very recently, infrared 'images' of celestial objects were really 'maps', not pictures. Each map was built up piecemeal by laboriously scanning the infrared scene point-by-point, back and forth in a sweeping pattern, called a **raster**, with a single sensitive detector having a small angular field of view on the sky. Usually, the field of view was a circle no smaller than 2 or 3 arcseconds. This is rather like using a photomultiplier tube, which is a single detector, to map out a scene which could be imaged in an instant with a snapshot by a CCD camera or a photographic plate! Alternatively, imagine using just *one* pixel in a CCD array of 500×500 pixels and then moving the image so that a different part of the scene fell on your pixel! Scanning was done by moving the entire telescope or some major optical component such as the secondary mirror. As a result, errors due to small mis-pointing of the telescope, imprecise registration of individual parts of the map, and slow changes in the background signal level — due to temperature changes, variation in the behaviour of the atmosphere and variations in the detector system — all introduce errors and artifacts, Moreover, even if the circular patches of sky are carefully overlapped in the mapping process it is impossible to record fine detail on the scale of the 'seeing' which is about 1 arcsecond.

Infrared astronomy has *never* had the equivalent of the photographic plate or the TV tube. Little wonder then that when infrared-sensitive, solid-state semiconductor analogues of the CCD came along, infrared astronomy received a staggering boost.

9.1.3 The infrared array revolution — *déjà vu*

Some infrared arrays make use of CCD technology but, as we shall see, several of the most sensitive devices currently are not actually CCDs and do not make use of the 'charge-coupling' idea. To generate an infrared image an infrared array detector must:

- convert radiation into electrical charge by means of the **internal photoelectric effect** — as with a CCD;
- store the electrical charge at the site of generation, i.e. in a picture element (pixel);
- transfer the charge on each pixel to a single (or a small number of) outlets; this is called **multiplexing**;
- enable the charges to be removed sequentially as a voltage which can then be converted or **digitised** into a number for storage in a computer.

Because of the fundamental atomic structure of silicon, ordinary CCDs cease to be sensitive at wavelengths longer than 1.1 μm, and so the infrared region of the spectrum is eliminated. Yet, each of the steps and stages in infrared detection are so similar to those employed in an optical silicon CCD that it was natural to attempt to manufacture an entire CCD from some of the other known semiconductor materials which, because they had smaller energy band-gaps between the valence and conduction bands, were sensitive to infrared radiation and, in principle at least, could be made into CCDs. Texas Instruments were among the first to try. An entire CCD was manufactured out of a material called cadmium–mercury–telluride (CMT) in the UK, but mer–cad–tel (HgCdTe) in the USA, as early as 1974. Other companies tried indium antimonide (InSb), and still others tried silicon which had been specially treated or doped, all with varying degrees of success. The problems of manufacture turned out to be very difficult due to the limited experience in processing and purifying these materials compared to the extensive technology base which had already grown-up around silicon. Finally, a different approach was taken . . . the **hybrid**.

In the hybrid infrared array, the functions of **detecting** infrared radiation and then **multiplexing** the resulting electrical signal were separated, with the latter task going, of course, to silicon. At present then, infrared arrays are like 'sandwiches' (as shown in Fig. 9.2) in which the upper slab is the infrared sensor (InSb, HgCdTe; or doped silicon — usually written as Si : XX with XX replaced by the symbol for the dopant actually used (see Table 9.2)), and the lower slab is a silicon multiplexer of some sort. Both slabs are provided with a grid of electrical connections in the form of tiny raised sections — referred to as 'bumps' — of a soft electrical conductor called indium. The slabs are pressed together to enable the bumps to mate. What to use as the multiplexer? CCDs would be quite acceptable since the potential wells provide the required charge storage. Charge injection devices (CIDs) can also be used or, alternatively, a special array of 'switches' made from metal–oxide-semiconductor field-effect transistors (MOSFETs) in microscopic form can be used to access the signal from each infrared detector (whether photodiode or photoconductor). Charge storage may occur on the junction capacitance of the infrared sensor itself or, in more recent designs, on a separate storage capacitor associated with the silicon circuitry.

Many forms of hybrid devices were constructed during the period 1974–1984, due mainly to the extreme importance of the infrared for military applications — heat seeking weaponry and battlefield smoke conditions etc. — but even by 1982 when I carried out a survey, very few infrared array devices were actually for sale and virtually none of them had the requisite performance expectations for low-back-

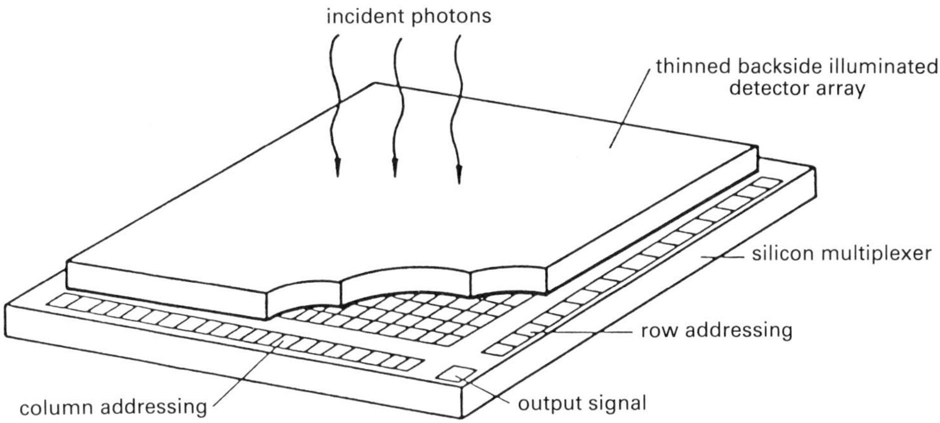

Fig. 9.2 — The hybrid or sandwich construction of infrared array sensors. On the upper face is an infrared semiconductor appropriate for the wavelength region of interest, while beneath is a silicon array — which may be a CCD — used simply to extract the photo-generated electronic charges from the infrared sensor.

Table 9.2 — Extrinsic semiconductors

Base material : Impurity	Long wavelength limit (μm)
Germanium (Ge) : Au	8.27
: Hg	13.8
: Cd	20.7
: Cu	30.2
: Zn	37.6
: B	119.6
Silicon (Si) : In	8.0
: Ga	17.1
: Bi	17.6
: Al	18.1
: As	23.1
: P	27.6
: B	28.2
: Sb	28.8

ground astronomical applications. Various workers in the field had reported technical evaluations of a few devices and some even obtained astronomical images.

Craig McCreight of the NASA Ames Research Center led in-house tests and coordinated a major programme involving a number of other groups with NASA financial support. Among the earliest champions of infrared arrays were John Arens and co-workers at the Goddard Space Flight Center where a 32×32 bismuth-doped silicon array at a wavelength of 10 μm (only a 2×10 pixel area was functioning) was tested as early as 1979. A 32×64 platinum silicide Schottky Barrier array was used by researchers at the Kitt Peak National Observatory and at the NASA Infrared Telescope Facility in collaboration with the US Airforce; those early results were disappointing but not disillusioning. Infrared CIDs for various wavelengths and long linear photodiode arrays using individual, switchable MOSFET multiplexers were tested and described in the technical literature. The most serious drawbacks with many of these devices were low sensitivity, high readout noise, and poor stability. The most hopeful sign came from tests of a 32×32 array of indium antimonide detectors hybridised to a silicon CCD. Testing of this device was done by Judith Pipher and Bill Forrest of the University of Rochester (USA) using a 'reject' device loaned to them by a former colleague who was now employed by Santa Barbara Research Center (SBRC) in California. This kind of situation occurred quite often. Experimental arrays developed with military applications in mind were loaned by manufacturers to just a few astronomical groups with good contacts!

In the UK, beginning officially in 1982 with my appointment as project scientist by the Astronomer Royal for Scotland Professor Malcolm S. Longair, the Royal Observatory, Edinburgh, acting on behalf of the UK/Dutch astronomical communities, undertook a major committment to stimulate work in this field and to produce a series of common-user infrared array systems (cameras and spectrometers) for astronomy. Our aim was to avoid the rut of laboratory testing of 'loaned' devices and to encourage commercial availability of infrared detectors at about the same level as had been achieved for silicon CCDs by then. Contact was established with Dr Alan Hoffmann of SBRC who had loaned the 32×32 CCD to Rochester. Realising how serious our approach was, he broached the subject of an astronomy programme with SBRC management. About one year later, two of SBRC's marketing personnel Carol Oania and Dick Brody, contacted us to initiate technical discussions and to learn more about the astronomy market. In March of 1984, with the magnificent background setting of a spectacular 12-mile lava flow erupting from Mauna Loa, SBRC and ROE representatives met in Hilo, Hawaii to define the technical specification and costs of a new infrared array device suitable for astronomy. During the months that followed, agreements were reached, funding approved and orders placed. At an infrared detector meeting in Wyoming in May of 1984, SBRC were able to extend their list of potential customers. At about the same time, through contracts being developed for second-generation space telescope instrumentation, several groups of US astronomers (in Arizona, Chicago and Hawaii) had obtained access to new mer–cad–tel arrays from Jon Rode of Rockwell International. French and German sources of infrared array technology, for space applications in particular, were also recognised. So excitement was high!

The first ROE infrared array instrument project commenced in June 1984 with the formation of a project team led by the author, and our first endeavour came to a climax in September 1986 when a cryogenic infrared imaging system, named IRCAM, incorporating the first of a new generation of indium antimonide detector arrays from the SBRC, was delivered to the 3.8-m UKIRT in Hawaii. The SBRC detector has an array of 3596 pixels. On the morning of October 23, 1986, in broad daylight, IRCAM obtained its first infrared image of a cosmic source (see colour plates). In the following year IRCAM was used from January to August on a 'trial' basis by staff and visitors, and then in September 1987 it became available to the entire scientific community.

The new infrared camera system was the result to two years of hard work by a core team of 10 scientists and engineers (the engineering group was led by Timothy Chuter) covering a wide range of disciplines, including electronics, mechanics, cryogenics, optics and software. Once again we have a very good example of how major scientific advances are intimately associated with new technology. IRCAM was also the first of a new generation of infrared instruments in which much more emphasis than ever before was placed on automation, computer-control, engineering reliability, and ease of use through 'user-friendly' software.

The advantages of the new array detectors can be summarised as follows:

- large numbers of elements compared to the old single-element detectors, therefore very time-efficient;
- high resolution, usually seeing-limited, but new techniques permitting diffraction-limited resolution have also been successful;
- no scanning, therefore much less prone to errors of measurement due to mis-registration of scans;
- very high sensitivity, partly the result of the small detector size.

By March 1987, the first astronomical results from many commercially available infrared arrays had begun to pour in. Short-wavelength mer–cad–tel was emerging as the means for optical telescopes to extend their capability to 2.5 μm and longer wavelength arrays were proving better than anyone dared hope. Typical formats were only 64×64, but plans for 128×128 and even 256×256 arrays were being laid (Fig. 9.3). A key moment in infrared astronomy was a 'workshop' on infrared array detectors organised by Eric Becklin and Gareth Wynn-Williams and held in Hilo, Hawaii in March 1987. That workshop attracted over 200 participants, and the 62×58 InSb array manufactured by SBRC emerged then as the most sensitive device (Fig. 9.4). To some of those attendees who had straddled the apparent divide between optical and infrared astronomy such as me, it was like history repeating itself. The euphoria was similar to the Harvard–Smithsonian meeting on optical CCDs only six years before.

Experience gained in the use of silicon CCDs has been of immense value in accelerating the introduction of infrared arrays. Electronic drive systems and noise reduction schemes are similar, device physics is closely related, and flat-fielding strategies, data reduction steps, image processing techniques and software are all identical. The equations used for signal-to-noise calculations in Appendix 4 are also valid for infrared arrays.

Examples of recent images obtained with infrared arrays are shown in Figs 9.5 and 9.6 and in the colour section. Perhaps the most astonishing result is the 'depth'

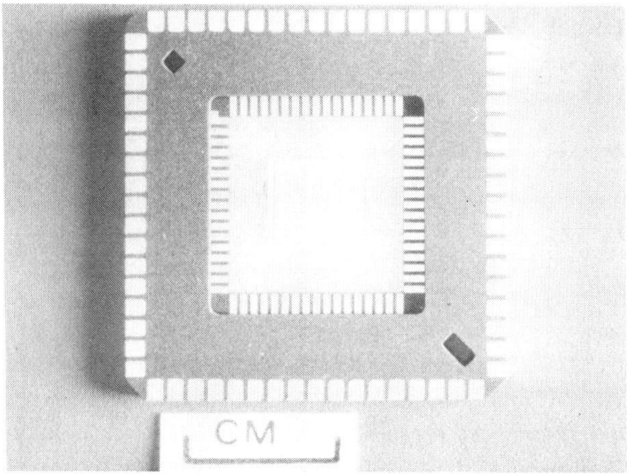

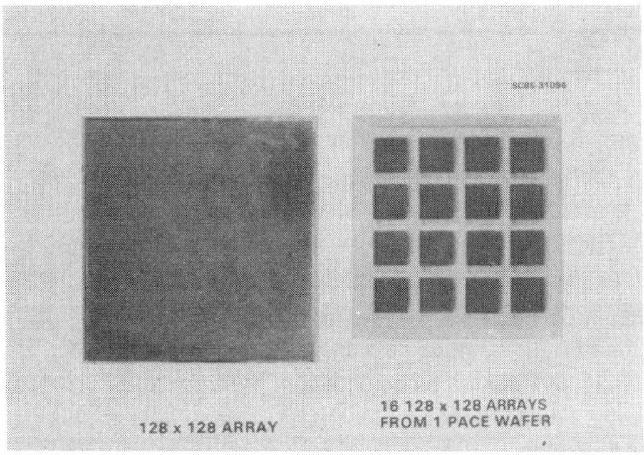

Fig. 9.3 — A photo of the SBRC 62×58 InSb array and a photo of the Rockwell International 128×128 HgCdTe array.

already being reached by infrared images. In a survey search for protogalaxies at high redshifts with the 3.8-m UKIRT, objects (distant galaxies) as faint as 21st magnitude at the infrared wavelength of 2.2 μm were detected (see Fig. 9.7), which corresponds to visual magnitudes fainter than 25.0! Even though the new infrared arrays are still relatively small — just like the first silicon CCDs in the mid-seventies — it is already clear that infrared astronomy too will benefit from the Semiconductor Age.

9.2 ULTRAVIOLET

Wavelengths shorter than the violet end of the optical spectrum (0.35 μm) are called **ultraviolet** (UV). Unlike the circumstances at infrared wavelengths, the Earth's

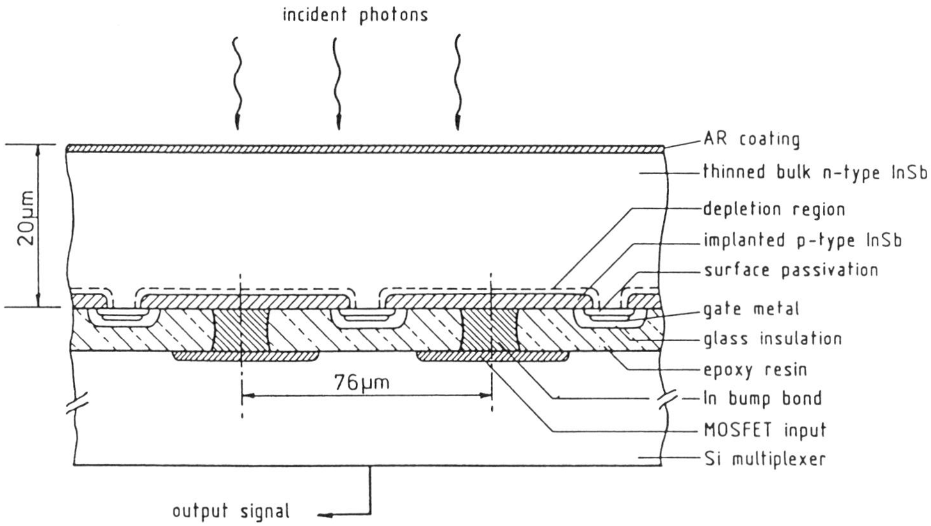

Fig. 9.4 — Cross-section of the SBRC InSb infrared array.

atmosphere is completely opaque for wavelengths shorter than 0.30 μm — even for high altitude sites like Mauna Kea, Hawaii. True UV observations, therefore, must be carried out from sounding rockets or spacecraft (see section 10.2). Several excellent and highly successful UV space programs such as Copernicus and IUE (International Ultraviolet Explorer) have already opened up this vast field. The next major advance is expected with the Hubble Space Telescope. Due to the considerable progress on thinning, coating and QE-pinning of CCDs, and with the 1-electron readout noise CCD firmly on the horizon, CCD detectors are becoming very important for UV astronomy, and future UV space observatories will almost certainly include UV-sensitive CCDs.

There are, however, a variety of observing modes, such as high-dispersion spectroscopy in the violet and UV, for which the finite readout noise of currently available CCDs, the presently limited UV response, and the slower data access time of CCDs makes them less satisfactory than **photon-counting detectors (PCDs)** employing good UV photo-emissive photocathodes and some form of light intensification.

The developers of the first generation of panoramic PCD systems, of which the Boksenberg IPCS is an example, employed large, and often heavy, magnetically-focussed image-tube chains and electron-beam-scanning TV tubes for detecting and reading out the 'events' or light emissions from the output phosphor screen of the image tube. Many of the limitations of these early systems, and hence one of the main gains for the simpler solid-state CCD-type detectors, were due to these massive components. For instance, mechanical instabilities due to size and weight, distortion and jitter of the TV image due to **beam-pulling** caused by the charges associated with a bright image, and the varying influence of the Earth's magnetic field on the electro-optical image-tube stages as the instrument moved with the telescope.

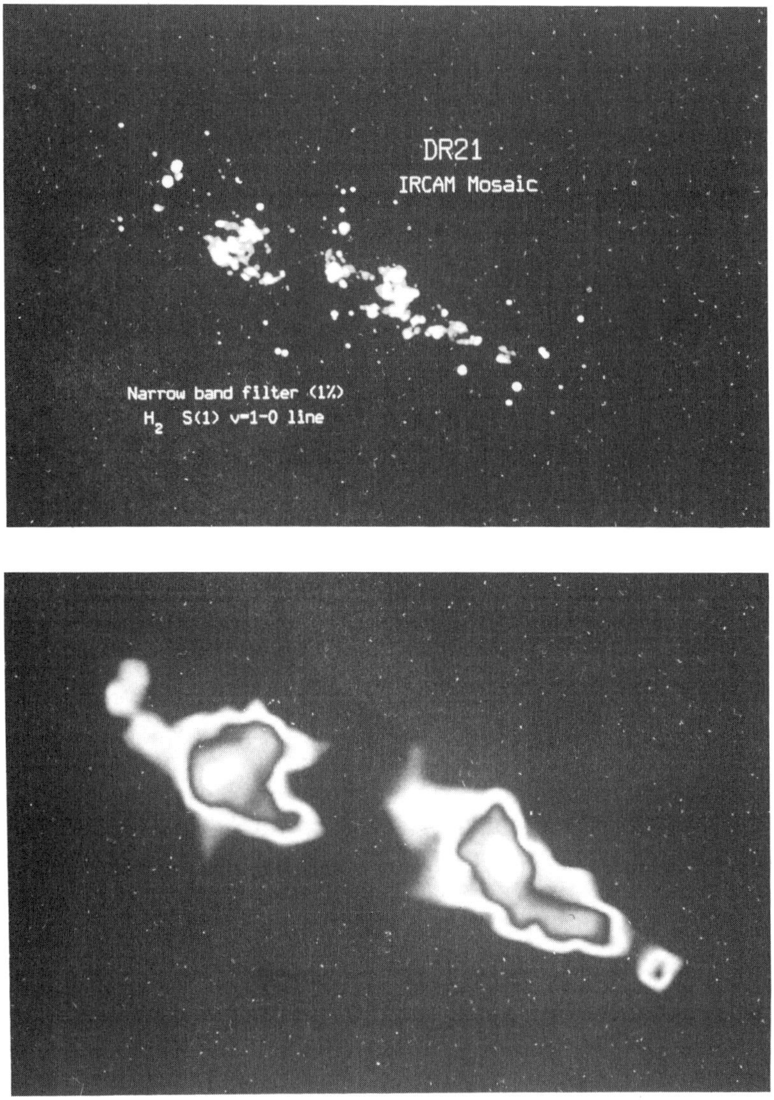

Fig. 9.5 — An infrared image of a region of star formation obtained by the old 'mapping' technique compared to one obtained with the IRCAM infrared camera. UK Infrared Telescope photo, courtesy R. Garden.

The current generation of PCDs have begun to address these problems and now employ compact and mechanically stable readout schemes — using CCDs of course — and modern methods of intensifying signals such as the **microchannel plate** (MCP) shown in Fig. 9.8. Table 9.3 summarises some of the principal panoramic photon-counting detector systems currently under development.

The reader is referred to other sources such as the *Proceedings of the 1987 Santa Cruz Workshop on Instrumentation for Ground-based Optical Astronomy* and

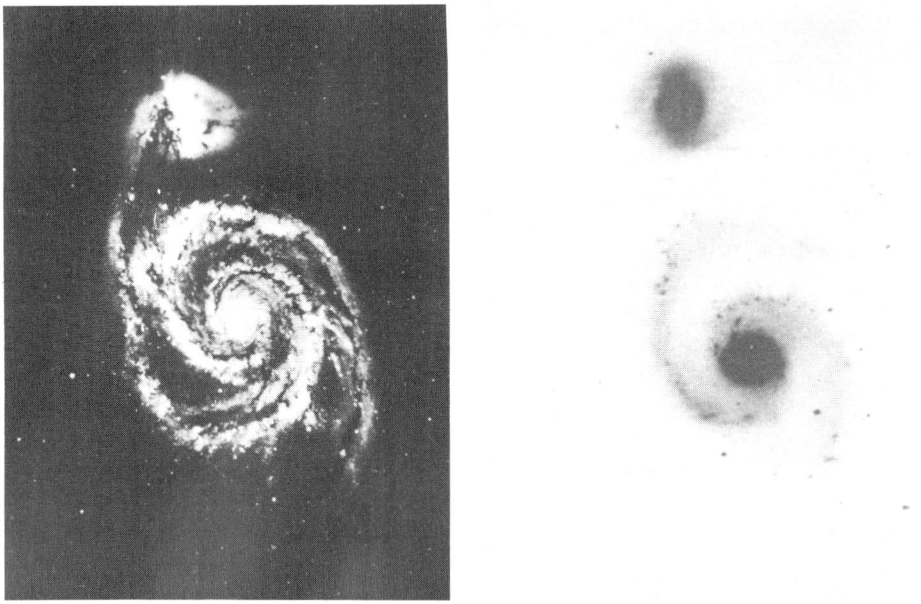

Fig. 9.6 — An optical (left) and an infrared image of the galaxy M51 obtained by Gillian Wright, Mark Casali and Dolores Walther with the UKIRT infrared camera.

Vol. 74 of *Advances in Electronics and Electron Physics* for reviews of the state of the art in PCDs. Very few of the systems mentioned in Table 9.3 have been used seriously at a telescope, but it is fair to say that they have a niche, especially at ultraviolet wavelengths.

Large 40-mm MCP intensifiers optically coupled by a demagnifying fibre-optic boule to a CCD detector are under development in the UK with the aim of replacing the bulky, four-stage magnetically focussed cascade intensifier and Plumbicon TV camera of the original IPCS. The new machine, which has already undergone some astronomical trials, is so compact (Fig. 9.8) as to bear little resemblance to its famous progenitor!

An alternative approach to a UV imaging detector is to accelerate the electrons released from the UV-sensitive photocathode — and certain opaque photocathodes can actually have QEs of 50% in the far UV — to high energies and allow them to impact a CCD. Such a structure is known as an **electron-bombarded** CCD or EBCCD, as shown in Fig. 9.9(a). Among the proponents of this approach is John Lowrance of Princeton Scientific Instruments and George Carruthers of the Naval Research Laboratories. Examples of far UV laboratory spectra with the EBCCD are shown in Fig. 9.9(b).

Since the output of a PCD is stored in computer memory and on magnetic disk as a two-dimensional image, data analysis can utilise much of the same software as a CCD. Indeed, software packages such as FIGARO, mentioned in Chapter 8, were developed from photon-counting systems and then extended to CCDs.

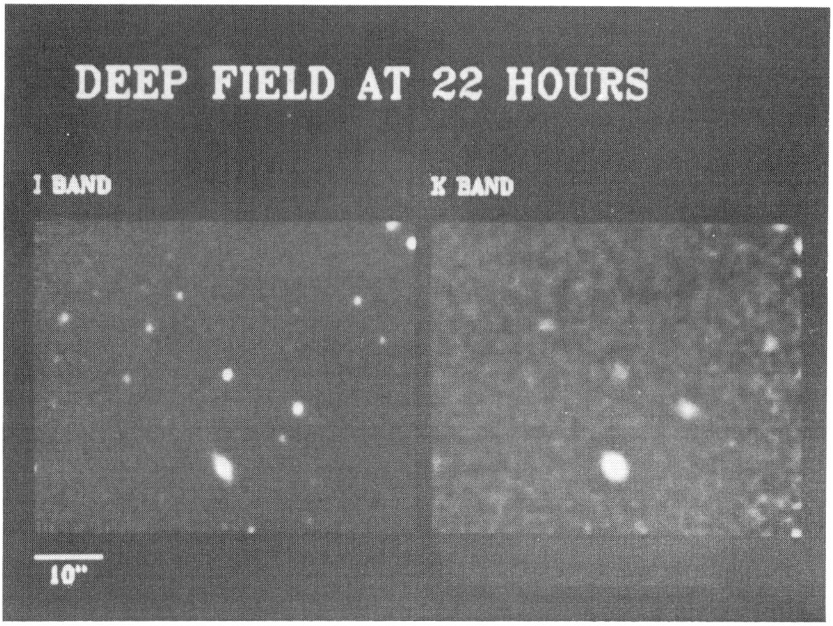

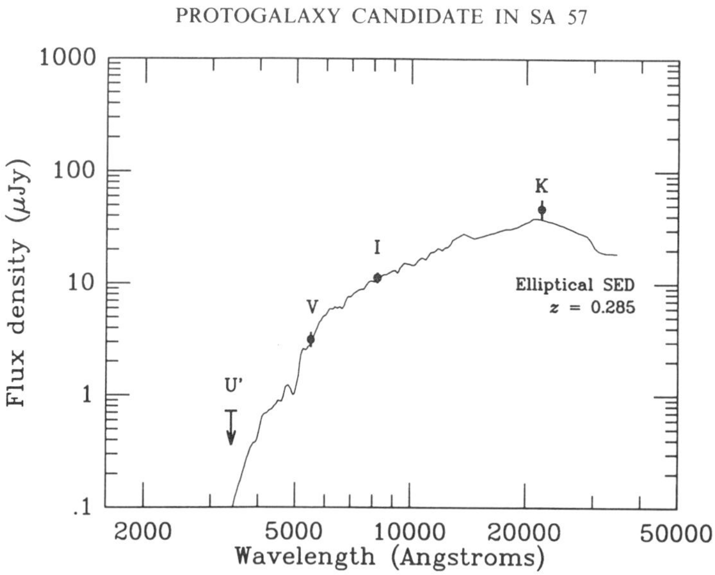

Fig. 9.7 — Extremely deep optical CCD I-band and infrared IRCAM K-band images. The noise limit on the infrared frames is about K=22, some 20 times better than previous studies and with much greater area coverage on the sky. This survey (by Cowie, Lilly, Gardner and McLean) led to the discovery of a new class of very young galaxies. Courtesy Lennox Cowie. Lower diagram shows optical CCD and IRCAM photometry of a protogalaxy candidate reported by Elston, Rieke & Rieke which turns out to be an elliptical galaxy at a redshift of 0.285.

(a)

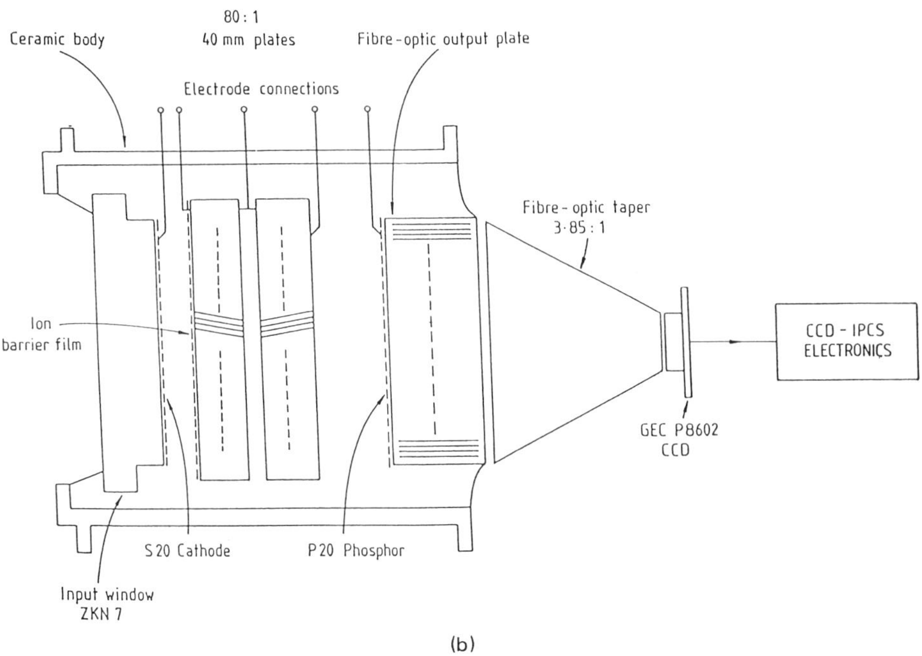

(b)

Fig. 9.8 — A typical microchannel plate (MCP) image intensifier. (a) Fully assembled; (b) schematic. Courtesy John Fordham.

Table 9.3 — Types of photon counting detectors

Intensification	Readout scheme
Stacked, straight-channel MCPs	Resistive anode
Stacked, straight-channel MCPs	Wedge-and-strip
Curved-channel MCP	Discrete anode-array (MAMA)
Stacked, straight-channel MCPs, with fibre-optic coupling to readout	CCD
Image tube intensifier, fibre-optic coupling	CCD
Image tube intensifier with coded apertures	Optical position encoding and photomultiplier array (PAPA)

In summary, the role of image intensifiers and other photocathode-based image sensors in astronomy is largely restricted to very low light levels, fast time response and certain wavelengths where CCDs are not yet competitive. CCDs find a place, however, in many photon-counting detector systems as the readout device.

9.3 X-RAYS

X-ray astronomy is a space-based rather than a ground-based discipline. Nevertheless, I bring up the subject because it is worth mentioning that, yet again, the ubiquitous CCD has a role to play. At JPL, Penn State University and MIT, programmes are under way to develop and utilize CCDs to obtain images of astronomical sources in the X-ray region of the spectrum. The idea at these wavelengths is to provide images which are spectrally resolved or, in other words, X-ray photons of different energies are separately identified. The best known form of X-ray imager is, of course, an emulsion film, but this is neither practical nor efficient for astronomy applications. Most approaches to X-ray imaging are via phosphor screens which emit visible light. However, a direct CCD-based X-ray imaging spectrometer is under review for the proposed NASA Advanced X-ray Astrophysics Facility mission (AXAF), and evaluation of existing CCDs has led to considerable insights in device design of future UV and X-ray CCDs. To date, the achievements have been CCD response over the unprecedented wavelength range of 1/10 000 to 1 μm (or 1 to 10 000 Å) — from X-rays to the near-infrared! Until recently, the best results were obtained only for selected areas of existing CCDs such as the Texas Instruments devices available to JPL, but several manufacturers have already converged on the recipe to make CCDs which perform like this over the entire array. Fig. 9.10 shows the approximate efficiency with which photons are converted to electronic charges over this immense spectral range for a backside-illuminated and UV-flooded CCD. The measured data points agree closely with theoretical expectations, lending credence to the as yet unmeasured portions of the curve.

The photoelectric absorption of a 'soft' X-ray photon in silicon ejects an electron

(a)

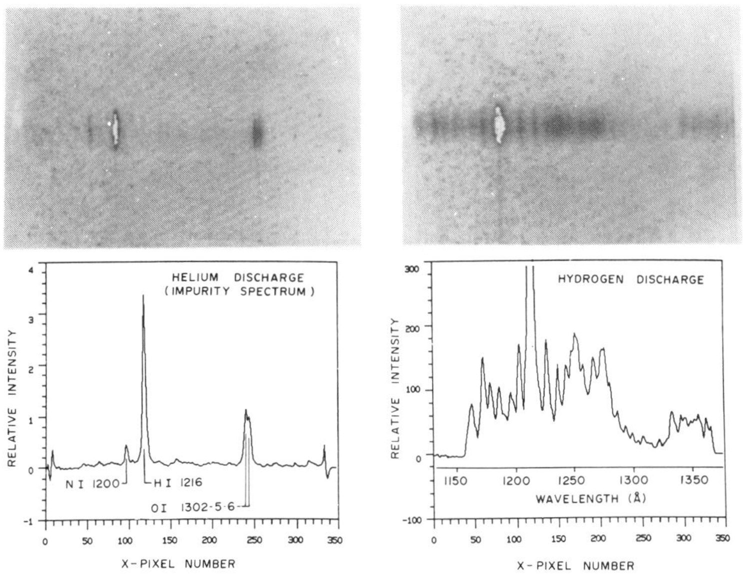

(b)

Fig. 9.9 — (a) An electron-bombarded CCD (EBCCD) camera for UV imaging. UV light is first converted to electrons in a vacuum tube; these are then focussed and accelerated to impact the CCD. (b) A UV spectrum obtained with the EBCCD. Courtesy George Carruthers.

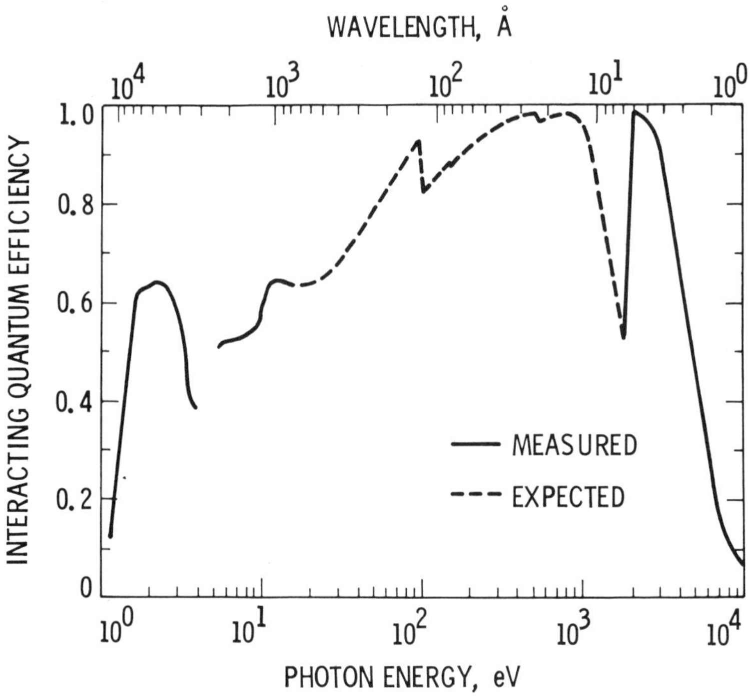

Fig. 9.10 — The QE of a thinned backside-illuminated CCD over the entire X-ray, UV, and optical region of the spectrum. Courtesy of James Janesick.

with energy $E-b$ were $E=h\nu$ and is the energy of the X-ray and b is the energy with which it is bound to the atom (typically 1780 electron volts or 1.78 keV). When free, this energetic electron produces a trail of electron–hole pairs by collisions with orbital electrons. On average it takes 3.65 eV of energy to produce a single electron–hole pair in silicon. If all the energy of an X-ray photon was used to produce electron–hole pairs directly, there would be no statistical variation in the amount of charge generated by the X-ray event. However, a small but significant amount of energy is transferred into the silicon crystal lattice structure, giving a small statistical difference in the number of electron–hole pairs actually formed. This uncertainty is characterised by the 'Fano factor'†. The Fano factor (F) reflects the ultimate limit of detector energy resolution and relates the measured width or spread in energy (δE) and the average energy (E) needed to produce electron–hole pairs in silicon by

$$\delta E = 2.355(FE/3.65)^{1/2}$$

where δE is the full width at half maximum energy of the observed spread (in electrons), E is the photon energy in electron volts and F is the Fano factor. The Fano

†The Fano factor was originally formulated by U. Fano in 1947 to describe the uncertainty of the number of ion pairs produced in a volume of gas following the absorption of ionizing radiation.

factor is approximately 0.1 for silicon, and Fano-noise-limited CCDs are now a reality over the entire soft X-ray regime from 100 to 10000 eV. To achieve this performance the CCD readout noise must be less than two electrons (rms) and any 'fat zero' or 'preflash' level needed to eliminate any form of charge-trapping must be less than one electron! Incredibly, several CCD manufacturers, notably Ford Aeronutronics and Lincoln Laboratory, now produce devices with such properties, and several others including Thomson-CSF, Tektronix and EG&G Reticon appear to be nearing this performance level.

Besides X-ray astronomy, there are several potential applications in high-temperature plasma diagnostic instrumentation and X-ray experiments on synchrotron sources. An example of the CCD as an X-ray sensor is shown in Fig. 9.11 for a

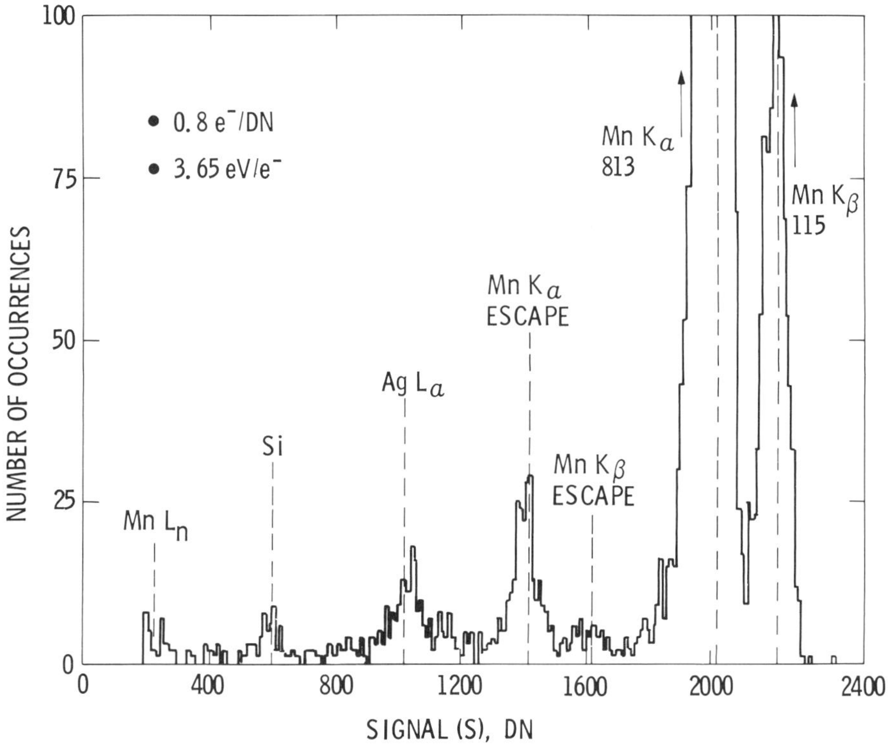

Fig. 9.11 — A histogram of X-ray events recorded from a CCD camera, revealing its ability to distinguish X-rays of different energies from different atomic species. Courtesy of James Janesick.

laboratory energy source of Fe^{55}, an isotope of iron; notice that Ag (silver) and Mn (manganese) are also detected in the source.

Jim Janesick and colleagues at JPL use the X-ray sensitivity of CCDs both as an analyser of X-ray emission and as a diagnostic of CCD performance. In particular, X-rays can be used to determine the CTE of CCDs because a known charge is deposited in a pixel by the X-ray event. For example, the Fe^{55} source yields 5.9-keV X-ray photons (the K-alpha line) which deposit 1620 electrons each in the CCD. Several references to the technical literature are given at the end of this chapter and two examples shown in Fig. 9.12(a),(b).

Undoubtedly, this is a field which will be receiving more attention in the years ahead.

REFERENCES

Allen, D. A. (1975) *INFRARED, the new astronomy*, Keith Reid Ltd. Shaldon, Devon.

Janesick, J., Elliott, T., Collins, S., Daud, T., Campbell, D., and Garmire, G. (1987) Charge-coupled device advances for X-ray scientific applications in 1986, *Optical Engineering*, Vol. 26 , No. 2.

Janesick, J., Elliott, T., Bredthauer, R., Chandler, C., and Burke, B. (1988) Fano-noise-limited CCDs, Optical and Optoelectronic Applied Science and Engineering Symposium; *X-ray Instrumentation in Astronomy* to be published by SPIE.

Keyes, R. J. (ed.) (1977) Optical and infrared detectors. *Topics in Applied Physics*, Vol. 19, Springer-Verlag, New York.

Kingston, R. H. (1978) Detection of optical and infrared radiation, *Springer Series in Optical Sciences*, Vol. 10, Springer-Verlag.

Lilly, S. J. (1988) Discovery of a radio galaxy at a redshift of 3.395, *Astrophysical Journal*, Letters, October.

McCaughrean, M. J. (1988) The astronomical applications of infrared array detectors, Ph.D. Thesis, University of Edinburgh, Scotland.

McLean, I. S. (1988) Infrared astronomy's new image, *Sky & Telescope*, March.

Morgan, B. L. (1988) *Advances in electronics and electron physics*, Vol. 74, Symposium on photo-electronic image devices, Academic Press Limited, London.

Robinson, L. B. (ed.) (1988) *Instrumentation for ground-based optical astronomy; present and future*, Springer-Verlag, New York.

Stern, R. A., Liewer, K., and Janesick, J. R. (1983) Evaluation of a virtual phase charge-coupled device as an imaging x-ray spectrometer, *Review of Scientific Instruments*, Vol. 54, 198–205.

Wolfe, W. L. and Zissis, G. J. (eds.) (1978) *The IR Handbook*. Office of Naval Research, Department of the Navy, USA.

Wynn-Williams, C. G. and Becklin, E. E. (eds.) (1987) *Infrared astronomy with arrays*, Institute for Astronomy, University of Hawaii.

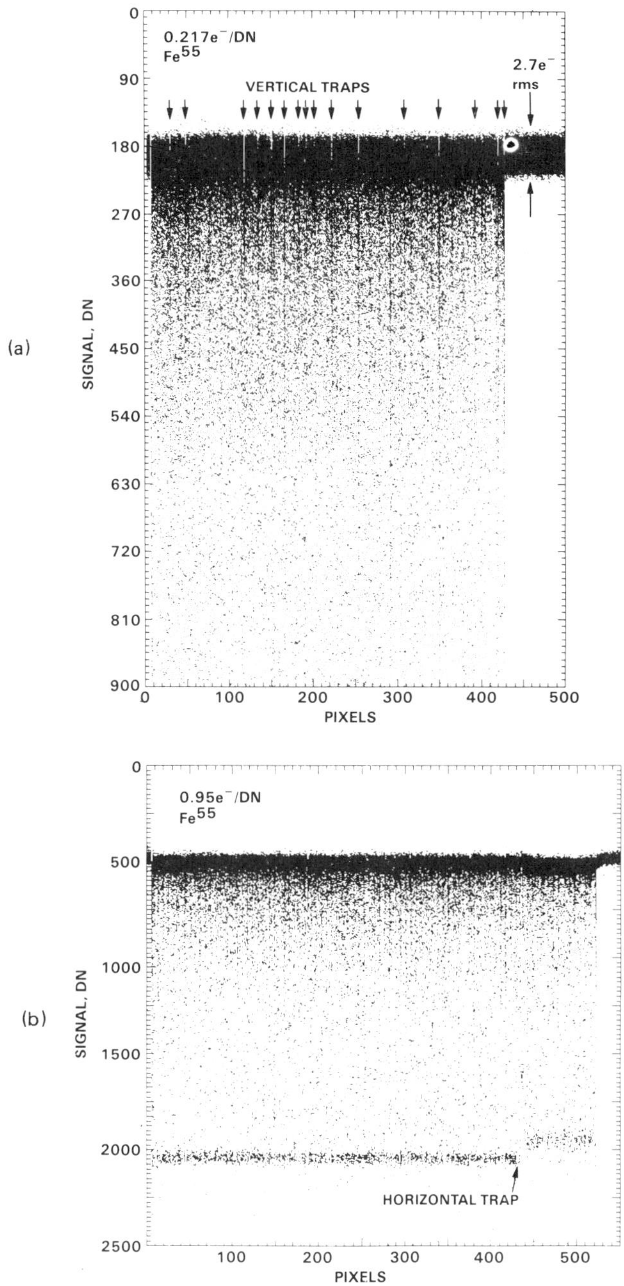

Fig. 9.12 — (a) An Fe-55 X-ray line-trace response of a Lincoln CCD which exhibits Fano-noise-limited performance at the 2.7 e^- read-noise level; vertical single-pixel traps are easily distinguished in this type of trace. (b) An Fe-55 X-ray line-trace response for a Ford CCD containing a small single-pixel trap in its horizontal register. The trap displaces about 100 e^- into deferred charge. The dark line at about 2100 e^- is due to the 1620 e^- single-pixel events from the Fe-55 source. Partial events and events split between pixels result in the grey-scale gradation of signal values. Courtesy Jim Janesick.

10

The future

The past 15 years have been a remarkable period of success for almost all branches of astronomy, and this despite budget cut-backs and delays at various times. As might be expected, X-ray, UV and radio astronomy have all blossomed. Ground-based optical and infrared astronomy, far from consolidating their traditional positions, have undergone a remarkable expansion. In this book I have tried to show that the underlying reasons for these developments are largely technological. The advent of microelectronics yielding new imaging devices of amazing sensitivity and reliability — such as the CCD and related infrared detectors — compact inexpensive computers of immense power for control of scientific instruments and analysis of data, easy international telecommunications and improvements in the manufacture of optical components have all played a role. With the aid of electronics and computers, the world's 4-m class telescopes, conceived in the 1960s, will enter the 1990s with a suite of scientific instruments which will permit near-perfect performance. Moreover, the advances in detector technology alone have revitalised many smaller telescopes, enabling them to be used for projects which only a few years earlier would have required a much larger telescope.

All of our knowledge of the universe revolves around the quality of our measuring apparatus and in the past decade we have seen ample evidence of the benefits of the technological innovations described in this book. For example, the incredible CCD images of the nucleus of comet Halley as it passed close to the Earth or the discovery of distant galaxies at redshifts greater than many quasars, or from new insights into the formation of stars to a new understanding of the three-dimensional structure of the entire universe.

What lies in the future? At the very beginning of this book it was mentioned that new techniques help scientists answer the old questions but, inevitably, new puzzles are also revealed. There must be further progress.

10.1 DETECTOR DEVELOPMENTS

Sensitive electronic detectors of radiation are central to the future of astronomy at all wavelengths. X-ray, UV, optical, infrared and even millimetre-wave astronomy all stand to benefit immediately by improvements in solid-state array detectors such as the CCD. Astronomers are concerned, however, that the next generation of CCD sensors for TV applications may be less well-suited to astronomical and other scientific uses, since, for example, the demands for colour TV sensors favour interline transfer and *X-Y*-addressable architectures, whereas it is certainly the frame-transfer construction which is most suited to slow-scan operation at very low light levels and high spatial resolution. Moreover, astronomers want CCDs with larger sizes, lower readout noise, and better uniformity!

The answer may lie in the concept of the 'silicon foundry'. To procure the 'perfect device', we need to separate the roles of device fabrication and camera vendor. Beginning with the bold move by Photometrics Ltd. to subcontract Ford Aeronutronics for the fabrication of a 516×516 CCD, followed by the JPL/Lincoln Laboratory development and the Lick Observatory/EG&G Reticon project, the first steps in this new approach are being taken. To be completely successful, however, a high level of inter-institute cooperation is essential if only to secure sufficient funding for such an endeavour. It is also encouraging to note a renewed committment by Tektronix, Thomson-CSF and EEV. Once again, the key element is funding, preferably on an international basis. Similar developments are underway for infrared arrays involving various astronomical institutes and companies such as Santa Barbara Research Center, Rockwell International, Cincinatti Electronics, Honeywell Electro-optics, Sofradir and Mullards. As indicated in Chapter 9, the potential in that field is tremendous.

How much larger can CCDs become? The Tektronix 2048×2048 CCD with a pixel size of 27 μm and an imaging area of 55.3×55.3 mm^2 occupies an entire 100-mm (4-inch) silicon wafer (Fig. 10.1). This device represents true 'wafer-scale' inte gration. With a major development within the semiconductor industry — which may well occur, who knows, since 200-mm diameter Si crystals have recently been fabricated — this large size could be exceeded within a decade. There are, however, performance limitations on size. For example, as CCDs become larger the requirement for very good CTE becomes more stringent. By the most recent technology it is just possible to obtain CTE factors of 0.999999, which, if applied to a 2048×2048 format device for example, means that the pixel most remote from the readout amplifier will retain only 98% of its original signal charge! The remainder will have been lost in both horizontal and vertical trailing pixels Very recently Ford Aerospace has also constructed a 2048 CCD and a 4096×4096 device is under consideration. While this is a remarkable achievement, it is not yet commonplace and many designers believe that it may represent an upper limit for the forseeable future.

Probably the next step will be to form sizeable 'mosaics' composed of large CCDs butted together, rather than to await unexpected technology breakthroughs. Tony Tyson of Bell Labs has already proposed the construction of a large-area CCD camera which would employ four Tek2048 chips in a 2×2 arrangement; he calls the mosaic the LACCD. Fig. 10.2 compares the size of the LACCD with other CCDs, and with the field of view of the prime and Cassegrain foci of the Canada–France–Hawaii Telescope.

A special development at Thomson-CSF in collaboration with ESO and INSU (France) has led to the design of a large CCD with all electrical connections on one side only. With this construction a mosaic of the kind shown in Fig. 10.3 is possible. Clearly, the mosaic can be continued to produce a large rectangular array of 2×*n*

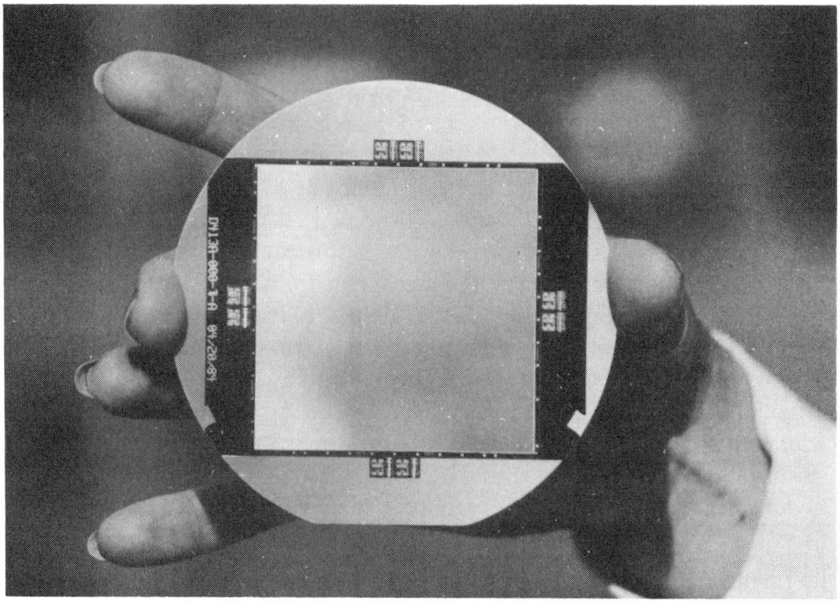

Fig. 10.1 — Wafer-sized silicon technology demonstrated in the Tektronix 2048×2048 CCD. Courtesy Morley Blouke, Tektronix Inc.

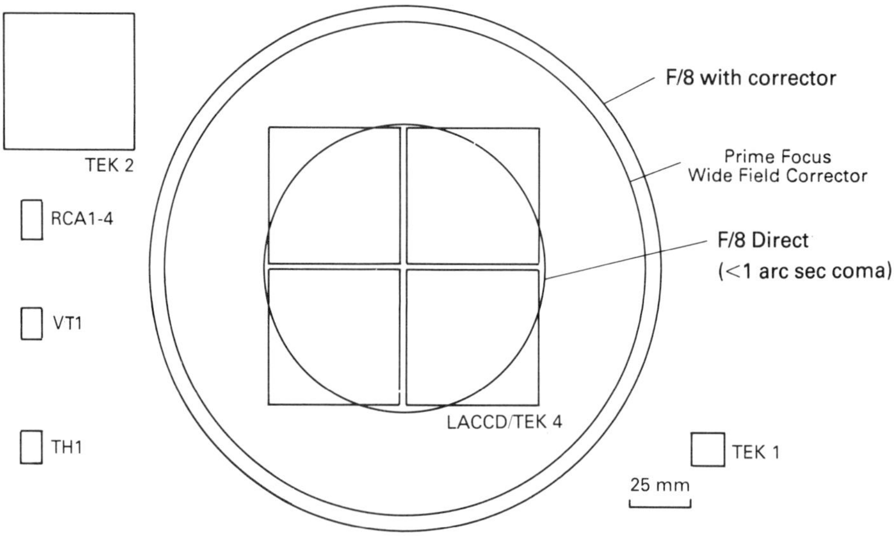

Fig. 10.2 — Tony Tyson's large-area CCD concept. Tek 4 is a special 2×2 mosaic of Tektronix 2048×2048 devices. For comparison other CCDs and the field of view on the CFHT are shown. Courtesy Tony Tyson and CFHT.

Fig. 10.3 — A 2×2 mosaic of specially designed CCDs by Thomson-CSF. These CCDs have all their electrical connections along one edge. Photo courtesy of Martin Cullum, ESO.

CCDs; for $n=3$ this gives 1148×1152 pixels. The Wide-Field/Planetary Camera pioneered by Jim Westphal for the Hubble Space Telescope employs eight of the Texas Instrument's 800×800 CCDs, although in that case an optical scheme involving a pyramid-shaped mirror is used to direct the split-field onto the arrays.

Nuclear physicists are interested in the use of silicon CCDs as vertex detectors in particle experiments where bubble chambers cannot be used and gaseous detectors have insufficient precision. Imaginative arrangements involving 100 CCDs have been published!

Mosaics of CCDs with an effective size of 10 000×10 000 elements, which is 100 million pixels, are likely in the next decade. This implies computer storage requirements of about 200 Mbytes per image or only 10 images per 2-Gbyte double-sided optical disk. By the time such mosaics are accomplished, however, it is likely that data storage devices will be 10 times more efficient. Image display systems will probably stabilise for a time at 1024×1024 and therefore one will display either a subsection of the full image or use a computer program to display the entire image at one-fifth or one-tenth of the maximum possible resolution.

How low can the readout noise go? Certain CCDs and their associated electronic control systems have achieved less than two electrons (rms) noise (see Fig. 10.4) and one-electron noise seems inevitable in the very near future. Remember that the noise is really a voltage variation which can be expressed as an 'equivalent number of electrons' by multiplying the voltage by C/e where C is the capacity of the output

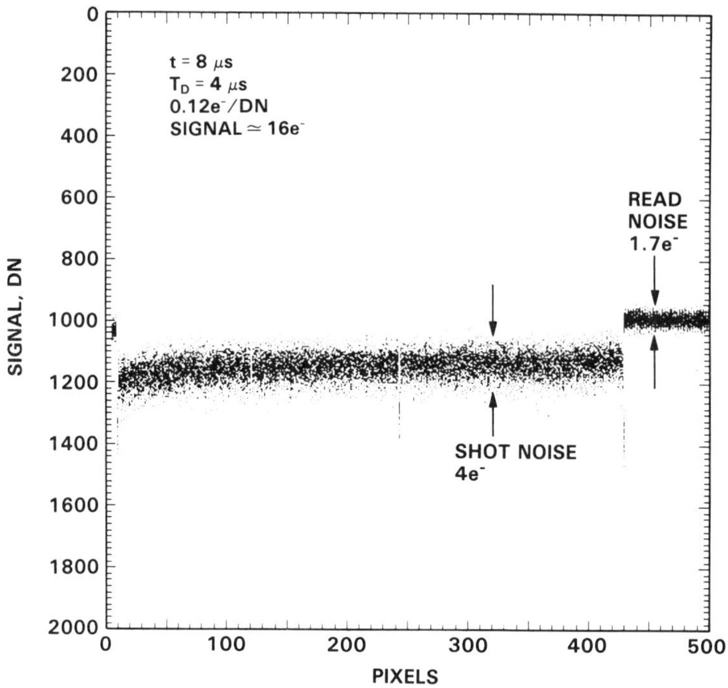

Fig. 10.4 — A raw line trace showing the shot noise of a 16 e^- flatfield signal. The shot noise of 4 e^- can be compared directly with the read noise of 1.7 e^- observed in the overscan region. The CCD is from Lincoln Labs. Courtesy of Jim Janesick, JPL.

node of the CCD to store charge and *e* is the charge on one electron. Many CCDs in operation at present do achieve a readout noise of about 10 electrons. If readout noise is the limiting noise source in the measurement then an improvement of 10 times in noise implies 100 times shorter exposures. Further improvements might be possible using new kinds of CCD outputs enabling non-destructive multiple sampling and these are under consideration. There is no known fundamental impediment to 'sub-electron' CCD readout noise; it should be possible to emulate photon counting systems and detect a single photoelectron from an optical photon. In fact, it is already possible to count X-ray photons with some of the latest CCDs.

In addition to making CCDs larger and less noisy, the X-ray, UV and even the optical response (or quantum efficiency) of silicon devices is expected to improve slightly through the use of coatings, surface treatments and specially doped fabrication. Coatings such as Photometrics' Metachrome II can raise the near-UV performance to two-thirds of the peak optical response, but the best breakthrough in sensitivity (times two at the most) will come when well-treated backside-illuminated CCDs reappear to replace the old RCA chips.

Further improvements in silicon technology and in CCD design will also improve the uniformity of CCDs and should reduce dark current to levels at which simple thermoelectric coolers instead of liquid nitrogen can be used. Apart from being operationally easier to use, such devices would have better charge-coupling efficiency because the silicon can be operated warmer.

Undoubtedly, the future will bring significant developments in infrared semiconductor technology. Infrared array detectors will get much larger and less noisy, and we can even expect the development of fully multiplexed array detectors for space applications which extend to wavelengths of 200 μm. Already 128×128 arrays of mer–cad–tel (HgCdTe) and 256×256 Schottky Barrier arrays are available for the very near infrared — just beyond (redder than) the cut-off wavelength of silicon CCDs, and several companies are working on arrays of up to 512×512. Because the sky is always bright at infrared wavelengths some special developments are needed to cope with large detector arrays. For example, the infrared array must store much more charge than a typical silicon CCD, the array must be read out much more frequently than a silicon CCD and much faster electronic and computer processing are needed. One approach is to split the infrared array into multiple subsections which can be electronically accessed simultaneously and then use the new technology of parallel processing computers called 'transputers' to handle the several channels of high-speed digital data. This field in particular promises to be very exciting over the next decade.

Astronomical instrumentation will also evolve. Multiple solid-state camera systems not unlike the Palomar four-shooter are likely to be developed so that several wavelengths can be imaged simultaneously. Optical fibres and very efficient faint-object spectrographs will be the norm at optical wavelengths with very low noise CCDs and totally cryogenic infrared spectrometers using diamond-turned optical components and the latest infrared array detectors will appear.

Space astronomy too will benefit from detector and instrument technology improvements.

10.2 THE HUBBLE SPACE TELESCOPE

Were it not for the tragic loss of the space shuttle 'Challenger' in January 1986, the Hubble Space Telescope (HST) (Fig. 10.5) would already be in orbit. Following the successful launch of the shuttle 'Discovery' on September 29, 1988 there is renewed hope that the HST will be launched before the end of 1989. The HST — one of the most ambitious payloads ever — will carry a battery of complex instruments, including CCD cameras, which are expected to produce a colossal flood of high-quality images and spectra — especially in the UV — free of contamination by the Earth's atmosphere. Although capable of collecting and focussing light over an immense span in wavelengths from 0.115 μm in the far UV to at least 10 μm in the infrared, the initial complement of instruments will contain no infrared detectors since, at the time of instrument selection, CCDs themselves had only just been invented! Because the telescope is warm, unlike the European Space Agency's ISO, it is not well-suited to infrared observations beyond about 2.5 μm. In the next generation of HST instruments, solid-state near-infrared arrays will be available and instrument definition is already under way.

The remote-controlled 2.4-m (94-inch) telescope and its scientific instrumentation (Fig. 10.6) will, of course, rely heavily on electronic detectors and computers.

There are five main instruments:

Fig. 10.5 — Artist's impression of the Hubble Space Tescope in orbit — a fully remote-controlled observatory with eight 800×800 CCDs and a host of other sensitive UV detectors. Two international data centres are already set up to receive the huge data flow. Courtesy of Terence Facey, Perkin–Elmer Corporation.

Wide-field camera/planetary camera (WF/PC)
Faint-object camera (FOC)
Faint-object spectrograph (FOS)
High-resolution spectrograph (HRS)
High-speed photometer/polarimeter (HSP)

The **wide-field/planetary camera** (the abbreviation WF/PC is commonly pronounced 'wiff-pick') can be used as a wide-field camera for survey observations or as a high-resolution camera capable of exceptionally fine detail imaging of, among other things, the planets of the solar system. Light coming along the optical axis of the telescope is 'picked-off' by a small mirror at an angle of 45°; only the central part of the beam is diverted and the remainder of the light goes on to the other instruments. In the wide-field mode a set of four TI 800×800 CCDs — arranged in a mosaic — 'see' a square area of 2.67 arcminutes and each pixel subtends an angle of 0.1 arcseconds, whereas in the planetary mode the field of view is only 68.7 arcseconds and an individual pixel corresponds to a mere 0.043 arcseconds — some

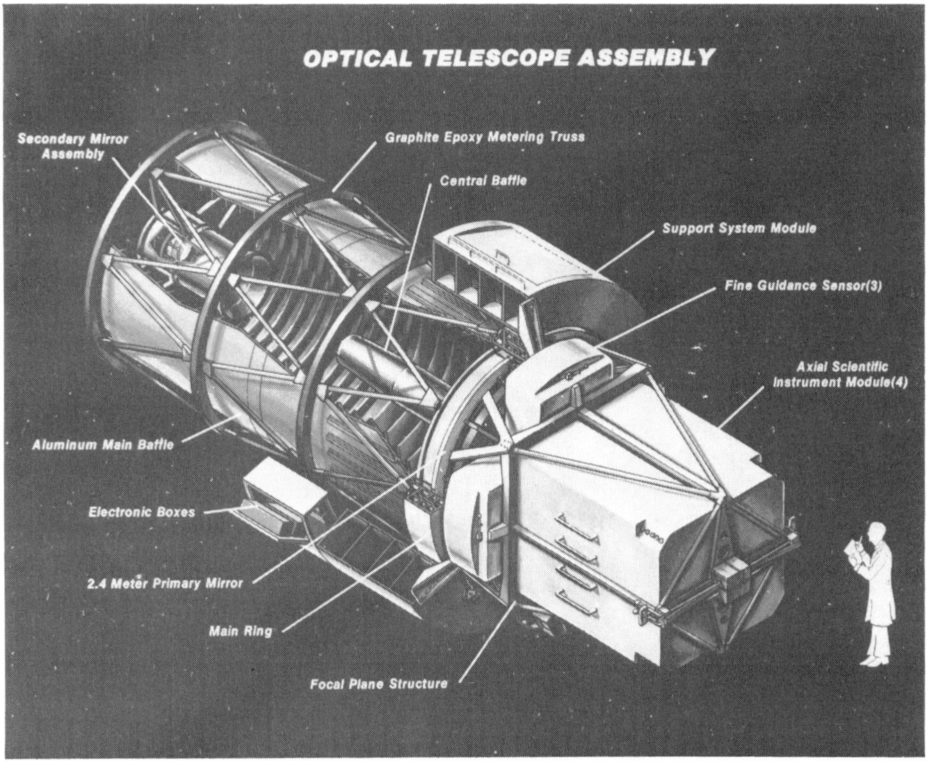

Fig. 10.6 — The HST showing the location of important components. Photo courtesy of Terence Facey, Perkin–Elmer.

10 times smaller than typically used with ground-based telescopes. The planetary camera takes advantage of the absence of atmospheric blurring and the superb optical quality of the telescope optics.

Good UV response from the TI CCDs is to be obtained by the use of a coating on the CCD surface of the phosphor coronene. After construction of the spacecraft and instrument it was found that it was necessary to illuminate the CCD with a strong UV light — a UV flood — to prepare the coronene and ensure good UV quantum efficiency (see section 5.8). To solve this problem a hole had to be drilled in the side of the telescope and a 'light-pipe' installed to feed UV light from the sun to the CCD chamber!

Analysis of CCD data from the WF/PC will be no easy task and, in some respects, ground-based experience has enabled the astronomical community to 'get prepared' for the enormous quantity of data. For example, in a ground thermal-vacuum test at JPL in February and March of 1988 10 Gbytes (or 10 000 Mbytes) of test data were accumulated. Jim Westphal also remarked that the system has run for 8000 hours now in ground tests.

The **faint-object camera** is complementary to the WF/PC. It is designed to have higher spatial resolution than the planetary camera and better UV sensitivity, especially for extremely faint objects. In the FOC, two similar but independent systems are provided to form an image. One system has a very small field of view of just 11 arcseconds with tiny pixels of only 0.022 arcseconds; the other system gives twice the field and pixel size. In each system the detector consists of an Image Photon Counting System of the Boksenberg type, i.e., an image intensifier tube with a UV-sensitive photocathode followed by a TV camera for readout and 'event-centroiding' electronics to locate individual photon impacts on the detector. As mentioned in the text, this kind of detector saturates very quickly, but it has no readout noise and so it really is intended for *faint* objects. The FOC also has a spectroscopic capability and a means of measuring polarization. F. Macchetto of the European Space Agency is principal investigator.

The **faint-object spectrograph** will be equipped with two detector systems, one blue-sensitive and one red-sensitive. Both detectors are of the Digicon-type, that is, a photoelectric photocathode as in a photomultiplier tube, with a surrounding magnetic focussing coil which collects and directs the emitted photoelectrons onto a linear array of 512 silicon photodiodes by which means they are detected. Associated with each diode is a discrete amplifier. This is a complex type of detector but it has many of the advantages of the photon-counting detectors and some of the advantages of silicon devices such as a large dynamic range. With its two channels, the faint object spectrograph will be sensitive to radiation ranging in wavelength from about 0.115 to 0.8 μm, and in addition it will have a superb polarisation-measuring capacity and it will be responsive to extremely fast time-varying astrophysical phenomena. During the first half of 1988 the FOS was refurbished at the Lockheed plant with two new Digicon detectors developed by principal investigator Richard Harms of the University of California at San Diego.

The **high-resolution spectrograph** proposed by John Brandt of Goddard also employs Digicon detectors and differs from the FOS largely in terms of the fine detail or resolution which it will achieve within the spectrum of relatively brighter sources.

The **high-speed photometer** by Robert Bless and colleagues at the University of Wisconsin at Madison is the simplest instrument in the sense that it has no moving parts. The detectors are called 'image dissector scanners' and are essentially photomultiplier tubes except that by means of magnetic focussing they respond to photoelectrons coming from the small region of the photocathode on which the light is actually focussed. In principle, the HSP will be able to detect time variations in light level which occur as fast as one every 10 μs.

A major component of the HST is the set of 'fine guidance' sensors. The telescope requires accurate target positions relative to an acquired guide star so that light from the target can be fed accurately and continuously into the entrance aperture of the appropriate HST instrument. To do this, a very extensive guide-star selection system was developed at the Space Telescope Science Institute in Baltimore. Basically, an image and a catalogue archive has been developed on optical disks using digitised photographic plate surveys from several different Schmidt telescopes (see section 2.1.1). Typically, each survey plate corresponds to an array of 14 000×14 000 pixels, with one pixel being equivalent to 1.7 arcseconds on the sky. Four plates (1600 Mbytes) are accommodated on each optical disk and up to 80 disks may be placed in a 'juke box', which gives about 20% of the sky 'on-line' for image display at any time. In addition, a catalogue with information on approximately 20 million individual objects is available.

Recently, the Space Telescope Science Institute in Baltimore released the Space Telescope Science Data Analysis System (STSDAS). This immense software package includes both calibration and analysis software and is fully portable to any host computer that runs the IRAF software system (see Chapter 7) from the NOAO; another form is available for non-IRAF systems.

In reading through the above account of the HST instruments one might be surprised that CCDs feature so little. As already mentioned however, this state of affairs is not so surprising when one considers the incredibly long 'lead-time' needed for space ventures of this scale. The idea of a large space telescope was first mooted in 1962, the first group of scientists selected by NASA in 1973 and construction finally authorised in 1977. The first astronomical images with a CCD — see the picture of Uranus obtained by Brad Smith, Jim Janesick and colleagues with the TI 400×400 chip — were obtained in 1976, and Jim Westphal and Jim Gunn proposed the WF/PC shortly after. Additionally, the space telescope was intended to provide excellent coverage in the UV and photoemissive vacuum-encapsulated TV-type tubes were at least known to be suitable at these wavelengths, and possibly at their peak of development.

Finally, it is important to realise the advantage of the HST over ground-based telescopes despite its size. Particularly in the blue and UV, the improved resolution resulting from the absence of atmospheric blurring leads to sharper images and much more light per pixel, which in turn means that the HST can image to very faint limits. On the other hand, the HST is very expensive indeed compared to ground-based telescopes, even very large ones.

10.3 VERY LARGE TELESCOPES (VLTS)

We can already improve our efficiency in spectroscopy by the use of optical fibres to observe many objects simultaneously. As CCDs get larger we can cover more area on the sky or more spectrum. But carefully designed optical cameras and spectrometers have transmission factors and efficiencies which are virtually impossible to improve due to fundamental effects associated with the refractive and reflective properties of materials and the quantum efficiency of the detector itself. Some improvements are no doubt possible, and should certainly be made even if the improvement is by no more than a factor of two. Already however, astronomers are beginning to wonder how to make astronomical measurements 10 times, rather than two times, more efficiently. How can we probe to the most distant objects in space and the earliest moments in time?

Once the quantum limits of sensitivity in detectors and instruments have been reached, the only way to gain large factors in efficiency is to construct even larger ground-based telescopes and to develop space observatories.

Table 10.1 lists the world's largest telescopes. Intriguingly, after construction of the 5-m (200-inch) Hale Telescope on Palomar Mountain in 1949, the trend was for smaller telescopes in what has become known by the imprecise term '4-metre class'; in this definition all telescopes with primary mirror diameters between 3.6 m and 4.5 m are lumped together. The exceptions were the Soviet BTA and the Smithso-

Table 10.1 — The largest telescopes

Telescope/observatory	Diameter (m)	f. ratio	Type
Bol'shoi Teleskop Azimutal'nyi Mt Pastukhov, (USSR)	6.0	f/4.0	Alt–az.
George Ellery Hale Telescope Mt Palomar, California (USA)	5.08	f/3.3	Equat.
Multiple Mirror Telescope Mt Hopkins, Arizona (USA)	4.5	f/2.7	Alt–az.
William Herschel Telescope Roque des los Muchachos, Canary Islands (UK, NL, Spain)	4.2	f/2	Alt–az.
Cerro Tololo Interamerican Obs. Cerro Tololo, Chile (USA)	4.0	f/2.8	Equat.
Anglo-Australian Telescope Siding Spring, NSW (UK, Australia)	3.9	f/3.3	Equat.
Nicholas Mayall Telescope Kitt Peak, Arizona (USA)	3.8	f/2.8	Equat.
United Kingdom Infrared Telescope Mauna Kea, Hawaii (UK, NL)	3.8(thin)	f/2.5	Equat.
Canada–France–Hawaii Telescope Mauna Kea, Hawaii (Can, Fr, USA)	3.6	f/3.8	Equat.
European Southern Observatory Cerro La Silla, Chile (ESO)	3.6	f/3.0	Equat.

nian MMT, both of which were unique in their design, and both of which may have been somewhat ahead of their time. What happened, of course, was that detectors and instrumentation improved greatly — as we have outlined in this book. Now, things have come full-circle, and astronomers are once again looking at ways of constructing large telescopes.

There are essentially three fundamental issues, namely (1) how to achieve a very large collecting aperture (mirror) of the required optical performance; (2) how to support and control in the optimum way such a potentially very heavy mechanical structure; (3) how to enclose a very large telescope in a cost-effective way with negligible degradation on image quality due to vibration, air disturbance or inadequate environmental protection (wind, dust). In addition, new telescopes must be designed to capitalise on the very best 'seeing' conditions at the world's best sites, and must be designed with 'remote control' in mind.

Without doubt, the new generation of telescopes will not be equatorially-mounted in the traditional way. Instead they will be 'altitude–azimuth' **(alt–az.)**-mounted like radio telescopes and they will be computer-controlled and computer-modelled to achieve exceptionally fine image quality and fine pointing and tracking. They will be designed using the latest computer-aided engineering tools, and will be equipped with electronic imaging spectrometers and cameras.

Several nations are currently engaged in telescope designs. The most advanced group is the 10-m W. M. Keck Observatory project managed by CARA — the California Association for Research in Astronomy — which comprises CalTech, the Lawrence Berkeley Lab and all of the University of California institutes. The Keck Telescope, which is under construction on the summit of Mauna Kea, Hawaii, will have an effective aperture of 10 m but it will possess a unique primary mirror concept championed by project scientist Jerry Nelson. The primary mirror will not be a single piece of glass, instead it will be composed of 36 hexagonal 'segments', each side of each hexagon being 0.9 m (i.e. they are 1.8 m across) and each segment being only 7.5 cm thick (Fig. 10.7). If this approach, which works well at longer (radio)

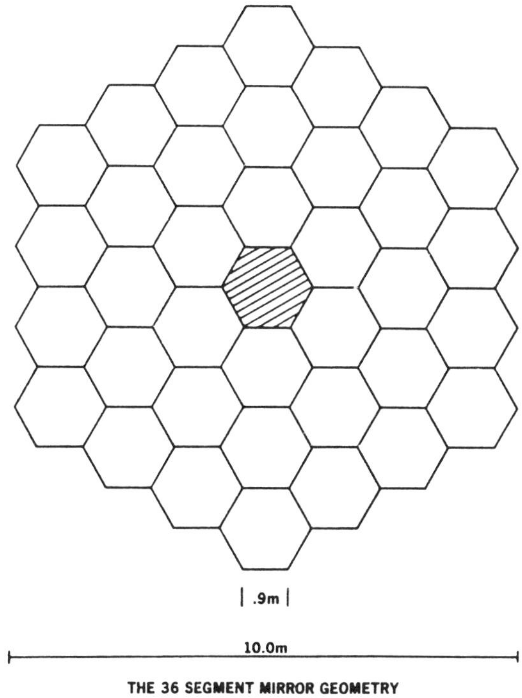

Fig. 10.7 — The 36-hexagonal segmented mirror design of the W. M. Keck 10-m telescope. Courtesy Jerry Nelson.

wavelengths as demonstrated by the UK's James Clerk Maxwell Millimetre Telescope completed in 1987, can be perfected at optical wavelengths then it will herald a new era in telescope design. What is needed is a complex scheme of adjusters attached to each mirror segment which, under computer control, can be manipulated remotely to bring all of the mirror segments into perfect alignment as if they were indeed part of a single giant surface. While the mechanical and optical aspects of this design are challenging enough, it would be impossible to manually set and control this segmented mirror surface; a computer is essential. Two position sensors are located along every inter-segment edge; a total of 168 sensors. A computer program uses the sensor readings to update 108 position actuators several times per second to keep the mirror in shape as the telescope tracks.

Since about 50% of the cost of an observatory resides with the enclosure or dome, considerable effort is expended to keep the overall length of the telescope to a minimum. In practice this means keeping the focal length of the primary mirror as short as possible despite its diameter or, in other words, making the focal ratio of the mirror as 'fast' as possible; Table 10.1 shows that traditionally this ratio is about 3. The primary mirror of the Keck telescope will be an f/1.75 hyperbola (rather than the usual parabola) with a resulting focal length of 17.5 m.

Equally daring is the proposal announced in 1988 by the European Southern Observatory (ESO) to build a 16-m telescope in the form of a 'linear array' of four 8-m telescopes (Fig. 10.8). Each of the 'unit' telescopes would be alt–az, and each

Fig. 10.8 — Artist's impression of the European Southern Observatory's very large telescope; four 8-m telescopes in a linear array with an equivalent aperture for combined images of 16 m. Photo courtesy of European Southern Observatory.

would contain a large monolithic glass mirror which would be so thin for its size that the mirror would require 'active' rather than 'passive' mechanical support. That is, hundreds of supporting adjustable 'pistons' would be placed against the back of the mirror to exert small forces to compensate for distortions of the figure of the mirror due to gravitational stress, thermal effects, wind-buffeting and slowly varying atmospheric blurring. Forces supplied by the mirror support would be updated a few times per second and the entire arrangement would be under computer control. ESO have already succeeded with laboratory trials of a smaller prototype active control system (Fig. 10.9). It would be hard to contemplate such an active control system

Fig. 10.9 — Numerous computer-controlled pistons and actuators used for active control of the surface shape of the next generation of telescopes. European Southern Observatory photo.

without the advances in speed and miniaturisation of electronic computers.

In the USA a major innovation in telescope mirror design has been pioneered by Roger Angel of Steward Observatory. To understand his idea we need to recall that for a mirror to focus the light of a distant object it must possess a curved surface — either a spherical or a parabolic surface. The deeper the curve, the shorter the focal length of the primary mirror and therefore the smaller the overall length and the total weight of the telescope. Most classical telescopes have a ratio of focal length to diameter of about 3 to 4, which leads to large structures and large buildings (domes). It would be much more economical and provide a stiffer mechanical design if shorter focal lengths could be manufactured. The problem is the depth of the curve and therefore the huge amount of glass which must be 'ground and polished away' from the original flat-surface blank. Roger's idea was the following: *melt* the glass in a rotating furnace and allow it to cool while spinning. During the molten stage the liquid glass will form a deep curve in the shape of a parabola due to the centrifugal force produced by the rotating furnace, and will keep this shape as it cools while spinning. The term for this technology is called 'spin-casting'. Roger and his team have recently cast their first 3.5-m blank and are aiming at a set of 7.5-m mirrors.

Although Roger started this amazing endeavour in a backyard furnace, there is little argument that it is now a large, sophisticated operation which relies heavily on computer-modelling, computer-simulation and computer-control of the furnace environment.

Interestingly, two of the major optical companies in the world have now shown considerable interest in the manufacture of large thin mirrors with short focal lengths for the next generation of astronomical telescopes.

10.4 ADVANCES IN DATA HANDLING

Computers, of course, will continue to get faster and smaller, data storage media will increase in capacity and decline in cost, and image display systems will improve. The most likely innovations in data handling will be the introduction of

(1) optical laser disks for data archives,
(2) 'transputers' for high-speed processing of CCD data in real time,
(3) much faster 'small' computers using 'parallel' processing for off-line analysis,
(4) an increase in the use of data base systems.

Astronomical instrumentation will increase in size and complexity with more attention to good engineering. Modularity will be the means of obtaining flexibility — through computer-controlled functions — and most instrument designers will opt for 'local intelligence' in their equipment to ensure that it operates quickly and independently of the the 'observatory' computer.

A development which this author would like to see, and there have been a few initial attempts, is 'instrument simulators' in the spirit of 'flight simulators'. The complexity and the common-user status of many astronomical instruments makes it difficult for potential users, which includes graduate students, to become thoroughly familiar with any particular system during their often all-to-brief observing sessions. Good documentation helps, of course, but most people learn more rapidly by 'hands-on' experience. The answer is to simulate the performance of the instrument in question by computer programs which not only simulate the required 'control commands' that the visiting astronomer will need, but will also simulate physical actions, such as filter selection or grating angle, and include data-taking, storage and display of a plausible image.

10.5 REMOTE OBSERVING

With the availability of high-speed, high-density intercontinental links by satellite telecommunications, the possibility for 'remote observing' at ground-based observatories, in a manner analogous to remote observing using space facilities, is receiving considerable attention. One of the first demonstrations of remote observing was carried out by the Royal Observatory, Edinburgh, Scotland. Using satellite telephone links, the UKIRT in Hawaii and an infrared photometer was controlled and data returned to Edinburgh.

The difficulties with full remote observing are several:

(a) *All* functions must be under computer control, for if even *one* item requires human presence then the operation is not fully remote. This places very considerable constraints on the observatory and its equipment. Safety of the telescope and instrumentation become a major design issue. There is likely to be a lot more expense arising from the additional equipment needed, and the only way to partially recover such costs is to eliminate staff.
(b) The reliability and cost of intercontinental links, and their ability to handle large quantities of data such as CCD images. This situation is by no means perfect at the moment, but it is an area of great commercial importance and improvements are likely. Various methods of data compression can be used.
(c) Proper assessment of climatic conditions at the remote site.
(d) Development of a 'mission control centre'.
(e) Organisation of maintenance, repair and installation at the remote site.

While these problems can all be overcome in time, given sufficient funding, most national ground-based facilities have opted for a reduced form of remote observing. The UK SERC (Science and Engineering Research Council) establishments in Hawaii, La Palma and Australia have pioneered the concepts of

- **remote eavesdropping,**
- **service observing.**

These techniques are largely self-explanatory. With remote eavesdropping, the distant participant can take part in the observing session by viewing 'mimic' displays showing telescope and instrument status, receive data after a short delay and by means of a computer console or telephone can give advice. The actual observing is performed by a qualified telescope operator and/or a staff scientist. If the guest observers form a large team then one of them can go to the overseas site to represent the team while the others contribute by remote eavesdropping. The service observing mode is similar except that the proposers of the scientific experiment are not expected to send a representative, nor are they expected to eavesdrop. The observations are carried out by a qualified staff scientist, the data are transferred to magnetic tape in a standard form and mailed to the proposer. Electronic computer mail systems are used to solicit scientific proposals for service observing, the proposals are vetted by peer review in the normal way, there is a limit of one to two hours per proposal, and staff scientists arrange an optimum observing schedule and maintain the selected proposals on-file until completed. Usually the deadline for submission of service proposals is only a month in advance of the allotted nights.

These two forms of remote observing are on the increase, and more and more effort will be directed toward fully remote-controlled instruments so that, for example, service observing can be 'pre-programmed' well in advance of the observing run and on the night the operator will simply execute a 'command file' or a series of these files.

10.6 TAKING STOCK

Astronomers of the 1980s have experienced a transitional phase in which many aspects of the subject have grown rapidly as the result of technological innovations. In some cases, astronomy has actually stimulated the technology. Optical and infrared astronomy from the ground, UV and X-ray astronomy from space, and even radio astronomy, which from its very inception was a discipline relying heavily on the skills of electronic engineers, have all benefitted from the Semiconductor Age.

CCDs, infrared array detectors, optical fibres, advances in computer-aided design of optical, electronic and mechanical components, high-speed 'single-board' microprocessors and powerful minicomputers, high-resolution colour image displays, intercontinental telecommunications and other innovations have blown a significant wind of change through astronomy in a mere 15 years. What now?

Astronomers of the 1990s will rarely find themselves perched on step-ladders, wrapped up in warm parkas and woollen hats for protection aginst the cold mountain air as they peer through the eyepiece of some light-measuring instrument and manually operate its controls. Instead, they will remain by a computer-console and image display system in a warm control room while electronic eyes and computer brains manipulate the flux of weak radiation arriving at the Earth from distant objects in the universe. Construction will begin on large telescopes on the ground and in space. Measurements of the intensity, the spectrum and the polarisation will be made and stored together with calibration data, fast computer processors will perform 'reductions' of those data with the minimum of interaction by the astronomer, and 'near final form' results will appear on computer screens to allow the observational experiment to be carefully monitored and the quality of the data assessed. Supercomputers will produce theoretical models to try to match observations and the cycle of theory and experiment established so long ago by Galileo will continue. Galileo's telescope was over 1000 times smaller than current telescopes, his detector (the human eye) was about 50 times less sensitive than a CCD and with no capacity to 'integrate' on faint sources, and his awareness of the universe was confined to a tiny part of the energy spectrum. Incredibly distant objects whose light began its journey to us more than 10 billion years ago, and objects 100 million times fainter than any that Galileo could have seen, have been confidently detected using a technology invested to improve a device which itself would have astounded Galileo — the telephone!

And CCDs? Well, they can only get better.

Exercises

The following set of questions may be of use to astronomy and physics tutors who have (rightly) chosen to include a description of CCDs and other electronic imaging devices in their courses.

1. Describe with the aid of diagrams a three-phase surface channel CCD and show how photogenerated charges are collected and stored and transferred to the output. (Hint: section 4.2.2.)
2. Describe with the aid of a diagram what is meant by a 'virtual phase' CCD. (Hint: section 4.2.6.)
3. Draw a diagram of a buried-channel CCD and cite its advantages over a surface-channel CCD. Explain what is meant by 'inverted' operation. Why is inverted operation advantageous? (Hint: sections 4.2.5, 5.2.)
4. List the advantages of a CCD camera for astronomy and compare these to any kind of vidicon TV tube. (Hint: sections 1.4, 2.1.2.)
5. What is meant by charge transfer efficiency? If a 2048×2048 CCD is quoted as having a CTE of 'five nines', what fraction of the original charge in the last pixel will remain when it reaches the output? (Hint: section 5.4; answer: 0.960.)
6. What are the main sources of 'noise' in a buried-channel CCD with a standard floating diffusion output? Calculate the value of 'kTC-noise', in electrons, for a CCD at 150 K if it has an output capacitance of 0.5 picofarads. (Hint: section 5.5; answer: 200 e.)
7. Describe in detail how the reset or kTC noise can be removed by the technique of correlated double sampling. (Hint: section 6.3.1.)
8. Why must CCDs be cooled for astronomical use? (Hint: section 5.2.)
9. What is meant by 'slow-scan'? Calculate the 'pixel time' and the total 'readout time' for an EEV/GEC CCD having a format of 576×385 pixels if the readout rate is 10 kHz. (Hint: section 6.3.1; answers: 100 μs, 22.2 s.)
10. Explain what is meant by 'flat-fielding' a CCD. Describe some practical strategies for flat-fielding when at the telescope. (Hint: sections 5.6, 8.3.)
11. Calculate the required amplifier gain to match a CCD with an output capacitance of 0.1 pF and a source follower gain of 0.7 to a 10 V 16-bit A/D in order to achieve 5 e/DN. (Hint: section 8.1.2; answer: 27.2.)

12. What is meant by the photon transfer curve of a CCD? Explain how certain CCD system parameters can be derived from such a curve. (Hint: section 8.1.2.)
13. Explain the term 'charge-trapping' and describe a method for overcoming the problem. What are the consequences of the solution? (Hint: section 5.4.)
14. A certain CCD requires a 'fat-zero' or 'preflash' of 400 electrons. If it has a readout noise of 15 electrons what will be the final noise when the preflash is used? (Answer: 25 e.)
15. (a) Give a simple formula for calculating the signal-to-noise ratio of an object spread over 3×3 pixels on a CCD in the background-limited regime. (Hint: Appendix 4.)

(b) What is the total noise in electrons for a pixel in a CCD spectrograph which yields 49.33 e/min of sky background, a dark of 4 e/min and a readout noise of 10 electrons if the integration time is 15 min? (Answer: 30 electrons.)
16. What gives rise to the phenomenon of 'fringing' in some CCDs? (Hint: section 8.4.)
17. A certain CCD camera system seems to be showing a faint but regular pattern of near-horizontal, slightly sloping lines on bias frames. What could be wrong? (Hint: section 5.1.)
18. If a CCD camera system is using a 16-bit A/D how many levels of digitisation are there? If the wire carrying bit 14 breaks and the line sits 'high', what will be the consequences? (Hint: section 8.1; Answer: 16384 counts will be added to everything.)
19. (a) If a CCD system is read-noise limited, is it better to integrate twice as long 'on-chip' to improve signal-to-noise performance or is it better to repeat the measurement a second time? (Hint: Appendix 4.)

(b) A useful figure-of-merit for a detector is the exposure time required to become background-limited. If the only sources of noise are readout noise R and the background photoelectron rate B (e/s), give the expression for t, the time required for both noise values to be equal. (Hint: Appendix 4.)

(c) If $B=100$ e/s/pixel and $R=10$ electrons, how long should the integration be in the above (b) example? (Answer: 1 s.)
20. Why are CCDs not sensitive to light with a wavelength longer than 1.1 μm? What materials can be used to make array detectors for long wavelengths? What are the attractions of imaging in the near-infrared for astronomy? (Hint: section 4.2.1; Table 4.2; Chapter 9.)
21. Explain what is meant by 'thermal' background in the infrared. (Hint: section 9.1.1; Fig. 9.1.)
22. What is meant by 'Fano-noise-limited' CCDs? (Hint: section 9.3.)
23. A CCD camera uses a 516×516 pixel sensor, digitises pixel values to 16 bits and makes 100 images per night, how many megabytes of data does this correspond to? Why would this number be an underestimate of the actual amount of data storage required? (Answer: 53.25 Mbytes; Hint: section 7.2.5.)
24. What is meant by the term 'false-colour'? An image display is described as having '8 bit-planes', how many intensity levels does this correspond to? If the range in brightness in the data is 1 to 10000 DN, how many data numbers correspond to one intensity level in the display assuming a linear relationship? (Answers: 256; 39 DN.)

Glossary

aberrations Effects associated with the performance of optical components which give rise to imperfect optical images.

absolute zero A temperature of −273°C or 0 Kelvin (K). The temperature at which an atom would have no energy of motion.

alt–azimuth mounting A form of mounting similar to that of a radar which allows the telescope tube to be moved horizontally (by rotation in azimuth or compass direction) and vertically (by rotation in altitude or elevation). To follow a star the telescope must be adjusted simultaneously in both axes.

anti-reflection coating (AR coating) A layer of material of lower refractive index of just the right thickness (1/4 wave) is deposited on the optical surface to be coated. More complex coatings are possible which cover a large wavelength range.

arcminute A measure of angular distance equal to 1/60th of a degree. There are 90 degrees in a right angle and 360 degrees in a circle.

arcsecond A measure of angular distance equal to 1/60th of an arcminute or 1/3600th of a degree.

aspherical An optical surface designed with departures in shape from a perfect sphere in order to cancel optical imperfections or aberrations.

astrometry The accurate determination of the positions and motions of celestial objects.

back focal length The distance between the last surface of a compound optical system and the focal plane of the system. This distance may be quite different from the actual focal length.

band gap *See* **forbidden energy gap.**

baud rate The number of high/low transitions per second on a line.

binning On-chip binning. Combining data from a rectangular group of pixels on the CCD itself by appropriate clocking. An improvement in signal-to-noise ratio is achieved at the loss of some picture detail (or spatial resolution).

birefringent *See* **refractive index.**

blackbody This is the name given to an object which absorbs *all* radiation incident on it. Such a body is also the most efficient emitter of radiation. The simplest model of a blackbody is a small hole in a matt-black, heated hollow chamber. Since radiation entering the hole has minimal chance of re-emerging, radiation leaving the hole will be characterised by the temperature of the chamber. The amount of radiant energy emitted by a blackbody at any wavelength depends only on the absolute temperature of the body, and is given mathematically by the **Planck function**. Certain astronomical sources are almost blackbody emitters, but most objects are not quite black enough and therefore have an **emissivity** of less than 100% of ideal.

buried channel A construction in a silicon CCD which results in a collection zone for photogenerated electrons which is buried well below the surface of the silicon. The zone is produced by a combination of a pn junction within the silicon and a metal electrode on the surface of the silicon.

bus The general term for hardware for dealing with the input–output pathway and backplane of a computer. There are many types e.g. IEEE-488 bus, Q-bus, VME-bus.

byte A group of eight 'bits' or binary digits. Two bytes (16 bits) make a word. A **megabyte** is one million bytes and a **gigabyte** is one thousand megabytes.

calibration A general term used to describe the process of converting initial measurements made with a piece of apparatus into final form on an absolute scale or on an agreed 'standard' scale of reference. For example, converting the number of photoelectrons per second in a CCD image to astronomical magnitudes.

Cassegrain focus Refers to a compact design of reflecting telescope in which the light collected and focussed by the large concave primary mirror is refocussed by a smaller convex secondary mirror on the same axis as the primary. The refocussed beam passes through a central hole cut into the primary mirror and emerges behind the primary.

CCD Charge-coupled device. A small photoelectronic imaging device (typically 1.5 cm square) made from a crystal of semiconductor silicon in which numerous (at least 250 000) individual light-sensitive picture elements (pixels) have been constructed. Each tiny pixel (less than 0.03 mm in size) is capable of storing electronic charges created by the absorption of light. The name derives from the method of extracting the locally stored charges from each pixel which is done by transferring or 'coupling' charges from one pixel to the next by the controlled collapse and growth of adjacent storage sites or 'potential wells'. Each 'well' is formed inside the silicon crystal by the electric field generated by voltages applied to tiny, semi-transparent metallic electrodes on the CCD surface.

CDS Correlated double sampling. A technique used with CCDs to remove an unwanted electrical signal, associated with resetting of the tiny 'on-chip' CCD output amplifier, which would otherwise compromise the performance of the detector. It involves making a double measurement of the output voltage before and after a charge transfer and forming a difference to eliminate electrical signals which were the same, i.e. correlated.

Cepheid variables A class of stars named after Delta Cephei which vary in brightness over a regular period of time (typically a few days). The period of change is directly related to the true average brightness or luminosity of the star. Once the period is known the true brightness can be calculated and the distance estimated by observing the 'apparent' brightness of the object as seen from Earth.

channel stops Narrow, heavily doped strips in a silicon CCD which act like walls to prevent sideways movement of charge in a pixel.

charge-coupled device *See* **CCD.**

charge traps *See* **deferred charge.**

clock frequency The rate at which a CCD is clocked or read out. It is the reciprocal of (or one divided by) the pixel time; e.g. 40 microseconds (μs) per pixel corresponds to 25 kiloHertz (kHz).

clocking The process of raising and lowering the voltages between two levels — high and low — on the electrodes or gates of a CCD in order to move charges from one pixel to the next.

coaxial cable A type of electrical wiring. It consists of a copper wire surrounded by an insulator which in turn is surrounded by a braided copper shield which is encased in another plastic insulator.

common-user A shared facility or common resource, such as a CCD spectrograph, a computer data reduction program or even a telescope, which has been carefully designed to meet the needs of many research programmes. Common-user facilities are expected to be very reliable and well-supported.

conduction band The combined unoccupied higher energy levels or orbits of all the atoms in a semiconductor crystal.

corrector plates Thin lens-like optical pieces which remove certain optical aberrations.

coronene The first ultraviolet phosphor to be tried on the surface of a CCD.

CPU Central processing unit. The part of a digital computer responsible for interpreting and executing instructions.

critical path A term used in project planning to indicate a segment of the proposed work which if not completed on time will result in one or more other segments being delayed with serious 'knock-on' effects for the project.

cryostat A vacuum chamber containing a large reservoir of some liquid cryogenic material, such as liquid nitrogen or liquid helium. Any components attached to the cold face of the reservoir will be cooled down.

CTE Charge transfer efficiency. A term used to characterise the amount of charge successfully moved from pixel to pixel in a CCD. It is usually expressed as a fraction such as 0.99999X per pixel transfer, meaning that 99.999X% of the charge is moved on each time; X stands for a number less than nine.

dark current Signals generated in a detector such as a CCD merely by the heat energy of the atoms which, through random jostling, results in the production of free electrons. Dark current is greatly reduced by physically cooling the detector.

data sheet A manufacturer's document giving electrical and other specifications for the operation of a dcvicc such as a CCD.

dc bias Direct current (dc) fixed voltages.

deferred charge There are a number of phenomena in CCDs which result in charges becoming 'trapped' during the charge-coupling process. Some of these traps are due to design faults, others are due to imperfections in the raw materials or difficulties introduced during fabrication. The global effect of charge trapping is poor charge transfer efficiency. Traps often release captured charge at a later time leading to the presence of charge in trailing pixels which should have been empty, that is, the readout of that charge has been delayed or deferred.

depletion The region of a semiconductor in which an electric field has swept out any free charge carriers such as electrons. The field can be applied by a voltage to a metal electrode on the surface of the semiconductor.

diffraction grating An optical device containing thousands of very fine parallel grooves which produce interference patterns in a way which separates out all the components of the light into a spectrum.

diffraction of light A wave-like property of light which allows it to curl around obstacles whose size is about that of the wavelength of the light. The disturbed waves then interfere to produce ripple-patterns.

DMA Direct memory access. An efficient electronic method of transferring digital (numerical) information or data to a computer.

DN Data number. One count or one bit; the smallest unit of an analogue to digital conversion system. For example, a 10-V 16-bit system converts the range 0–10 V into 65 536 levels or data numbers.

doping The process of selectively adding known amounts of foreign atoms into the silicon (or other semiconductor) crystal to subtly change its electrical properties.

Doppler effect The apparent change of frequency or wavelength of radiation from an object due to its motion toward or away from us. If the object is receding the frequency is decreased and the wavelength is increased, i.e. becomes red-shifted.

down-converter *See* **fluorescence.**

down-loading The process of passing a microprogram from a main computer to a microprocessor.

DQE Detected quantum efficiency. The quantum efficiency of an idealised system with no readout noise but which yields the same signal-to-noise-ratio as the real detector.

drift-scan *See* **TDI.**

dynamic range The ratio of the maximum possible signal, the saturation level, to the readout noise.

EBCCD Electron-bombarded CCD. An imaging device containing a thin photosensitive target which emits electrons when illuminated and then magnetically focusses these electrons to impact onto a silicon CCD where they generate a large charge.

echelle This is a type of diffraction grating with groove angles of 90°. With the grating at an angle of 45° the grooves resemble a staircase.

electrodes Also called **gates** or **phases**. Small electrically conducting plates connected to a voltage source (battery or power supply) and arranged in strip patterns to define the picture elements or pixels of a CCD. The plates create an electric field within the semiconductor which therefore forms a storage site for photogenerated charges.

electron A subatomic particle of very small mass and negative electric charge which 'orbits' the nucleus of an atom.
electron–hole pairs When a photon is absorbed in silicon its energy causes an electron in the valence band to be ejected into the conduction band leaving a (positively charged) vacancy or hole in the valence band.
energy levels The allowed energy states for electrons orbiting the nucleus of an atom.
epitaxial A thin layer of differently doped semiconductor used in the construction of solid-state devices such as the CCD.
EPROM Erasable programmable read only memory. A small silicon chip containing thousands of individual locations which can be set to either a low or a high voltage level; a 0 or a 1. The settings can be erased by exposure to ultraviolet light.
Ethernet A single coaxial cable serial link.
expansion of the universe There are billions of other galaxies and all except the closest ones (the Local Group) are receding from us as deduced from the Doppler red-shift effect of their spectra. Thus the universe appears to be expanding. Moreover, the greater the distance the faster the speed of recession.
extinction The combined effects of absorption and scattering of light, usually by interstellar dust, which dims our view of an object.

false colour The use of colours, instead of shades of grey, on a computer image display screen to represent different brightness levels and highlight very small differences in a dramatic way. For example, in an ordinary black-and-white image, objects which differ only slightly in brightness appear as almost the same shade of grey and are hard to distinguish. If instead the numerical brightness values are assigned carefully chosen colours then two objects of almost equal brightness will be strongly distinguished when the image is displayed.
fast interface states *See* **traps.**
fat zero *See* **preflash.**
FET Field-effect transistor. A form of semiconductor amplifier which behaves much like a variable resistor which is controlled by the electric 'fields' associated with a junction between two differently doped regions.
fibre-optic A long, thin strand of glass capable of excellent transmission of light over large distances.
field of view The patch of sky or of any image scene which can be seen by an optical system or by one picture element (pixel) of a detector system. Usually expressed in angular measure.
FIFO First-in first-out buffer. Used in many microprocessor-controlled systems to smooth the flow of data.
flash gate An ultra-thin transparent electrode across the entire back surface of the TI 800×800 CCDs used to control the charge on the back surface and hence the quantum efficiency for blue and ultraviolet light.
flat-field An extensive, very uniformly illuminated source of light used to determine the relative sensitivity of an imaging detector composed of many picture elements (pixels) each with a slightly different response to light.
fluorescence The emission of light at one wavelength, the green say, following

absorption of light with a much shorter wavelength such as the ultraviolet. The ultraviolet photon parts with its energy by ejecting the electron into a high energy level from which it cascades back down, releasing photons of lower energy and therefore longer wavelengths. A material which has this property is called a **phosphor.**

focal plane scale The relationship between angles on the sky, in seconds of arc, and millimetres of size at the focus of the telescope; i.e. the number of arcsecs per mm.

focal reducer An optical component or system for changing the image scale of a telescope to achieve a better match between the seeing disk and the pixel size. *See* **optical matching.**

forbidden energy gap The unoccupiable interval of electron energy levels which forms in a crystalline substance (that is one having a periodic atomic formation) between the valance band (bound electrons) and the conduction band (free electrons). The forbidden energy gap is very large in insulators, non-existent in metals, and small but finite in a semiconductor.

forward bias A term applied to an electronic device known as a diode — usually formed by a junction of p-type and n-type semiconductor material — in which current flows easily if the externally applied voltage has the correct polarity or direction. If the opposite polarity is applied there is almost no current flow and the device is said to be **reverse biased**.

frame transfer A CCD construction in which one half of the imaging area of the device is purposely covered with a mask opaque to light to provide a temporary charge-storage section.

frequency Used to characterise electromagnetic radiation, it is the number of waves or cycles per second passing a given point. The product of the frequency and the wavelength is the speed of light.

gain The amplification factor.

galaxy A colossal system of billions of stars and the raw gaseous material (mostly hydrogen) from which the stars are formed. The Solar System is part of a galaxy of 200 billion stars. This system is known simply as the Galaxy.

getter A chemical absorption method of removing (pumping) gas from a chamber by tieing up molecules on a surface.

grism This is a right-angled glass prism with a transmission diffraction grating deposited on the hypotenuse surface. The spectrum produced by the grating is deflected to remain along the optical axis and the apex angle of the prism is chosen to get a certain wavelength in the centre.

ground-loop A condition in which two pieces of apparatus are connected together while each has a separate earth connection and these are not at identical potentials. A current will flow and small signals will be affected by fluctuations, called noise.

hold-time The time taken to use up all the liquid cryogens, like liquid nitrogen in a cooled CCD cryostat.

host computer The main or master computer in an instrumentation system. The computer responsible for interacting with the user.

hybrid array A device in which the roles of radiation (infrared mostly) detector and signal multiplexer are separated. The device is a sandwich of two slabs.

image display A high-resolution TV monitor device attached to a computer for visualising digital data such as CCD images.

image intensifier An electronic device for increasing the brightness of a faint optical image. The image is first formed on a thin metallic surface called a photocathode from which electrons are then ejected. The stream of electrons is accelerated and focussed onto a phosphorescent screen which glows brightly as a result of the impact.

infrared The region of the electromagnetic spectrum from a wavelength of about 1 μm to 200 μm.

InSb Indium antimonide. A compound semiconductor used as an infrared photoelectric detector.

integrated circuit A small electronic component made of semiconductor silicon on which an entire electronic circuit of numerous microscopic transistor amplifiers, diodes and resistors has been constructed. The basis of this technology is that these electronic components can actually be made in miniaturised form by selectively doping tiny regions of the silicon crystal.

integration time The interval of time used to collect photons of light on a detector and build-up a strong signal.

interline transfer A CCD construction consisting of vertical strips which are alternately opaque and light-sensitivie. The opaque strips conceal charge-transfer registers.

interstellar dust Small grains or particles in the interstellar medium.

inverse square law The intensity of light falls as $1/r^2$, where r is the distance from the source.

IPCS Image photon-counting system. A form of very low light level detector used in astronomy. By means of an image intensifier the IPCS is capable of counting individual photons of light.

IR Infrared radiation.

IRAS Infrared Astronomical satellite.

JHKLMN Designations for parts of the infrared waveband transmitted by the Earth's atmosphere in the wavelength range 0.001 to 0.020 mm.

kTC noise *See* **reset noise.**

LAN Local area network. A means of interlinking computers.

LHe The symbol for liquid helium. The temperature of liquid helium is normally 4°K, that is, 4 degrees above absolute zero.

line spectrum A pattern of lines, each corresponding to an image of the entrance slit of the spectrometer, seen when light is either emitted by or interrupted by a hot rarified gas such as hydrogen. The pattern is characteristic of the gas and the wavelength at which the features are observed to occur is indicative of the velocity of the object.

LN_2 The symbol for liquid nitrogen. The temperature of liquid nitrogen is normally 77K, that is, 77 degrees above absolute zero.

LSB Least significant bit. *See also* **DN**.

luminescence Light emission caused by currents through faulty parts of, for example, a CCD structure which behave like a light-emitting diode.

magnitude A logarithmic brightness scale for astronomical objects such that each difference of one magnitude corresponds to a ratio of brightness of 2.5. Therefore a ratio of 100 in brightness is equivalent to a difference of 5 magnitudes, a ratio of 250 corresponds to 6 magnitudes and so on. Fainter stars have numerically larger magnitudes. The starting point of the scale is such that the star Vega is approximately 'zero magnitudes' in brightness. Objects brighter than this such as Sirius, or the nearest planets, have negative magnitudes.

median A mathematical estimator of the true value of a set of values when some of these values are contaminated, i.e. known to be much larger than average.

megaFLOPS Millions of floating-point operations per second. A computer benchmark.

mer–cad–tel Mercury–cadmium–telluride (HgCdTe). A semiconductor alloy useful as an infrared photoelectric detector.

Metachrome A chemical coating developed by Photometrics Ltd. which can be applied to the surface of a CCD by vacuum deposition to enhance its ultraviolet sensitivity by fluorescence.

microchannel plate A compact electrostatic high-voltage multiplier attached to a photocathode.

micrometre (μm) One millionth part of 1 metre (m); one thousandth part of 1 millimetre (mm).

MIPS Millions of instructions per second. A computer benchmark.

MOS Metal–oxide semiconductor. A construction used to fabricate microelectronic components, including CCDs, which consists of three layers, namely a metal conductor, an insulating layer usually made from an oxide of silicon, and a semiconductor such as silicon.

native oxide The silicon dioxide layer which grows in air on the exposed backside surface of a thinned CCD.

noise A term used to describe unwanted electronic signals, sometimes random sometimes systematic, which contaminate the weak signal from an astronomical source. Types of noise include, $1/f$ noise which get smaller at high frequencies, white noise which is independent of frequency, and reset noise which can be removed by correlated double sampling.

observing Experimental astronomy, at all wavelengths, is a passive act — unlike in a physics lab where conditions can be changed — and is therefore called 'observing'. Large, well-maintained or innovative astronomical facilities are expensive and astronomers must compete through peer review evaluations of their proposals in order to win 'observing time'.

optical matching The use of lenses or other optical devices to match the size of the image of the seeing disk, as it appears in the focal plane of the telescope, to the physical size of the CCD pixels. If the telescope yields 10 arcseconds per mm and the seeing is 1 arcsecond then the image is 0.1 mm in size. But a typical CCD pixel is 0.022 mm, five times smaller.

out-gassing The absorbed gases released from the interior walls and components of a vacuum chamber which has already been 'roughed-out'.

overscan Additional clock pulses in both the horizontal and vertical directions in excess of the actual number of real pixels in order to sample the electronic offsets in the system.

phosphor *See* **fluorescence.**

photocathode A thin metallic plate housed inside an evacuated tube capable of releasing electrons through the 'photoelectric effect' when illuminated by light. These surfaces are best for optical and ultraviolet light.

photoconductivity Absorption of light increases the number of charge carriers.

photoelectric effect The forced ejection of electrons from certain metals due to the transfer of energy from incident 'photons' of light. A demonstration (by Einstein) of the quantum effect. If the ejected electrons are collected as a charge or a current then the intensity of the incident light can be measured. Almost all modern detectors rely on some variant of the photoelectric effect.

photometry The quantitative measurement of the intensity of light.

photomultiplier tube A vacuum-encapsulated photocathode from which electrons are ejected by the photoelectric effect followed by multiple cathodes from which many additional electrons are emitted in a cascade. When finally collected, the original single electron may have generated a pulse of over 1 million electrons.

photon The term used most often to describe the basic package or 'quantum' of electromagnetic energy (light). Photomultiplier tubes are capable of detecting single photons.

photovoltaic effect Absorption of a photon leads to the production of a voltage across a junction.

polarisation modulator An optical device sensitive to the plane of vibration of electromagnetic waves, i.e. to their polarisation. It is used to convert the polarisation of light into a measurable brightness change.

polysilicon A non-crystalline form of silicon with a high conductivity like a metal; preferred in CCD manufacture to the use of metals because it keeps the entire process in silicon and is more transparent to visible light.

preamplifier A low noise amplifier designed to be located very close to the source of weak electronic signals, but capable of delivering amplified signals down tens of metres of cable.

preflash The technique of illuminating the CCD with a low light level flash before beginning a long exposure in order to 'fill up' any charge traps.

prime focus The focal point of the large primary reflecting mirror in astronomical telescopes when the light source is extremely distant. This focus actually falls at a point just within the upper structure of the telescope itself and is therefore accessible to CCD cameras and other instruments; it provides a large field of view.

printed circuit A compact double-sided circuit board with no wires but instead fine tracks, etched on a copper-clad board, perform the same function.

PROMS *See* **EPROMS.**

PSF Point spread function. The size and shape of the actual image of a point source as a result of the combined effects of atmosphere, optics, and guiding.

QEH Quantum efficiency hysteresis. An increase in quantum efficiency after exposure to light.

QE pinned The quantum efficiency of thinned TI CCDs can be driven to the maximum by ultraviolet flooding and pinned there by immediate cooling.

quadrature A means of combining estimates of errors or unwanted signals — noise — by summing the 'square' of the values and then taking the 'square root'.

quanta Light can carry energy only in specific amounts, proportional to the frequency, as though it came in packets. The term quanta was given to these discrete packets of electromagnetic energy by Max Planck.

quantum efficiency (QE) The ratio of the number of photoelectrons released for each incident photon of light absorbed by a detector. This ratio cannot exceed unity.

quasars or **QSOs** Compact-looking objects, often radio sources, with emission lines in their spectra which are displaced by very large amounts towards the red. These redshifts correspond to velocities which are a large fraction of the speed of light, and hence these objects are believed to lie at great distances.

RAM Random access memory. A silicon microchip capable of temporary storage of information in the form of binary digits, either 0 or 1, and enabling rapid access to any part of its storage area.

ray tracing Computer simulation of light ray paths through an optical system.

refractive index A number which characterises the properties of a material when light is transmitted through it. A vacuum has a refractive index of 1.0; air is fairly close to this also. As the refractive index rises, light waves travel more slowly in the material, more light is reflected and the bending or refraction of light is greater. Some materials have a different refractive index depending on the plane of vibration of the electromagnetic wave, they are called **birefringent.**

reset noise The unwanted and uncertain electrical signal transmitted to the output pin of a CCD during the process of recharging, via the **reset transistor**, the output storage capacitor to its preset value in readiness for the next pixel charge.

reset transistor *See* **reset noise.**

resolution The ability of an optical system, including detector, to separate two adjacent objects — this is called 'spatial resolution' — or two adjacent wavelengths in a spectrometer — this is called 'spectral resolution'.

seeing A term used to describe the blurring of a stellar (point-like) image due to turbulence in the Earth's atmosphere, both at high altitudes and within the telescope dome. Seeing estimates are often given in terms of the full width in arcseconds of the image at the points where the intensity has fallen to half its peak value. The typical value at a good site is a little better than 1 arcsecond.

semiconductor A material like silicon or germanium in which the valence band and the conduction band are separated by a small (forbidden) energy gap. Such materials have some of the properties of a good electrical conductor — in which the energy gap is zero — and some of the properties of an insulator — in which the gap is very large.

serial (horizontal) register The final row of a CCD in which the controlling electrodes are arranged at right angles to those on the rest of the CCD. This enables charges coupled onto this row to be transferred in single-file through the CCD output amplifier.

sequencer That part of an electronic system responsible for the accurate phasing of time-critical events such as CCD clocking and readout.

slew The relatively rapid motion of a telescope (under computer control) as it moves to point at a new position in the sky. Once at the new position the motion of the telescope returns to that required to cancel the effect of the Earth's rotation relative to the stars — the sidereal rate.

software A general term used to describe the sets of instructions or programs which cause an electronic computer to perform some action, and to distinguish this process from a purely physical or hardware cause.

speckle A technique of obtaining rapid snapshots of a point-like source like a star in order to essentially freeze for an instant the turbulent blurring of the atmosphere.

spectroscopy The quantitative study of the spectral or energy content in a beam of light. Usually achieved by spreading or dispersing the light into its constituent wavelengths by means of a diffraction grating and recording the image with a CCD or other electronic imaging detector. The patterns, wavelengths, intensities and shapes of the spectral features yield considerable physical information.

surface channel A semiconductor device construction in which the electron charges are held or moved near the surface of the silicon crystal.

TDI Time delay and integration. **Drift-scan.** Methods for averaging the response of a CCD along columns by reading out at the same rate as a mechanical motion is shifting the optical image along the CCD.

three-phase A CCD construction in which three overlapping metal electrodes are used to define a pixel and effect the transfer of charge, in either direction along a column, by the charge-coupling method. If only two electrodes are used then the device is **two-phase.**

throughput A measure of the efficiency of an optical system, formally given by the product of the solid angle it can accept multiplied by its effective aperture.

timing waveform A diagram showing the time sequence and voltage levels of a stream of pulses required, for instance, to perform charge-coupling in a CCD.

traps Poorly understood irregularities in the silicon crystal lattice which can absorb free charges created in the semiconductor by, for instance, the absorption of light.

twisted-pair A form of wiring consisting of two strands of single wire twisted together to form a transmission line. One strand carries the signal and the other is grounded to the single 'star' ground point of the system.

UBVRI Designation for parts of the optical waveband, isolated by means of special glass filters which eliminate the unwanted regions, and used for standard astronomical intensity measurements.

UV Ultraviolet radiation. That part of the electromagnetic spectrum beyond the violet end of the optical spectrum with wavelengths from 0.35 μm (millionths of a metre) down to 0.01 μm. Beyond this is the extreme ultraviolet (EUV) and beyond that at wavelengths of less than 1/10 000th of a micrometre is the X-ray region.

valence Valence band or valence electrons; the electrons in the outermost orbit.

virtual phase A type of CCD in which only one electrode is physically outside the silicon and is such as to obscure only half of the pixel. A specially doped layer under the transparent part acts as another or virtual electrode.

wavelength The effective distance between intensity peaks in the electromagnetic waves of light. The product of the frequency of the waves and the wavelength is the wave-speed — the speed of light. For single energy packets of light (individual photons) rather than beams, it is still possible to define a wavelength by noting, as Einstein did, that the energy of the photon E is related to the frequency ν by the simple equation $E=h\nu$, where h is a universal constant known as Planck's constant.

white dwarf Faint very compact stars at the end of their life.

workstation A single-user image processing set-up with a powerful small computer, a colour image display and access to a data storage system (a magnetic disk).

X-rays A large band of electromagnetic radiation with wavelengths smaller than extreme ultraviolet light. A typical X-ray photon has over one thousand times as much energy as a photon of visible light.

zeropoint The magnitude corresponding to one data number per second from the CCD for a star of zero colour term (like Vega) corrected for absorption in the Earth's atmosphere.

Appendix 1: Physical and astronomical constants

Constant	Symbol	Value
Speed of light in vacuum	c	2.9979×10^5 km/s
Charge on the electron	e	1.6022×10^{-19} C
Planck constant	h	6.6262×10^{-34} J s
Boltzmann constant	k	1.3807×10^{-23} J/K
Mean Earth–Sun distance = 1 Atronomical Unit (AU)		1.4960×10^8 km
1 Light year	ly	9.4605×10^{12} km
1 Parsec	pc	3.0857×10^{13} km = 3.2616 ly
Radius of the Sun	$R_\odot$	6.96×10^5 km
Mass of the Sun	$M_\odot$	1.99×10^{30} kg
Luminosity of the Sun	$L_\odot$	3.83×10^{26} J/s

Appendix 2: Units of measurement and powers of 10 notation

Quantity		Units	
Time		seconds, s	
Mass		kilograms, kg	
Length		metres, m	
Charge		Coulombs, C	
Temperature		Kelvins, K	
Derived units			
Energy	E	Joules, J	
		Electron volts, eV	
Power	P	Watts, W	1 W=1 J/s
Current	I	Amperes, A	1 A=1 C/s
Potential	V	Volts, V	1 V=1 J/C
Capacitance	C	Farads, F	1 F=1 C/V
Resistance	R	Ohms, Ω	1 Ω=1 V/A
Frequency	v	Hertz, Hz	1 Hz=1 cycle/s

Notes:

(1) One **electron volt** is the energy gained by an electron of charge e after being accelerated by a potential (V) of 1 V, that is $1\ \text{eV}=e\times 1\ \text{V}=1.6022\times 10^{-19}\ \text{J}$.

(2) The product RC of a resistance R and a capacitance C, has the units of **time,** and is known as the 'RC time constant'.

(3) The product of the wavelength of light (λ) and the frequency (ν) is the speed of light, that is, $\lambda\ \nu=c$.

(4) The energy of a photon of electromagnetic radiation (E) is the product of the Planck constant (h) and the frequency (ν), that is

$$E=h\,\nu.$$

But since $\nu=c/\lambda$ we also have $E=hc/\lambda$. The constant hc has the value 1.99×10^{-28} J km or if instead we express 1 km as 10^9 μm (micrometres) then

$$hc=1.99\times10^{-19}\ \text{J}\ \mu\text{m}$$

POWERS OF 10 NOTATION

When writing down numbers which are very large or very small it is useful to introduce a shorthand notation called 'powers of 10'. Thus, instead of writing 1 000 000 for one million we write 1×10^6 and understand this to mean 1 followed by six zeros. Similarly, instead of writing 0.000 001 for one millionth we write 1×10^{-6} where the minus sign tells us that this number is one *divided* by one million.

Names and symbols are given to the most frequently used powers of 10 as shown below. These names and symbols can be prefixed to any of the units of measurements given above to infer that the basic unit is to be multiplied by that power of ten.

Name	Symbol	Power of 10
femto	f	10^{-15}
pico	p	10^{-12}
nano	n	10^{-9}
micro	μ	10^{-6}
milli	m	10^{-3}
centi	c	10^{-2}
kilo	k	10^{3}
mega	M	10^{6}
giga	G	10^{9}
tera	T	10^{12}

Commonly used Greek letters are

alpha α	beta β	delta δ	epsilon ε	eta η
theta θ	lambda λ	mu μ	nu ν	pi π
sigma σ,Σ	phi ϕ	omega Ω		

Appendix 3: Astronomical magnitude scale and its relation to lux and watts

The range of brightnesses of astronomical objects is so great that it is useful to find a way to 'compress' that range. This can be done by using a **logarithmic scale**. Astronomers call their logarithmic scale for brightness a **magnitude scale**. The magnitude m of a star of brightness b is given by

$$m = \text{constant} - 2.5 \log b$$

where the 'constant' has a value which determines the scale. Notice that the fainter the object the larger the value of m. When we have two stars of different brightnesses b_1 and b_2 then we can obtain their magnitude difference $m_1 - m_2$ from

$$m_1 - m_2 = 2.5 \log (b_2/b_1)$$

In these equations log represents the logarithm to the base 10; a logarithm of a number to the base 10 is the **power of 10** which exactly yields the number. The magnitude difference corresponding to a brightness ratio of 100 is therefore 2.5 $\log(100) = 2.5 \times 2 = 5$. To find a brightness ratio from a known magnitude difference we can turn the equation around to give

$$b_2/b_1 = 10^{0.4(m_1 - m_2)}$$

so that by raising 10 to the power of $0.4 \times (m_1 - m_2)$ we get the ratio b_2/b_1.

The logarithm or the **anti-logarithm** — i.e. the value corresponding to 10 to the power of log(number) — can be obtained from mathematical tables and from scientific pocket calculators which usually have 'push-button' evaluation of the log and anti-log of any entered number.

To determine a magnitude from a brightness we need to establish either a relative scale or an absolute scale or both. Astronomers use a relative scale such that the magnitude of the star Vega (Alpha Lyrae) is nearly zero at all wavelengths. On that scale stars fainter than sixth visual magnitude ($m = 6$) cannot be seen by the unaided eye, the brightness of 1 square arcsecond of dark (moonless) sky at an excellent mountain site in the visible (V) waveband is about $m = 21$; CCDs have enabled stars and faint galaxies to be detected to levels fainter than $m = 27$, which is 63 billion times fainter than Vega.

It is also possible to establish an absolute scale. It has been determined by careful experiments that at a wavelength of 0.54 μm in the middle of the visible spectrum that an $m = 0$ star of the same type as Vega produces 3.92×10^{-12} watts per square centimetre per micrometre interval of wavelength, just outside the Earth's atmosphere. This is the **irradiance per unit wavelength**. Notice that the dimensions here are energy per unit time per unit area per unit wavelength interval, and the units are well-known and fundamental. A few other useful terms are as follows, the total amount of energy emitted by a light source per second is its **radiant power**, measured in watts, and the power emitted in a cone-shape about a particular direction (that is, per unit solid angle) is called the **radiant intensity** (watts per steradian).

In the electro-optical industry the term **illuminance** is used to describe the amount of light received per unit surface area and it is measured in a unit called **lux**. One lux is the illuminance produced by a standard light source of one **candela** at a distance of one metre and 60 candelas is the luminous intensity of a 1-cm^2 blackbody at the temperature of melting platinum (2,042 K). One lux is therefore also equivalent to 1 **lumen** per square metre. Such units are complicated and are never used in astronomy or scientific research in general; they are often found in reference to CCD-based cameras intended for low light level television applications, however. An illuminance of 1 lux is approximately equivalent to a star of visual magnitude -14. Note the minus sign which indicates that this source is very much brighter than Vega, in fact it is roughly equivalent to the full moon! The brightness in lux of any source in the 'visual' waveband is

$$B(\text{lux}) = 10^{-0.4(m_v + 14)}$$

Again note the minus sign. The illuminance of Vega ($m_v = 0$) is only 0.000 002 lux. Ordinary hand-held commercial TV 'camcorders', even those with CCDs, operate at frame-rates of 1/60th second and are typically designed with one-inch lenses to image scenes with light levels greater than a few lux (bright moonlight). Some intensified CCD systems are capable of imaging with illumination levels as weak as 0.02 lux. Of course, while a 2-lux CCD camcorder cannot actually take images in a room where the illumination level is equivalent to that from the star Vega, it could actually image the star itself if its standard lens was removed and it was attached to a modest telescope. The reason is simply that the telescope collects more light and, more importantly, focusses all of the available light onto a single (or small number) of pixels so that the illuminance for that pixel is increased by several hundred thousand times. If the CCD is cooled to allow integration times of a few seconds, instead of 1/60th second, then objects can be recorded which are several hundred times fainter still. Cooled, slow-scanned CCD cameras with very long integration times on the world's largest telescopes are therefore very sensitive indeed!

Appendix 4:
Calculating the signal-to-noise ratio of a CCD measurement

A detailed calculation of the signal-to-noise ratio (S/N) of a CCD-type detector is quite complicated and depends on how closely one wishes to 'model' the CCD behaviour. Here we will assume that the CCD electronics system is perfect and so there are no other sources of unwanted noise save the irreducible minimum readout noise. The simplest approach to analysing a CCD image is to construct a 'final frame' by performing the following steps to subtract dark current (including bias) and normalise with a flat-field derived from images of blank sky:

$$\text{Final frame} = \frac{\text{Object frame} - \text{Dark frame}}{\text{Sky frame} - \text{Dark frame}}$$

We assume that the only sources of 'noise' are

(1) readout noise (R electrons)
(2) photon noise on the signal (S) from the object,
(3) photon noise on the signal (B) from the sky background,
(4) noise on the dark current signal (D).

where the signals S, B and D in electrons per second are derived from observed counts in data numbers (DN) by multiplying by the transfer factor g electrons/DN, and we assume that the noise sources associated with the signals S, B and D all behave according to Poisson statistics so that the noise is equal to the square root of the number of electrons. To find the original photon arrival rate for S or B we divide by the quantum efficiency (QE). To find the number of counts recorded by the A/D system (DN or data numbers) we divide by the conversion factor g electrons per DN.

The noise terms can be added in quadrature (i.e. by squaring and adding), and we must allow for the noise sources in each of the two calibration frames—sky and dark. We can further assume that the number of object, sky and dark frames combined or 'co-added' to form the final frame is not necessarily equal, but for simplicity we will assume that the normal practice pertains of keeping the exposure time (t) the same. Suppose we have N_S object frames, N_B sky (background) frames, and N_D dark frames.

Let,

$T = t\,N_S$ be the total integration time accumulated on the object frame

$f = S/B$ be the ratio of the source signal to that of the background signal per pixel

then, for a source covering n pixels on the CCD, we can estimate a total signal-to-noise ratio of

$$S/N = \mathbf{S}T^{1/2}/[u_r^2 + \mathbf{S} + \sum_{i=1}^{n}\{(B + D + R^2/t) + \epsilon_D(D + R^2/t) + (1+f)^2\epsilon_B(B + D + R^2/t) + (1+f)^2\epsilon_D(D + R^2/t)\}]^{1/2} \quad \text{(A4.1)}$$

where $\mathbf{S}$ is the total signal summed over n pixels, i.e. $\mathbf{S} = \Sigma\,(S)$ and u_r represents the average, over n pixels, of any residual error due to failure in the flat-field. Note that B, D and f are not the same from pixel to pixel.

We need to distinguish the cases of point sources and extended objects. A stellar object has a seeing disk of diameter b arcseconds corresponding to the full width at half maximum intensity (FWHM), whereas each CCD pixel is a square a arcseconds on a side. If $\pi b^2/4$ is less than a^2 then almost all of the light falls on a single pixel, otherwise a star image will occupy about

$$n = \frac{\pi}{4}(b/a)^2 \text{ pixels}$$

and the sums (Σ) must be taken over this area to estimate the S/N ratio. For a small but extended object such as a faint galaxy of size x arcseconds, the sums will be over

$$n = \frac{\pi}{4}\frac{(b^2 + x^2)}{a^2} \text{ pixels.}$$

However, if x is much larger than the seeing diameter b, then it is more convenient to express m as a 'surface brightness' in magnitudes per square arcsecond (as is done for the sky) and so $n = 1$ and each pixel is treated separately.

In the ideal case, we would have very accurate calibration data so that ϵ_B and ϵ_D become very small indeed, and u_r is negligible. Then,

$$S/N = \mathbf{S}T^{1/2}/[\mathbf{S} + \sum_{i=1}^{n}(B + D + R^2/t)]^{1/2} \quad \text{(A4.2)}$$

There are two further simplifying cases:

1. B very much larger than $(D + R^2/t)$, and S is much weaker than B; this is called the 'background-limited' or 'sky-limited' case and so

$$S/N = ST^{1/2}/[\sum_{i=1}^{n} (B)]^{1/2} \tag{A4.3}$$

if we restrict the object to just one pixel and set $S = fB$ then

$$S/N = f(BT)^{1/2}$$

In this case, the accuracy of the measurment scales only as the square root of the integration time—four times as long gives two-times improvement. Also, the signal-to-noise scales only as the diameter of the telescope used because a larger telescope gives more background B in proportion to the *area* of its mirror ($\pi D_{TEL}^2/4$).

Suppose $f = 0.001$, that is the source is only 0.1% of the brightness of the sky, then to just barely detect it requires that $S/N = 1$ and therefore $BT = 1\,000\,000$ electrons. Suppose the sky background gives about 300 e/s/pixel, then this observation will take over 3000 s. Note that for R^2/t to be negligible (say 1% of B or 3 e/s), then R^2 must be less than $3t$, or R less than 95 electrons if the entire exposure is made in one go ($t = T$). However, if the exposure is made in five separate $t = 600$ s blocks then R^2 must be less than 3×600, that is R less than 42 electrons. In practice t is limited by the 'full-well' capacity of a pixel.

2. If R^2/t is very much larger than $(B + D)$, and the signal S is also weak in comparison, then the dominant source of noise is the readout noise and this is known as the 'readnoise-limited' case.

$$S/N = ST^{1/2}/[\sum_{i=1}^{n} (R^2/t)]^{1/2} \tag{A4.4}$$

if we restrict the object to just one pixel then

$$S/N = \frac{S(N_S)^{1/2}t}{R}$$

which shows that the signal-to-noise ratio for a given source S, and detector noise R, scales as the on-chip integration time and as the square root of the number of repetitions. In a spectrometer the background may be less than 0.3 e/s per pixel and so for $t = 600$ s we are formally readnoise-limited if $R^2 > 0.3 \times 600$, i.e. if R is larger than 13.4 electrons.

The above simplified treatment shows that the best detections and therefore the highest signal-to-noise ratios are obtained when

(a) sufficiently accurate calibration frames are obtained,
(b) the readout noise and dark current are as small as possible,
(c) the on-chip integration time is as long as possible.

The power collected by a telescope of area A cm^2 in a wavelength interval of $\Delta\lambda$ μm from a source of apparent magnitude m (after allowance for extinction by the

Earth's atmosphere), transmitted by an optical system of efficiency τ onto a CCD detector pixel of quantum efficiency q is

$$P = \tau(\lambda)q(\lambda)A\Delta\lambda F_\lambda(0) \times 10^{-0.4m} \mathrm{W}$$

where $F_\lambda(0)$ is the flux from a zero magnitude standard star above the atmosphere.

Since the energy of a single photon of wavelength λ is just hc/λ W s/photon, then the photoelectron detection rate is

$$S = (hc)^{-1}\tau(\lambda)q(\lambda)A\lambda\Delta\lambda F_\lambda(0) \times 10^{-0.4m} \text{electrons/s}$$

By dividing S by g electrons per DN we get the observed signal S_0 in A/D counts. If we set $S_0 = 1$ DN and allow for the atmospheric extinction in the waveband of interest then we can derive the corresponding magnitude m_{ZP} which is the 'zero-point' of the instrumental scale. Corrections for extinction by the atmosphere are expressed as magnitudes per air mass δm; the 'air mass' (X) is a computed function of the zenith angle of the star. That is

$$m = m_{\mathrm{obs}} - \delta m\, X$$

where X is usually in the range 1–3 and δm varies from about 1.0 in the UV to 0.05 in the infrared and must be determined experimentally at each observatory.

The signal from the sky or background (which includes thermal emission from the telescope at infrared wavelengths) depends on wavelength, local conditions and moonlight for any given telescope.

$$B = (hc)^{-1}\tau(\lambda)q(\lambda)A\lambda\Delta\lambda a^2 F_\lambda(0) \times 10^{-0.4m\mathrm{sky}}$$

where m_{sky} is the magnitude per square arcsecond of sky relative to a star of zero magnitude above the atmosphere and the pixel size is in arcseconds. Since B is proportional to Aa^2 or $(\pi D_{\mathrm{tel}}^2/4)(206265\, d_{\mathrm{pix}}/F)^2$, where d_{pix} is the effective pixel size in the telescope focal plane and F is the focal length, then B is inversely proportional to the square of the telescope focal ratio $(F/D)^2$. Many optical telescopes have focal ratios of about f/7.5 whereas infrared telescopes use f/36 or larger to greatly reduce the background. At a good site with no moon in the sky, m_{sky} is about 22 in the blue, 21 in the visible and red, about 20 or brighter (and variable) near 1 μm and about 13 or brighter at 2.2 μm; the standard passbands are assumed.

As an example, consider a CCD camera with $\tau = 0.5$ and $q = 0.5$ at a wavelength of 0.54 μm in the middle of the visible spectrum with a filter having a passband of 0.1 μm on a 4-m class telescope, i.e., a 4-m mirror with a 1-m Cassegrain hole in its centre. With the appropriate value of $F_\lambda(0)$ from Table A4.1, we get

$$S = 3.136 \times 10^{10} \times 10^{-0.4m} \text{ e/s}$$

so that for $m = 17$ — a faint object but brighter than the sky — we obtain 5000 e/s, whereas for the extremely faint sources seen on deep CCD images at $m = 27$ we detect only 0.5 electrons per second. (In practice the situation is more complex since star images are not sharp-edged, filter transmissions are not simply represented by a wavelength interval, and the stellar energy spectrum will not be the same as that of Vega.)

It is sometimes convenient to express the limiting precision of a measurement in terms of an amount of power equivalent to the noise, this is the **Noise equivalent power (NEP)**. This term was traditionally used with infrared sensors before the advent of array detectors. To convert the total noise N (in electrons averaged over the integration time) to power in watts we divide by the quantum efficiency to find the number of photons and then multiply by the energy per photon

$$\text{NEP} = \frac{hc}{\lambda}\frac{[S + \Sigma(B + D + R^2/t)]^{1/2}}{q T^{1/2}} \text{ watts}$$

where the simpler expression for the noise used in equation (A4.2) has been substituted.

To convert NEP to noise equivalent flux density (NEFD) in Jansky we have

$$\text{NEFD} = \frac{\lambda \, \text{NEP} \, R}{\tau A \, 3 \times 10^{-16}}$$

where the wavelength is expressed in micrometres and the telescope collecting area (A) is in square centimetres; R is the resolving power $\lambda/\Delta\lambda$.

At infrared wavelengths the 'sky' background (B) becomes progressively dominated by thermal emission from the sky and the telescope optics. To estimate F_λ we use the Planck function

$$F_\lambda = \frac{2\pi hc^2}{\lambda^5} \cdot \frac{1}{e^{hc/\lambda kT} - 1} \quad \text{W cm}^{-2}\,\mu\text{m}^{-1}$$

where T is the absolutetemperature.

Table A4.1 — Absolute flux from a zero-magnitude star like Vega

Symbol	λ (μm)	ν (Hz)	F_λ (W cm^{-2} μm^{-1})	F_ν (Jy)[a]
U	0.36	8.3×10^{14}	4.35×10^{-12}	1880
B	0.43	7.0×10^{14}	7.20×10^{-12}	4440
V	0.54	5.6×10^{14}	3.92×10^{-12}	3810
R	0.70	4.3×10^{14}	1.76×10^{-12}	2880
I	0.80	3.7×10^{14}	1.20×10^{-12}	2500
J	1.25	2.4×10^{14}	2.90×10^{-13}	1520
H	1.65	1.8×10^{14}	1.08×10^{-13}	980
K	2.2	1.4×10^{14}	3.80×10^{-14}	620

[a] These units are called Janskys. One Jy equals 10^{-26} W per square metre per Hertz or $(3\times10^{-16}/\lambda^2)$ watts per square cm per micrometre if λ is expresed in micrometres.

Appendix 5: Typical CCD clock voltages

A tabulation of CCD clock voltages for a wide range of devices in use at many different observatories. The values given here could be used as a guide for setting up a CCD system.

Table A5.1 — RCA CCD (30-μm pixels) clocks

Clock	RCA3[a] CFHT 501	RCA-1[a] KPNO 501	RCA-2[a] KPNO 501	RCA-3[a] KPNO	RCA4[a] CTIO 501	RCA5[a] CTIO 501	RCA2[a] CTIO 53612	RCA#5[a] AAO 501
V_{HH}	+2.0	3.0	3.0	5.0	+8.0	+7.0	0.0	+1.0
V_{HL1}	−11.0	−8.0	−8.0	−10.1	−10.0	−12.0	−9.0	−12.0
V_{HL2}	−11.0	−8.0	−8.0	−10.1	−10.0	−12.0	−9.0	−12.0
V_{HL3}	−11.0	−8.0	−8.0	−10.1	−10.0	−12.0	−9.0	−12.0
V_{VH}	0.0	0.5	0.5	0.5	2.0	0.0	0.0	−3.0
V_{VL}	−9.0	−9.5	−9.5	−8.5	−10.0	−12.0	−8.75	−8.0
V_{RH}	0.0	4.0	4.0	5.0	5.0	5.0	5.9	+0.5
V_{RL}	−3.0	−1.0	−1.0	−10.0	−10.0	−10.0	−3.5	−5.0
V_{RD}		14.0	12.0	12.5	17.32	16.65	14.0	17.50
V_{OD}		14.0	14.0	14.0	17.08	17.55	14.0	17.50
V_{BD}		—	—	—	9.01	9.0	9.0	6.0
$V_{OG}H$		1.0	3.0	5.0	−6.01	−6.24	0.0	$H_{\phi 1}$
$V_{OG}L$		0.0	−8.0	−10.0				
V_{OPW}		14.0	14.0	—				
IOS					1.0	1.0	1.0	
ILS					0.75	0.75	0.75	
$R_n(e^-)$	54	79	72	42	45	45	45	36.2
CTE	good	fair	poor	good				Exc
$FW(e^-)$	6×10^5	1.1×10^6	1.2×10^6	8×10^5	$\sim10^6$	$\sim10^6$	$\sim10^6$	$\sim10^6$
prefl(e^-)	—	—	—	—	—	—	—	—

[a]'inverted' clocks (recommended for 3ϕ).
$V_{BB}=V_{SUB}=$substrate$=0V$.

Table A5.2 — More RCA CCD (30 μm pixels) clocks

Clock	CCD#3[a] ESO 501	CCD#5[a] ESO 501	A SO	B SO	C SO
V_{HH}	~5.55	~4.25	7.0	4.0	10.0
V_{HL1}	−8.50	−9.70	−5.0	−5.0	−3.0
V_{HL2}	−8.47	−9.85	−5.0	−5.0	−3.0
V_{HL3}	−8.45	−9.85	−5.0	−5.0	−3.0
V_{VH}	+1.00	+1.10	0.0	3.0	5.0
V_{VL}	−9.00	−8.90	−9.5	−6.0	−9.0
V_{RH}	+5.00	+5.00	5.0	5.0	5.0
V_{RL}	−5.00	−5.0	−5.0	−5.0	−5.0
V_{RD}	20.00	15.60	14.5	13.0	14.5
V_{OD}	15.00	14.00	12.5	13.0	12.5
V_{BD}	11.76	8.00			
V_{OG}	−0.15	−2.00	0	0	0
$R_n(e^-)$	40	49	37	38	65
CTE	Vg	Vg	fair	good	fair
$FW(e^-)$	10^6	10^6	3×10^5	2×10^5	3×10^5
$prefl(^-)$	—	—	—	—	—

[a]'inverted' clocks (recommended for 3ϕ).

Table A5.3 — RCA CCD (15-μ pixels) clocks

Clock	RCA2[a] CFHT SID503	RCA4[a] CFHT 503	CCD#8 ESO 503	CCD#9 ESO 503
V_{HH}	+2.0	+3.0	+5.0	+4.88
V_{HL1}	−9.0	−9.7	−7.00	−8.15
V_{HL2}	−9.0	−9.7	−7.00	−8.10
V_{HL3}	−9.0	−9.7	−7.0	−8.07
V_{VH}	+0.0	+0.0	+0.0	+0.0
V_{VL}	−10.0	−11.0	−9.00	−9.05
V_{RH}	+4.0		+5.00	+5.00
V_{RL}	− 6.0		−5.00	−5.00
V_{RD}	18.5	18.0	17.00	19.08
V_{OD}	16.0	16.5	15.00	16.00
V_{BD}		10.0	6.49	9.00
V_{OG}	$V_{\phi H1}$	$V_{\phi H1}$	−4.00	−4.00
R_n	45	55	36	40
CTE	exc	exc	Vg	Vg
$FW(e^-)$	10^5	10^5	10^5	—
$prefl(e^-)$	300	300	—	—

[a]'inverted'.
$V_{BB}=V_{SUB}$=substrate=0V.

Table A5.4 — TI (3φ) 800^2 CCD clocks

Clock	TI#1 CTIO	TI#2 CTIO	Cryocam KPNO	TI-2 KPNO	TI-3 KPNO	TI-4 KPNO	TI-5 KPNO	A SO	B SO
$V_{\phi H1-3S}$	+12.8	+9.0	12.0	12.0	12.0	12.0	12.0	11.0	10.5
$V_{\phi L1-3S}$	−2.0	−2.0	0.0	0.0	0.0	0.0	0.0	0.0	0.0
$V_{\phi H1(c)}$	+5.0	+5.0	7.0	7.0	7.0	9.0	9.0	9.0	9.5
$V_{\phi H2-3P(c)}$	+5.0	+5.0	8.4	8.4	7.0	9.0	9.0	9.0	9.5
$V_{\phi L1p(a)}$	−10.0	−8.0	−8.0	−8.0	−8.0	−7.9	−7.9	−7.0	−6.0
$V_{\phi L2-3P(c)}$	−5.0	−8.0	−1.5	−1.5	−1.5	−1.5	−1.5	−7.0	−6.0
V_{HOPW}	+12.8	+9.0	12.0	12.0	12.0	12.0	12.0	11.0	10.5
V_{LOPW}	−2.0	−2.0	0.0	0.0	0.0	0.0	0.0	0.0	0.0
V_{REF}	+16.4	+17.1	17.3	16.4	17.3	17.0	17.0	20.0	19.25
V_{DD}	+28.4	+28.7	27.7	29.8	29.8	29.8	25.0	30.0	31.25
V_{OPG}	+4.05	3.0	3.0	3.0	3.0	3.0	3.0	5.0	5.0
$V_{+\phi R}$			14.0	14.0	14.0	14.0	14.0		
V_{HOPG}	+10.0	+10.0	3.0	3.0	3.0	3.0	3.0		
$V_{\phi HT}$	+5.0	+9.0	4.0	4.0	4.0	9.1	0.0	11.0	9.0
$V_{\phi LT}$	−2.0	−2.0	−3.5	−3.5	−3.5	−3.4	−3.5	−11.0	−9.0
V_{HTL}			4.0	4.0	4.0	9.0	4.05		
$R^n(e^-)$	6	6	10.5	8.0	13.5	11.3	8.0	6	9
$FW(e^-)$	4×10^4	4×10^4	4×10^4	4×10^4	4×10^4	4×10^4	4×10^4	4×10^4	4×10^4
CTE	good	good	good	good	good	good	good	exc	exc
prefl(e^-)	30	30	50–100	50–100	50–100	50–100	50–100	—	—

Table A5.5 — Thomson CCD clocks

Clock	TH1 CFHT 7882	INSU 7882
V_{HH}	11.25	9.98
V_{HL}	3.0	0.403
V_{VH}	8.75	9.98
V_{VL}	1.1	0.50
V_{RH}	10.0	9.80
V_{RL}	0.0	0.50
V_{RD}	14	11.0
V_{OD}	16	14.80
V_{OG}	3	2.850
$R_n(e^-)$	13	
CTE	good	good
$FW(e^-)$	$>10^5$	1.5×10^5
prefl(e^-)	200	600

Table A5.6 — TEK CCD clocks

Clock		TEKI KPNO
$V_{+\phi}$	V_{HH}	14.0
$V_{-\phi 1}$	V_{HL1}	−1.5
$V_{-\phi 2}$	V_{HL2}	−2.0
$V_{-\phi 3}$	V_{HL3}	0.0
$V_{+\phi p}$	V_{VH}	8.0
$V_{-\phi p}$	V_{VL}	−1.0
$V_{+\phi sw}$	V_{RH}	14.0
$V_{-\phi sw}$	V_{RL}	−1.1
$V_{+\phi R}$	$V_{resetgate}$	12.0
V_{OD}	V_{RD}	14.3
V_{DDA}	V_{OD}	20.0
V_{SA}	V_{shield}	14.9
$V_{+\phi L}$	V_{OG}	2.4
$V_{+\phi TA}$	$V_{\phi HT}$	4.8
$V_{+\phi TA}$	$V_{\phi LT}$	−6.6
$R^n(e^-)$		7
$FW(e^-)$		3×10^6
CTE		good
prefl(e^-)		—

Table A5.7 — EEV/GEC CCD clocks

Clock	P8603
V_{HH}	−1.0
V_{HL}	−12.0
V_{VH}	−2.0
V_{VL}	−12.0
V_{RH}	−2.0
V_{RL}	−12.0
V_{RD}	+4.5
V_{OD}	+13.5
V_{OG}	−1.0
V_{SS}	−6.5
I_{OS}	1.0 mA
R_n(e)	7
CTE	exc
FW(e)	10^5
prefl	—

Table A5.8 Nomenclature cross-reference

RCA (CFHT)	RCA (ESO)	TI (CTIO)	Description
V_{HH}	HAMP	V_{HS}	High horizontal/serial phase clocks
V_{HL}	HLO	V_{LS}	Low horizontal/serial phase clocks
V_{VH}	VAMP	V_{HP}	High vertical/parallel phase clocks
V_{VL}	VLO	V_{LP}	Low vertical/parallel phase clocks
V_{RH}	RAMP	V_{HOPW}	High reset transistor/output well
V_{RL}	RLO	V_{LOPW}	Low reset transistor/output well
V_{RD}	RD	V_{REF}	Reset drain
V_{OD}	OD	V_{DD}	Output drain
V_{BD}	BD	—	Barrier drain
V_{OG}	OG	V_{OPG}	Output gate
0	0	V_{SUB}	Substrate voltage
0	0	V_{IPG}	Input gate
—	—	V_T	Transfer gate
—	—	—	IOS, ILS constant current sources: CCD amp (AAO)

REFERENCES

ESO: CCD detectors available at La Silla: ESO technical report June 1986, Roland Reiss, personal communcation to C. Christian.

NOAO (Kitt Peak): Richard Reed, personal communication to C. Christian.

NOAO (CTIO): Alistair Walker, personal communication to C. Christian.

AAO: John Barton, personal communications to C. Christian and I. S. McLean.

JPL: Jim Janesick, personal communication to C. Christian.

Steward Obs.: Robert Leach, Optimising charge-coupled device detector operation for optical astronomy, *Optical Engineering,* October 1987, Vol. 26, 1061.

CFHT: Patrick Waddell, Christopher Clark, personal communication to C. Christian.

EEV/GEC: Craig Mackay, Johanthan Wright, Paul Jordan personal communication to I. S. McLean.

Appendix 6
CCD suppliers and camera manufacturers

A6.1 DEVICE SUPPLIERS

There are actually numerous CCD detectors for sale but manufacturers tend to sell isolated devices designed specifically for black-and-white or colour TV operation, and some CCDs are available only in complete systems. Many commercially available devices are not suited to astronomy; those that are and which are also available for sale now or in the near future are listed below.

1. EEV/GEC: 385×578 frontside-illuminated; phosphor coating for UV. Work on thinning and larger formats in progress.
   Device Marketing
   English Electric Valve Co., Ltd.
   106 Waterhouse Lane
   Chelmsford
   Essex CM1 2QU
   UK Tel. 0245-493493

2. Thomson-CSF: 384×576 frontside-illuminated; phosphor coating for UV. Formats up to 1024×1024 being developed and versions buttable on three sides being developed.
   Division Tubes Electronique
   38 Rue Vauthier
   B P 305
   92102 Bologne-Billancourt CEDEX
   FRANCE Tel. 604-8175

3. Photometrics: 516×516 frontside-illuminated; Metachrome coating for UV.
   Photometrics Ltd., 2010 N. Forbes Boulevard, Suite 103

Tucson
Arizona 85745-1418
USA Tel. (602) 623 8961

4. Tektronix: 512×512 front- and back-illuminated devices. Format as large as 2048×2048 possible.
   Charge Coupled Devices
   Integrated Circuits Operation
   Tektronix, Inc.
   P.O. Box 500
   Mail Station 59-420
   Beaverton, OR 97077
   USA Tel. (503) 627 2515

5. Reticon: 400×1200 frontside developed, backside planned.
   EG&G Reticon
   345 Potrero Avenue
   Sunnyvale
   California 94086
   USA Tel. (408) 738 4266

Excellent devices have been supplied to JPL by Lincoln Laboratory, Massachusetts Institute of Technology, P.O. Box 73, Lexington, Ma. 02173.

Infrared array detectors are commercially available from several companies including;

Santa Barbara Research Center
75 Coromar Drive
Santa Barbara
California 93117
USA

Rockwell International
Science Center
1049 Camino Dos Rios
P.O. Box 1085
Thousand Oaks
California 91360
USA

Cincinnati Electronics
2630 Glendale-Milford Road
Cincinnati
Ohio 45241

Sofradir
BP 21
38113 Veurey-Voroize

FRANCE

Mullard
Millbrook Industrial Estate
Southampton SO9 7BH
UK

A6.2 SLOW-SCAN CCD CAMERA MANUFACTURERS

Many companies now market TV and intensified TV cameras which employ CCD detectors, and these firms advertise in the electronics literature or in specialised literature for the market they are aimed at, such as underwater imaging, surveillance, medicine or astronomy. For instance, Hydro Products (Honeywell), 11777 Sorrento Valley Road, San Diego, Ca. 92121, utilises an Amperex NXA 1000 series frame transfer CCD with 610×490 pixels, and Astro Link, P.O. Box 1978, Spring Valley, Ca.92077, supply various cooled cameras with selected 572×485 Sanyo CCDs. Scanco Ltd., of Waimea, Hawaii offer astronomical real time TV systems employing TI virtual-phase CCDs.

Low-noise, slow-scan cryogenic CCD cameras for astronomy at national observatories are usually constructed 'in-house' to exacting standards, but there are now several commercial sources with demonstrated expertise. Four of these, two in Europe and two in the USA, are given below:

Photometrics Limited
2010 N. Forbes Boulevard, Suite 103
Tucson
Arizona 85745-1418
USA Tel. (602) 623 8961

Princeton Scientific Instruments Inc.
306 Alexander Street
Princeton
New Jersey 08540
USA Tel. (609) 921 6629

Astromed Limited
Innovation Centre
Cambridge Science Park
Milton Road
Cambridge CB4 4GS
UK Tel. 0223-316705

Wright Instruments
4 Chalkwell Park Avenue
Enfield
Middlesex EN1 2AJ
UK Tel 01-363-7242

Index

aberrations, 148, 261
absolute temperature; degrees Kelvin; absolute zero of temperature, 119, 261
advantages of,
  CCDs, 32, 41, 42
  infrared array detectors, 229
Aikens, Richard, 81, 86, 92, 111, 137
Albrecht, Rudolf, 184
Allen, David, 224
alt-az. mountings; telescopes, 252, 254, 261
Amelio, Gilbert, 32, 81, 86
amplifier,
  gain, 168, 197
  on-chip, 193
  pre-, 153, 156
analogue-to-digital unit (A/D); analogue to digital conversion, 43, 162, 168, 197
Angel, Roger, 59, 255, 256
Anglo Australian Observatory (AAT), 42, 56, 74, 75, 252
anti-reflection coating; AR coating, 145, 149, 261
aperture photometry with CCDs; software apertures, 211
Arens, John F., 228
aspherical optics, 149, 261
Astromed Ltd, 45, 87, 112, 292
astrometry, 39
atmospheric windows, 225
AT&T, *see* Bell Labs
automation, 34, 73
AXAF, 236

back focal length, 147, 149, 261
background-limited; BLIP; background signals,
  sky, 279–283
  thermal, 221, 281, 283
backside,
  charging, 138, 139
  illuminated, 109, 136, 227
  treatment of CCDs, 137, 138, 139
Baird, John Logie, 25
baud rate, 174, 261
Bell Telephone Laboratories, 27, 29, 81, 82, 111, 180, 188

bias, 171, 197, 199, 201
bidirectional charge transfer, 65
binning, on chip, 170, 216
black-body; black-body radiation, 194, 261
Blouke, Morley, 85, 91, 92, 99
Boksenberg, Alec, 27, 231
Boyle, Willard S., 29, 31, 81, 112
bucket analogy for CCDs, 99, 100
bump bonds, 226, 231
Burt, David, 87
bus,
  IEEE-488, 174, 262
  VME, 175, 262
byte; megabyte; gigabyte, 174, 186, 245, 262

CAD/CAM, 178
cadmium mercury telluride, 94, 97, 226, 230, 247, 268
calibrations, 193–219
calorific rays, 222
Caltech, 83, 136
Cambridge University, England, 86
Canada-France-Hawaii Telescope, CFHT, 42, 50, 51, 223
CASE; computer-aided software engineering, 185
Cassegrain focus, 42, 148, 262
CCDs
  advantages of, 32, 41, 42
  buried channel, 81, 106, 107, 262
  calibration of, 196, 198, 202, 205, 207, 215
  comparison of, 217, 218, 284
  cooling of, 118, 119, 149, 151
  first astronomical image, 83, 84
  frame transfer, 103
  future of, 242
  interline transfer, 103
  invention of, 29, 31, 81
  makers of, 290
  outputs, 104, 106, 131
  surface channel, 100, 106
Cepheid variables, 21, 262
Cerro-Tololo Inter-American Observatory, CTIO, 42, 252

channel stops, 102, 263
charge-coupling, 31, 98, 102
charge transfer efficiency, CTE, 124, 128, 131, 243
Christian, Carol, 115
clocks; clocking, 103, 263
  drivers, 153
  frequency, 119
  inverted, 121
  levels, 113, 284
  serial, parallel, 113, 114
coaxial cable, 118, 263
common-user instrumentation, 72
computers, 34, 37, 74
  data analysis, 69, 177, 179
  for control, 41, 60, 158
  personal computers (PC), 187
conduction band, 93, 263
conferences,
  Infrared Astronomy with Arrays, 229
  Instrumentation in Astronomy VI, 90, 139
  Solid State Imagers for Astronomy, 32, 229
control room, 17, 75
coronagraph, 67
coronene, 136, 249, 263
corrector plate; Wynne corrector, 149
correlated doubling sampling (CDS), 132, 163–167, 262
cosmic rays, 121
COSMOS, 40, 41, 188, 189
Crab Nebula, 66
critical path, 78, 263
cryostat; cryogenics, 119, 149, 151, 152, 154, 263
CTI; CCD Transit Instrument, 68, 70, 190, 191
Cullum, Martin, 137

DAOPHOT, 211, 216
dark current, 43, 118, 119, 120, 202
data
  archives, 186
  cube, 61
  management, 37, 186, 256
  rates, 21, 173
  reduction, 74, 188, 207
  sheet, 113
dc bias voltages, 113
deferred charge, 124, 126, 263
depletion region, 96, 99, 264
diamond-machining, 149
diffraction (of light),
  grating, 54, 264
  limited by, 70, 222
Digital Equipment Corporation (DEC), 179, 181
digitised, 23
direct memory access (DMA), 174
Disney, Mike, 180
DN (data number), 168, 197, 264
doping, 96, 97, 227, 264
Doppler shift, 55, 264
down-converter, *see* fluorescence
down-loading, 159, 264
DQE (detected quantum efficiency), 196, 264
Draper, Henry, 20
drift-scan, 204
dynamic range, 168

echelle, 87
EEV/GEC, 89, 92, 243
EG&G Reticon, 92, 243
Einstein, Albert, 25
electrodes; gates, 95
  polysilicon, 108
electromagnetic radiation, 23, 24
electron-bombarded CCD (EBCCD), 233
electron-hole pairs, 94
electrons per DN, 197, 198
electrons, 25
electronvolt, 94, 238, 274
Ellis, Richard, 59
energy levels, 92, 93, 98
EPER; extended pixel edge response, 127, 129
epitaxial layer, 121
EPROM, 161
Ethernet, 175
European Southern Observatory, ESO, 34, 254
expansion of the universe, 20
extinction, 222

Fabry–Perot interferometers, 61
Fairchild Semiconductor, 86, 88
false-colour, 75, 178
Fano-noise-limited, 238, 239
fast interface states, 132
fat zero, 124
Fermi level, 96
fibres, optical, 58, 59, 60
field of view, 148
filters, 211, 212, 225
FITS (flexible image transport system), 181
FLAIR, 61
flash gate, 139
flat-fields, 132, 202, 207
fluorescence, UV, 136
focal plane scale, 148
FOCAS (faint object classification system), 188
focal reducer, 148
forbidden band gap, 93, 97
Ford/Photometrics CCD, 130, 131, 169, 243
four-shooter, 46
Fowler, Albert, 86
Fraunhofer, Joseph, 20, 54
  lines, 20
fringes, 111, 136, 205, 207

Galileo Galilei, 19, 258
Geary, John, 32, 85
GEC (*see also* EEV/GEC), 87
germanium, 27, 227
Giotto CCD camera image of Comet Halley, 89, 90
GONG (global oscillation network), 68
Grandi, Steve, 179
gravitational lenses, 33
gravity waves, 23

grism, 64
ground-loops, 117, 118
guiding, 17
Gunn, Jim, 43, 85
Gursky, Herbert, 33

Hale, George Ellery, 20
hardwired, 157
Harvard-Smithsonian Observatory, 33, 89
Helium, liquid, *see* LHe
Herschel, Sir William, 222
Hill, John, 59
Hoffmann, Alan, 228
hold-time, 151, 152
holes, in the valence band, 94
Hooker telescope, 20
Horsehead Nebula, 47, 48
Hubble, Edwin, 20
Hubble Space Telescope, 247–250
Humason, Milton, 20
Huygens, Christiaan, 25
hybrid arrays, 226, 227

IBM PCs, 187
image display, 176, 178, 179, 183
image intensifier, 26
image processing, 177
indium antimonide, InSb, 48, 226, 228, 229, 230
infrared arrays, 225
infrared astronomy, 220
infrared detectors, 223
integrated circuit, 22, 27
integrating detector, 21
integration time, 16, 280
interstellar dust, 222
IPCS (image photon counting system), 27, 54, 55, 231
IRAF (image reduction and analysis facility), 183
IRAS (infrared astronomical satellite), 224
IRCAM (infrared camera) (UKIRT), 47, 229
ISIS, spectrograph for WHT, 57
ISP, the ROE Imaging Spectropolarimeter, 64
ISO (infrared space observatory), 224

Janesick, James R., 83, 85, 117, 139, 239, 251
Janskys, 283
Jet Propulsion Lab, JPL, 83, 139, 236
JHKLMN filters, 225

Keck Telescope, 253
Kelvin, degrees, *see* absolute temperature
Kitt Peak National Observatory, 33, 42, 68, 158
Kodak,
  CCDs, 11, 131
  emulsion, 25
kTC noise, 163

Landauer, Fred, 83
La Palma Observatory, 42
Latham, David, 32
Leavitt, Henrietta, 21
LHe, 119
Lick Observatory, 42, 92, 139
Lilly, Simon J., 223
Lincoln Laboratory, 130, 131, 169
linearity, 208
light emitting diode effects, 122
LN2, 119, 151, 152
local area networks, lans, 175
Longair, Malcolm S., 89, 228
Low, Frank, 224
Lowrance, John, 84, 85, 233
LSB (least significant bit), 197
luminescence, 122
Lunar and Planetary Lab, 83, 87
Lux, 277

M51, 233
Mackay, Craig, 86, 112, 121
magnitudes, 215, 276, 277
Mauna Kea, Hawaii, 42, 231
Maxwell, James Clerk, 23, 222, 253
McGraw, John, 68
median filtering, 205, 207
MEDUSA, 59
MegaFLOPS, 178
mer-cad-tel, 226
Metachrome, 92, 137
metal oxide semiconductor (MOS), 98, 100
microchannel plate, 232, 235
micro-chip, 22
microprocessor, 22, 158, 159
MIDAS, 181
MIPS, 178
Multiple-Mirror Telescope (MMT), 252
multi-pinned phase (MPP), 121

n-type semiconductor, 96
National Optical Astronomy Observatory (NOAO), (*see also* Kitt Peak National Observatory), 183, 194
native oxide, 138
Nelson, Jerry, 253
Neugebauer, Gary, 223
neutrinos, 23
Newton, Sir Isaac, 25
NGC1068, 88
night assistant, 15
nitrogen, liquid, *see* LN_2
noise, 27
  equivalent power, NEP, 283
  photon noise, 198, 199
  reset, 131, 132, 163
  white, 168
  $1/f$, 132

observing time, 15
Oke, Bev, 55
optical fibres, *see* fibres
optical matching, 145, 147
Orion nebula, 53
outgassing, 151
overscan, 127, 171, 172, 201

Palomar Observatory, 42
panoramic or multiplex advantage, 26
peer review, 15
Pettie, Donald, 89
PFUEI — 'foo-ee', 42
phase CCD, three, 100
two, 108
virtual, 108
photocathode, 26
photoconductivity, 223
photodiode, 95
photoelectric effect
in metals, 25
in semiconductors, 95
photography, birth of, 25
Photometrics Ltd, 48, 92, 111, 188, 243, 246 292
photometry, 208, 211, 215, 216
photomultiplier tube, 25, 26
photon counting, 27, 231
photon transfer curve, 199, 200
photovoltaic, 223
Pickering, Edward Charles, 21
Picturephone, 29
pixel, picture element, 29, 99
Planck function, 194
Planck, Max, 25
Plumbicon TV tube, 27
pn junction, 96
pockets (in CCDs), 91, 129
polarimeters, 63, 64
polarised light and its measurement, 63, 88
polysilicon, 108
preamplifier, *see* amplifier
preflash, 124, 201
prime focus, 42, 262
Princeton Scientific Instruments, 87, 233, 292
project planning of astronomical instruments, 77
PROMs, 158
PSF, point spread function, 211
p-type semiconductor, 96
pulsar, *see* Crab Nebula

QEH, quantum efficiency hysteresis, 136
QE pinning, 138
quanta, quantum, 25
quantum efficiency, 29, 193, 238
quasars, 45

radiation damage to CCDs, 85
RAM, random access memory, 81, 158
ray-tracing, 148
RCA, 85, 86
readout noise, 198, 246
redshift (*see also* Doppler shift), 222, 223
refractive index, 145
Reitsema, Harold, 88, 89
remote observing, 73, 257
reset transistor, 106
resolution (spatialand spectral), 148
Reticon, EG&G, 92
reverse bias, 96
Robinson, Lloyd, 42, 92, 139, 141, 159
Rochester, University of, 228
Rockwell International, 228
Royal Greenwich Observatory, RGO, 91, 159
Royal Observatory, Edinburgh, ROE, 40, 89, 228, 257

Santa Barbara Research Center (SBRC), 228
saturation, 124, 168
Savoye, Dick, 85
Schmidt telescope, 39
seeing disc, 42
semiconductor,
Age, 22, 230
devices, impact of, 32
materials, 95, 227
technology, 243, 244
sequencers, 157
Serkowski, Krystof, 67, 87
sidereal rate, 68
signal-to-noise ratio, 196, 279, 280, 281
silicon
chip, 22, 29
doped, 227
foundry, 243
nitride, 131
oxide, 131, 138
slew, 16
slow-scan, 162
Smith, Bradford, 83, 87, 251
Smith, George E., 29, 31, 81
Smith, Gerald M., 83
Smith, Malcolm G., 89
software, 37, 96, 180
speckle, 33, 70, 71
spectrographs,
echelle, 87
faint object, 55, 56, 58
grating, 54
multiple object, 56
spectroscope, prism; invention of, 20
spectrum, spectra, 20, 216, 217
STARLINK, 180
Stetson, Peter (*see also* DAOPHOT), 216
Steward Observatory, 59, 255
Stockman, H. S. (Peter), 64, 87, 89
Sun Microsystems Inc., 179, 183
supernova 1987a,
light echo, 34
spectrum, 59

TDI, *see* drift-scan
Tektronix Inc., 91, 130, 243, 244
Texas Instruments Inc., 83
thermal conductivity, 151
Thomson-CSF, 92, 243
Thorne, David, 91
throughput, 61
timing waveform, diagram, 153, 167
Tody, Doug, 183
Tompsett, Mike, 31, 32, 81
transistor, invention of, 27
traps, (*see also* pockets), 106, 124

twisted pair wire, 118
Tyson, Tony, 45, 88, 189, 190, 204, 243

UBVRI, *see* filters
UK Infrared Telescope (UKIRT), 47, 224, 229
ultraviolet (UV), 230
uniformity of CCDs, 132
University of Arizona, 87
UNIX, 62, 180, 183
Uranus, 83, 84

vacuum chambers and pumps, 119, 153
valence band, valence electrons, 92, 93
variance, 198
Vax computers, VMS, 178, 179
Vega, 277, 283
virtual phase CCDs, 108, 109
VLTs (very large telescopes), 251

Waddell, Patrick, 86
Wall Jasper, 42
Wampler scanner, 27
Watson, Fred, 61
wavelength,
  cutoff, 95
  definition of, 23
Wells, Don, 180
Westphal, Jim, 43, 83, 84, 85, 136, 251
Wide Field/Planetary Camera — 'wiff-pick', 248
workstations, 179
Wright Instruments, 187, 292
Wright, Jonathan, 87, 187

X-rays, CCD detection of, 236, 238, 239. 241

zenith park, 17
Zeolite molecular sieve, 152
zeropoint, 215, 216